Concise Concepts of Nanoscience and Nanomaterials

Dr. Narendra Kumar, former director & DRDO fellow (Min. of Defence, Govt. of India), obtained his Ph.D degree in 'Organometallic Chemistry' from Delhi University, he has a very rich experience of working on a variety of advanced materials including organometallics, liquid crystals, conducting polymers, nanomaterials together with evolving green synthetic routes for the preparation of metal salts by electrochemical, conducting polymers and nanomaterials in aqueous media and products based on some such materials even for defence applications. Dr. Kumar has published 110 research papers, including four review articles, in International Journals of repute in the areas of Nano Materials, Organometallics, Conducting Polymers, and Electrochemical Synthesis together with one Chapter in Encyclopedia on Nano Science & Technology and has 12 patents to his credit. He has co-authored two book "Nanotechnology and Nanomaterials in Treatment of Life Threatening Diseases" (Elsevier, USA, 2013) and 'Essentials in Nanoscience and Nanotechnology' (Wiley, USA, 2016). He also served as a visiting research associate of CSIR during 1992-95 and is a recognized supervisor of JNV University, Jodhpur for Ph.D. and Guided 6 students for the degree. He has delivered several invited talks on Conducting Polymers and Nanomaterials in several International and National Conferences/Seminars, Universities in Japan and USA. Recently, delivered 2nd Kalam Memorial Oration "Nanotechnology: Potential Applications in Military and for Rural Development" at DMSRDE, Kanpur.

He is recipient of 'DRDO Technology Award' for his pioneering research work on conducting polymers, DRDO Best Scientist of the Year Award, from the Prime Minister of India, for products based on conducting Polymers and nanomaterials for defence applications, and the national MRSI-ICSC Super Conductivity and Materials Science Annual prize by Materials Research Society of India. Dr. Kumar is the member of a number of scientific societies including the prestigious American Chemical Society and Material Research Society of India.

Sunita Kumbhat, Ph.D., Professor of Chemistry and Dean, Faculty of Science at J.N.V. University, Jodhpur, India, having teaching experience of more than 30 years with specialization in Analytical Chemistry and Nanoscience. Her research fields include electrochemistry, sensor/biosensor, SPR technology and nanomaterials. She has published more than 50 research paper, number of educational films, one patent to her credit and have supervised 15 doctoral students. Dr Kumbhat co-authored a book 'Essentials in Nanoscience and Nanotechnology' (Wiley, USA, 2016). Awards and recognition includes Commonwealth Academic Staff Fellowship (1994-95) at Oxford, UK., UGC National Associate (1997) at BARC, Mumbai and INSA-JSPS Visiting Scientist (2005) at Kyushu University, Fukuoka, Japan. Dr. Kumbhat is associated with various International and National Scientific Societies.

Concise Concepts of
Nanoscience and Nanomaterials

Narendra Kumar
Sunita Kumbhat

Published by
SCIENTIFIC PUBLISHERS (INDIA)

5 A, New Pali Road
P.O. Box 91
Jodhpur - 342 001 INDIA

ISBN: 978-93-86652-22-5

Cover illustration : Carbon nanotube, an engineered nanomaterial as background.
'Lotus Effect' i.e. self cleaning mechanism in lotus leaves, represent nanotechnology in Nature.
Mayan Painting made up of hybrid organic-inorganic nanocomposite of natural clay palygorskite and plant derived indigo dye, represent nanotechnology of ancient civilization.

Visit the Scientific Publishers (India) website at
http://www.scientificpub.com
Printed in India

Preface

'Nano' refers to a size scale as one billionth of a meter (10^{-9} meters) and nanoscience is the study of objects/particles broadly in the size range of 1-100 nm and their unique properties. Nanomaterials and nanotechnology has been always been existed in Nature in various forms both in living and non-living world. The fundamental building blocks of sustaining life, such as DNA, RNA, proteins, lipids, biomembranes are all nanostructured materials; bones, teeth, feathers, shells are beautiful examples of nature's nanocomposite showing extraordinary characteristics. Nanoparticles found in interstellar dust, sediments, minerals, etc. are some example of nanomaterials in non-living species.

Though we have many examples of nanomaterials/ nanocomposites such as pigments, colloidal suspensions, unknowingly created by earlier generations, however the potential of nanoscale was projected by Scientific thinker like Richard Feynman (1959) and Eric Drexler (1980s) followed by National Nanoinitiative (NNI) taken up by US Government in 2001. Landmark discoveries of Scanning Probe Microscopies (Binning *et al.,* 1980s); Fullerenes (Smelly *et al.,* 1985) and CNTs (S. Ijima, 1991) gave impetus to the R & D activities in Nanoscience and Nanotechnology.

Past three decades have witnessed unprecedented growth in terms of synthesis, characterization of a wide variety of nanomaterials showing novel physicochemical and biological properties. Such properties have led to the technological applications of newer nanomaterial in diverse area concern with better health, wealth and security of mankind. Nanotechnology along with biotechnology and information technology have proved to be the cutting edge technologies of twenty first century. This is an area where different disciplines of science and technology converges while breaking their boundaries. The significance of nanoscience and nanomaterials calls for the students, budding scientists and industrialists to remain educated on fundamentals of nanoscience and updates on the recent advances in development of nanomaterials.

Present text book is an attempt to embodied the fundamental aspect of nanoscience and nanomaterials while including the latest developments in the area. Book is divided in seven chapters; first chapter introduces the subject of nanoscience with various terms, definitions, nanoscience in nature, history and its transformation into nanotechnology. Second chapter gives comprehensive description of different categories of nanomaterials including inorganic, carbon based, organic, self-assembled supramolecular structures, nano-composites and nano-coatings. Third chapter of the book deals with the different synthetic strategies to prepare nanomaterials. Characterization of nanomaterials is spread over two chapters with an objective to generate a better understanding of newer developments in the area. Chapter four, deals with the spectroscopic and scattering techniques, whereas chapter five includes microscopic imaging techniques. Chapter six describes various physicochemical unique properties of nanomaterials evolved as a consequence of their nano size. Chapter seven deals with technological impact of nanoscience in diverse areas including consumer goods, health care, energy, information and computer technology; defence and security; environment and pollution control.

We believe that this book will generate and promote the basic understanding on the complex and revolutionary discipline of Nanoscience and Nanotechnology offered as core subjects for UG and PG courses in most of the academic Institution across the World.

Narendra Kumar
Sunita Kumbhat

March, 26, 2018
Jodhpur

Acknowledgement

Authors are thankful to their colleagues from Academic and Research Institutions for their reviews, comments and suggestions in formulating the text of the book. Dr. Kumbhat acknowledges with thanks the constant support of Vice Chancellor and colleagues from Chemistry Department, JNV University and her research scholars while Dr. Kumar to his former colleagues from Defence Laboratory, Jodhpur. Authors are grateful to their family members for the patience and hidden support during the course of writing this book.

Authors would like to thank Mr. Tanay Sharma, Scientific Publishers (India), Jodhpur for taking up this project with keen interest and timely execution to make this publication a reality for Indian and Overseas students.

Narendra Kumar
Sunita Kumbhat

Contents

Introduction to Nanoscience 1

1.1 GENERAL DEFINITIONS

Imagine a world where the unique phenomena at the molecular scale can lead to entirely new, innovative, and transformative product designs, done by harnessing properties of materials at the nanoscale size level. Nanoscale materials though recognized in last half of 20th century are not new to nature or in science. Nanoscience, nanotechnology, nanoparticles and nanostructures have now become common words not only in research publications but also in daily life.

The term **Nanoscience** refers to the study and understanding of the fundamental interactions of physical systems confined to nanoscale (10^{-9} m) dimensions and as a consequence evolving unique chemical, physical, and biological propcrtics. Important propcrtics of matcrials, such as thc chcmical, electrical, optical, thermal and mechanical properties, are determined by the way molecules and atoms assemble on the nanoscale into larger structures. Moreover, in nanometer size structures these properties are often different then on macroscale, because of nano scale, surface and quantum mechanical effects become predominant. Materials formed from nanoscale grains can exhibit dramatic improvement in properties. Nanoscience strives to understand these changes related to the 'characteristic length scale' which is driven intuitively, *e.g.* diameter for a spherical particle and thickness for a thin film. For more complex systems intuition can, however, lead to ambiguous results. It is therefore suggested to use the ratio of total volume, V total interfacial area A of a system as the characteristic length scale, λ $(\lambda = V/A)$. Nanoscience is thus the science of systems with λ in the range of 1 to 100 nm. It is noted that neither 1 nm nor 100 nm is a "bright line". Bucky balls are also included under the category of nanomaterials even though they have a size <1 nm.

Nanotechnology is the application of nanoscience leading to the use of new nanomaterials and nanosize components in useful products. Nanotechnology essentially provide us with the ability to design custom-made materials and products with new enhanced properties such as new nanoelectronics components, new types of "smart" medicines and sensors, and even interfaces between electronics and biological systems. Generally, nanotechnology can be understood as a technology, which allows, in controllable ways, not only to create nanomaterials but also to operate them, *i.e.* to influence them or to use them according to their intended purpose.

Nanomaterials are the materials having at least one dimension in the size regime between 1-100 nm. The term includes all types of nanosize materials, including engineered, incidental nanomaterials and other nanoobjects, like those that exist in nature. The nanomaterials which exist in nature are natural nanomaterials; incidental nanomaterials are those produced as a by-product of combustion, industrial manufacturing, and other human activities **(Fig. 1.1);** and engineered nanomaterials refer to those materials which are synthetically produced in zero-,one-, two-, or three dimensions.

Fig. 1.1. Few sources of incidental nanoparticles

Nanomateriasl must meet at least one of the following criteria:

- Consist of particles with one or more external dimensions in the nano size range
- 1-100 nm for more than 1% of their number.
- Internal or surface structures in one or more dimensions in the size range 1-100 nm.
- Specific surface area to volume ratio >60 m^2/cm^3, excluding materials consisting of particles with a size lower than 1 nm.

One nanometer is a billionth of a metre (1 nm = 10^{-9} m); for comparison few typical examples are given below and illustrated in **Fig. 1.2**.

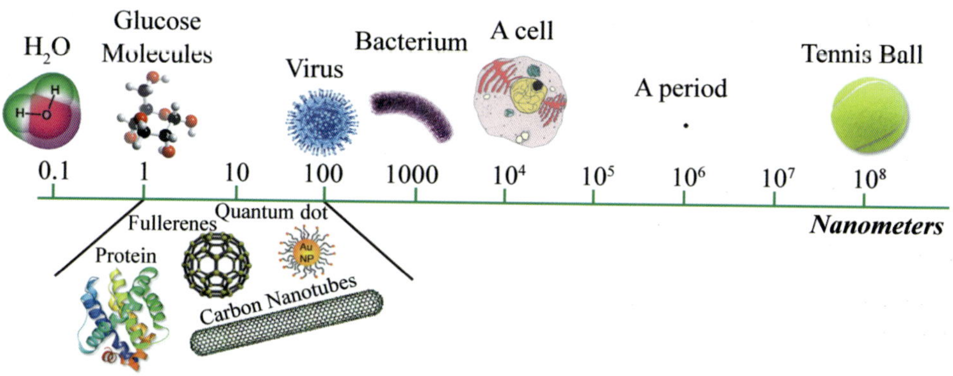

Fig. 1.2. Size comparison of different objects, synthetic nanomaterials and biomolecules

- Atom: ~0.1 nm.
- Atoms in a molecule: ~0.15 nm apart.
- DNA double-helix: ~2 nm in diameter.
- Typical protein: ~10 nm long.
- Typical bacteria: ~200 nm long.
- Virus: 50-60 nm.
- Computer transistor: ~100-200 nm wide.
- Human hair: 50,000–100,000 nm in diameter.
- One sheet of paper: ~100,000 nm thick

Nano-object*:* Material confined in one, two, or three dimensions at the nanoscale are called nano-object. This includes nanoparticles (all three dimensions in the nanoscale), nanofibers (two dimensions in the nanoscale) and nanoplates (one dimension in the nanoscale). Nanofibers are further divided into nanotubes (hollow nanofiber), nanorods (solid nanofiber) and nanowire (electrically conducting or semiconducting nanofiber). However, the nano-object term is not that popular.

Fig. 1.3. Lycurgus cups

Fig. 1.4. Ancient Maya fresco painting

Particle: A minute piece of matter with defined physical boundaries. A particle can move as a unit. This general particle definition applies to nano-objects.

Nanoparticle: Nano-object with all three external dimensions in the nanoscale. Nanoparticles can have amorphous or crystalline form and their surfaces can act as carriers for liquid droplets or gases.

Nanoparticulate matter: Refers to a collection of nanoparticles, emphasizing their collective behavior.

Agglomerate: A group of particles held together by weak forces such as van der Waals forces, some electrostatic forces and or surface tension. It should be noted that agglomerate will normally retain a high surface to volume ratio.

Aggregate: A group of particles held together by strong forces such as those associated with covalent or metallic bonds. It should be noted that an aggregate may retain a high surface to volume ratio.

Nanoscience and nanotechnology is a broad and interdisciplinary area of research and development activity that has grown exponentially, worldwide, during the past three decades. It has the potential for revolutionizing the ways in which materials and products are created and the range and nature of functionalities that can be accessed. It has already made a significant impact which promising super health-wealth, and security for the man kind with assuredly a very bright future. Nanoscience and nanotechnology constitute a new multidisciplinary area where the borders between different scientific disciplines such as physics, chemistry, biology, medicine, computer science and their respective sub- and intermediate disciplines calling nanotechnology in the concept of converging technologies is more appropriate.

1.2 HISTORICAL PERSPECTIVE

It's hard to say when human beings started making incidental nanoparticles, but probably as soon as people started taming fire. One can find nano-scale particles of soot in the embers and smoke. The earliest engineered nanoparticles are often attributed to the ancient Roman and Egyptian civilizations. Damascus steel was used to make blades from about 500 AD in Damascus. These blades are recognized for their resilience, sharpness, extreme strength and the aesthetic appeal of their unique surface pattern. Many researchers tried to reveal reason behind the special properties of Damascus steel and what they found was the presence of multi-walled carbon nanotube in the material. The Glass Lycurgus Cup is a Roman period cup, *i.e.*, fourth century AD, preserved in the British Museum. It appears green when observed in reflected light, for instance in daylight. However, it appears red when light is propagated through it **(Fig. 1.3)**.

Modern analytical methods have disclosed that the Glass Lycurgus Cup comprises silver and gold nanoparticles in the ratio of roughly 7:3, with a size of 70 nm diameter. The colors of Mayan paintings of 5th century AD **(Fig. 1.4)** also stem from the presence of metallic nanoparticles. The origin of Mayan color and its resistance to acids and bio-corrosion are still not understood but studies of authentic samples from Jaina Island show that the material is made of needle-shaped palygorskite (clay) crystals that form a super lattice with a period of 1.4 nm, with intercalates of amorphous silicate substrate, containing inclusions of magnesium metal nanoparticles.

The stained-glass windows of the great medieval cathedrals also contain metallic nanoparticles. Thousands of years BC ago, people knew and used natural fabrics such as flex, cotton, wool, silk, and processed them into products. What makes these fabrics so special that they developed a network of pores of size 1-20 nm, this farming of typical nanoporous materials. Due to their nanoporous structure, natural fabric possesses high utilitarian properties of absorbing sweat, quickly swell and get dried soon. Indian and Chinese art history is also filled with examples of nanotechnology. Photography, which was developed in the 18th and 19th Centuries, provides a more recent example of the use of silver nanoparticles. Nanostructured catalysts have also been investigated for over 70 years. By the early 1940's, precipitated and fumed silica nanoparticles were being manufactured and sold in the USA and Germany as substitutes for ultrafine carbon black for rubber reinforcements.

In 1857, Faraday had described the use of colloidal gold in his experiments and in his lecture at Royal Society, presented a purple color solution, stating that it contained reduced gold in exceedingly small particles. Faraday postulated correctly about the physical state of colloids; he also described how a gold colloid would change color (turning blue) on adding salt. Since then, colloidal science has evolved a lot. In the early 20th Century Gustav Mie presented the Mie theory, which is a mathematical treatment of light scattering that describe the relationship between metal colloid size and optical properties of solutions containing them.

Nobel Laureate Richard Feynman of Californian Institute of Technology dreamt of manipulating and controlling things on a small scale. Professor Richard Feynman in his now very famous speech *There's plenty of room at the bottom,* delivered to the American Physical Society in 1959, said that why not mimic nature and produce smaller and smaller functional materials, which will be highly efficient. He estimated that it should be possible to reduce the printing on a page by a factor of 25000 and if that were done, all the books ever printed could be copied in a space of about 35 ordinary pages by moving individual atoms around. He began speculating on the possibilities and potential of nanometric materials and on the fact that the manipulation of individual atoms could allow us to create very small structures whose properties would be very unique and different from larger structures with the same composition. Specifically, he suggested that as the scale got smaller and smaller, gravity would become more negligible, while Van Der Waals attraction and surface tension would become very important. In the end, Feynman's talk has been viewed as the first authentic talk that dealt with a main tenet of nanotechnology, the direct manipulation of individual atoms (molecular manufacturing) (Box 1). Although, he never explicitly mentioned "nanotechnology", however, he is now considered to be the father of nanotechnology.

Richard Feynman – speculated about atoms as quoted below in his own words…..

"Rearranging the atoms…

But I am not afraid to consider the final question as to whether, ultimately---in the great future---we can arrange the atoms the way we want; the very *atoms*, all the way down! What would happen if we could arrange the atoms one by one the way we want them (within reason, of course; you can't put them so that they are chemically unstable, for example).

Up to now, we have been content to dig in the ground to find minerals. We heat them and we do things on a large scale with them, and we hope to get a pure substance with just so much impurity, and so on. But we must always accept some atomic arrangement that nature gives us. We haven't got anything, say, with a "checkerboard" arrangement, with the impurity atoms exactly arranged 1000 angstroms apart, or in some other particular pattern.

What could we do with layered structures with just the right layers? What would the properties of materials be if we could really arrange the atoms the way we want them? They would be very interesting to investigate theoretically. I can't see exactly what would happen, but I can hardly doubt that when we have some *control* of the arrangement of things on a small scale we will get an enormously greater range of possible properties that substances can have, and of different things that we can do."

Box 1

The evolution of integrated chips and their subsequent miniaturization may also be viewed as a part of history of nanotechnology. The first transistor invented in 1947 was a bulky, macro object. To keep with the demand for miniaturization, the dimensions of the transistor have been reduced considerably in the last 70 years. In the year 2002, the nanosize was reached with the achieved size of a single transistor as 90 nm. As on today, a single transistor in an Intel Core 2 Quad processor is 45 nm. In order to keep pace with Moore's law, transistors should be as small as 9 nm by 2020. However, this dimension is below the fabrication capabilities of last generation tools used in the microelectronic industry. Numerous novel approaches such as quantum computing and molecular engineering are under investigation to achieve the workable transistor of this size.

The first use of the term "Nanotechnology" was by Norio Taniguchi in 1974 at the International Conference on Precision Engineering (ICPE). His definition referred to "production technology to get extra high accuracy and ultra fine dimensions, *i.e.* the preciseness and fineness on the order of 1 nm (nanometer), 10^{-9} m, in length". The development of nanotechnology has been enabled by the invention of two analytical tools that have revolutionized the imaging (and manipulation) of surfaces at the nanoscale. These are, the Scanning Tunneling

Micrroscope (STM) and the Atomic Force Microscope (AFM). The AFM and STM are capable of imaging surfaces at atomic resolution. Both the instruments were invented by Binning and his co-workers at IBM Zurich. Invention of these versatile tools, practically opened the doors of nanoworld to the scientists. With the advent of the STM scientist were given the tool not only to image surfaces with atomic resolution, but also to move indivisual atoms. The STM is the first step in realizing Feynman's vision of atom-by-atom fabrication.

In the 1980s the basic idea nanotechnology was explored in much more depth by Dr. Eric Drexler, who promoted the technological significance of nanoscale phenomena and devices through speeches and the books, *Engines of Creation: The Coming Era of Nanotechnology* (1986) and *Nanosystem: Molecular Machinery, Manufacturing, and Computation* and so the term acquired its current sense. These are assemblers that can be programmed to construct any stable atomic structure (Box 2).

Excerpt from Drexler's book ***Engines of creation***.........

Universal assemblers

These second-generation nanomachines - built of more than just proteins - will do all that proteins can do, and more. In particular, some will serve as improved devices for assembling molecular structures. Able to tolerate acid or vacuum, freezing or baking, depending on design, enzyme-like second-generation machines will be able to use as "tools" almost any of the reactive molecules used by chemists - but they will wield them with the precision of programmed machines. They will be able to bond atoms together in virtually any stable pattern, adding a few at a time to the surface of a workpiece until a complex structure is complete. Think of such nanomachines as assemblers.

Box 2

Birth of 'Cluster Science' in late 1980's followed by discoveries of new allotropes of carbon *i.e.* fullerene in 1986, CNT in 1991, and graphene in 2004 were another milestones in the development of nanoscience and nanotechnology. In another development, the studies on the synthesis and properties of metallic and semiconductor nanocrystals led to a fast increasing number of metal and metal oxide nanoparticles and quantum dots. Initiation of National Nanotechnology Initiative (NNI) programme by the United States and quick followup by European states, Japan, China, and India catalysed the research and development in this area. The milestones related to the evolution of nanoscience and nanotechnology, from pre-historic to modern era are given in **Table 1.1**.

Table 1.1. Milestones associated with evolution of Nanoscience and Nanotechnology

Year	Milestone
Origin of life	Cell: A magnificent nano machine
400 BC	Democritus of Abdera gave reasoning about atoms and matter
400 AD	Coloured Lycurgus Cups
500 AD	Damascus steel, Glaze artisan in Mesopotamia, Mayan Paintings
1857	M. Faraday synthesized nanosize gold colloids
1915	W. Ostwald, a famous chemist, wrote a book *World of Neglected Dimensions* (in German)
1931	E. Ruska and Max Knoll developed Electron microscope
1951	E. Müller developed the Field of Ion Microscope which enabled the imaging of atoms from the tip of metallic samples
1959	R. Feynman delivered his now very famous talk "There is Plenty of Room at the Bottom" pointing out to the scientists that reduced dimensionality of materials would create fascinating materials
1968	A.Y. Cho and J. Arthur developed Molecular Beam Epitaxy technique for layer by layer growth of materials
1970	L. Esaki demonstrated the quantum size effect in semiconductors
1974	N. Taniguchi used the term "nano-technology" for fabrication methods below 1μm.
1977	E. Drexler gave the concept of molecular nanotechnology
1981	G. Binnig and H. Rohrer developed the scanning tunnelling microscope (STM) capable of atomic size resolution. followed by a family of scanning probe microscopes. Received Nobel Prize of Physics in 1986
1985	R.F. Curl, H.W. Kroto and R.F. Smalley synthesized sixty atom carbon molecule, later named as 'Fullerene'. Received Nobel prize for Chemistry in 1996.
1986	i. G. Binning, C.F. Quate and Ch Gerber developed Atomic Force Microscopy (AFM). ii. Eric Drexler wrote the first book on nanotechnology: *Engines of Creation: The Coming Era of Nanotechnology* .
1989	D.M. Eigler wrote letters 'IBM' using xenon atoms
1991	S. Iijima discovered 'carbon nanotubes'
1998	C. Dekker, S. J. Tans and A. R.M. Verschueren developed the single carbon nanotube transistor
1999	C.A. Mirkin developed 'Dip Pen Lithography'
2000	D.M. Eigler devised 'Quantum Mirage' using Fe atoms on copper substrate. ii. S. Hell developed Stimulated Emission Depletion (STED)
2001	Intel Corporation surpassed Moor's Law by producing the world's fastest silicon transistorwhich switches on and off 1.5 trillion times per second.

Year	Milestone
2004	A. Giem and K. Novoselov discovered Graphene. Received Nobel Prize in 2010.
2004	Intel Corporation launched Pentium IV 'PRESCOFT' processor based on 90 nm technology
2006	E. Betzig and W. Moerner discovered Single-Molecule Microscopy (SMM). Based on STED and SMM techniques, super-resolved fluorescence microscopy has emerged to study synapses in Alzheimer's and Huntington's disease, and to gain a better understanding of protein development in embryos, leading to the Nobel Prize in Chemistry in 2014 jointly by E. Betzig, S, W. Hell and W. E. Moerner.

1.3 NANOSCIENCE IN NATURE

Mankind has always found inspiration from Mother Nature. Today, developing technologies allow us to probe and better understand many materials and processes related to nanoscience in nature. A closer look around us reveals the existence of innumerable nanomaterials and nanotechnology both in non-living and living matters and their impact in our lives including sustenance. Basically, understanding even natural phenomenon around us can lead to new discoveries and progress in science and technology.

1.3.1 Nanomaterials in living species

In the living species including human beings, the sizes of building blocks such as proteins, DNA, lipids, membranes having nanopores together with body's structural components such as bones and teeth all belong to nano regimes. Moreover, cells in living species performing various functions such as harnessing and storing energy, growth of the body, memory, sensing, movements, etc. are the most wonderful example of a nanomachine which is impossible to mimic so far. Biological systems have built upon inorganic-organic nanocomposite structures to improve the mechanical, optical, magnetic and chemical sensing properties in living species. Bones and teeth of varying degrees of mechanical characteristics, depending on the nature of the living species, are examples of nature's inorganic-organic nanocomposites. Nacre (mother-of-pearl) from the mollusc shell is a biologically formed lamellar ceramic, which exhibits structural robustness despite the brittle nature of its constituents. These systems have evolved and been optimized over millions of years into sophisticated and complex nanostructures. Some of the examples of nanotechnology driven sensors in living species are shown in **Fig. 1.5** which includes frog's inner ear, nose of a silk moth and an ant's eye. The inner ear of a frog has nanomechanical cantilevers that measure deflections as small as 3 nm due to sound. The capability of a male silk moth which can detect, with single-molecule precision, the pheromones of a female moth emitted even from a distance of 3 km. The eye of an ant and flies uses nanoscale features to greatly enhance their visual sensitivity.

Fig. 1.5. Nature's Nanotechnology: a. frog's inner ear; b. silk moth; c. eye of an housefly

Iridescent Colours of a Bird's feathers

The beautiful wings of a butterfly are because of nanoparticles that act as a photonic crystal. The glittering colors of a peacock feathers **(Fig. 1.6a)** arises as a consequence of ~ 500nm long barbules attached to each side of barbs. Electron microscopy reveals a highly ordered structure of melanin rods of high refractive index embedded in keratin of lower refractive index with an air tube between each square of melamine rod **(Fig. 1.6b,c)**. The whole array of melanin rods, keratin matrix and air holes acts like a photonic crystals to show vivid colurs.

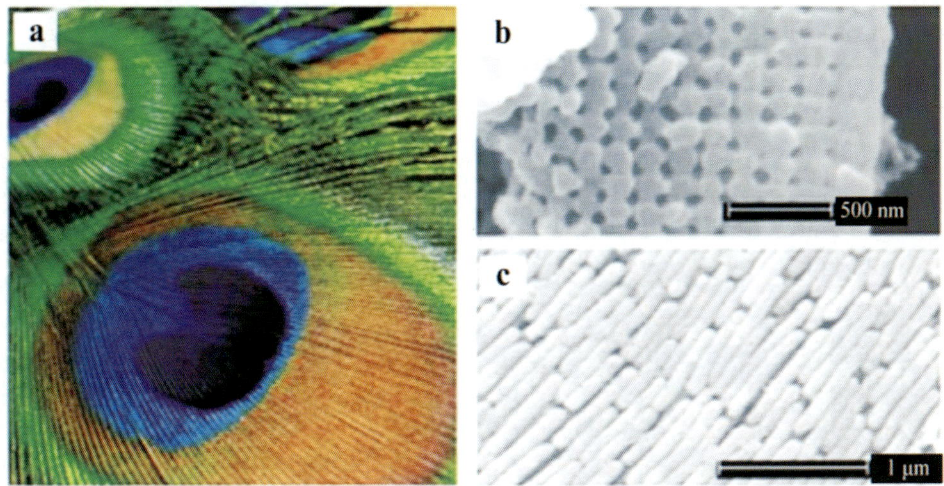

Fig. 1.6. (a) Glittering colors in barbules of peacock feather; (b, c) Electron micrograph of transverse and longitudinal sections of barbules embedded in keratin matrix showing photonic effect

Gecko Effect

Geckos are one of the few species of the lizard family in the animal kingdom that are known for sticky toes that allow them to climb up walls, even hang upside down on ceiling due to sticky toes. An interesting phenomenon observed in nature over centuries is so-called 'Gecko effect' or crawling of a lizard on a wall. How does a lizard balance its weight? Does it secrete any fluid to stick or some other phenomenon? It is the peculiar construction of the tips of its feet on which millions of nano hair about 200 nanometers in width and length exist on the sole of a gecko's toes, which help lizards in crawling **(Fig. 1.7).**

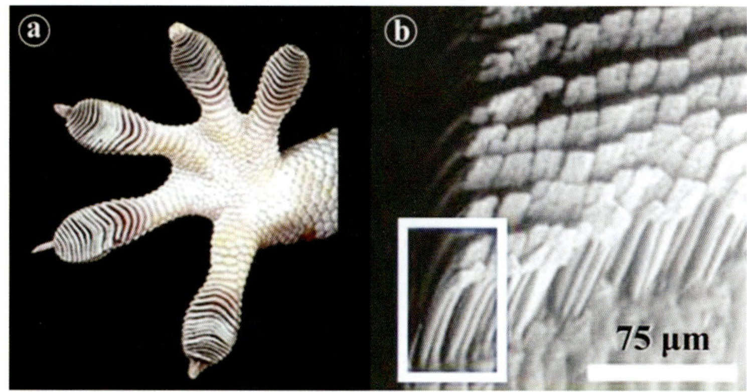

Fig. 1.7. Gecko's adhesive system: a. Ventral view of sole of the foot; b. Arrays of adhesive setae

Each hair strand exerts just weak Van der Waals force between the wall and itself is. However, due to millions of hairs total force is sufficient to balance its weight and remove the force in controlled manner to crawl. Today, this property is being exploited as **Geckos Technology** in producing super glues. This effect may be useful to make robots which would climb walls or do some scientific operations without manual aid. Some simple application of Gecko effect would be a stick tape, seals without nuts and bolts and so on.

Fig. 1.8. (a) Non-wettable lotus (*Nelumbo nucifera*) plant leaf; (b) Electron micrograph of convex papillae on the surface of lotus leaves coated with wax; (c) self-cleaning: a droplet picks up the dirt particles as it rolls off the leaf's surface

Lotus effect: Self-Cleaning

The water repellence of lotus had long been recognized, its scientific basis was understood only in 1997 when two botanists Wilhelm Berthelot and Christophe Neinhuis, examined leaf surfaces of lotus using a scanning electron microscope which resolves structures as small as 1–20 nm. **Figure 1.8** shows a non-wettable lotus plant leaf where a water droplet picks up the dirt particles as it rolls off the leaf's surface, showing self-cleaning process. The phenomenon of non-wettability in lotus leaf is also common in many other plant leaves, butterfly wings, skin of fish and many others in plant and animal kingdom.

Lotus effect can be understood in terms of hydrophobicity and the contact angle a water drop makes with the surface. Contact angle depends upon surface roughness, surface condition and surface material. Interaction between liquid and solid surface is very important. When the contact angle is $<20^o$ it is super-hydrophilic; $<90^o$ the surface is hydrophilic; $>90^o$ it would be hydrophobic and $>150^o$ then it is superhydrophobic. Super hydrophobicity in lotus leaf arises due to the convex papillae on the surface of leaves coated with wax crystals of nanoscale dimension: ~10 to ~100 nm. The papilla greatly reduces the contact area of water droplets with it. Every epidermal cell forms a micrometer-scale papilla and has a dense layer of epicuticular wax superimposed on it. Each of the papillae consists of branch like nanostructures on the surface for example of the lotus leaves, the almost spherical water droplets will not come to rest and simply roll off if the surface is tilted even slightly, which is now usually referred to as the "Lotus Effect". The self-cleaning effects of the surfaces of the lotus flower have been attributed to the combined micro-and nanostructure which in combination with hydrophobic groups give the surface a water and dirt repellent behavior. In the past few years, numerous companies have realized products such as paint, glass surface and ceramic tiles with dirt repellent properties resembling the surface morphology and chemistry of the lotus flower.

Some other examples of naturally occurring materials such as cotton, spider's silk and opals are also worth mentioning for their Nano features and unique properties. Cotton, have nano scale arrangement of cellulose fibers showing high strength, durability and absorbance. Spider silk showing five time higher strength than that of steel has natural supramolecular organization of fibroin at nano scale. Precious stone opal consists of spheres of silicon dioxide (150 to 300 nm diameter) in a hexagonal or cubic close-packed lattice. These ordered silica spheres produce the internal colors by causing the interference and diffraction of light passing through the microstructure of the opal. The realization that nature can provide the model for improved engineering has created a research field called "bio-mimicking" or bio-inspired material science. It has been possible to process several types of nanostructures inspired from biological nanomaterials **Table 1.2.** Despite all of the phenomenal and technological advancements in nanoscale synthesis and characterization, we are still far away to build nanotechnology-based devices that are quite common in nature.

Table 1.2. Bio-inspired unique properties

Natural system/materials	Bio-inspired properties
Sub-structure of nacre	Low density, high strength composites
Spider silk	High tensile strength fiber
Wood, ligaments and bone	High strength structural material
Eels and nervous system	Electrical conduction
Deep-sea fish and glow worms	Photoemission
Butterfly and bird wings	Photonic crystals
Moth eye	Antireflective
Lotus leaf, human skin, fish scales	Hydrophobic surfaces, self-cleaning
Shark skin	Drag reducing
Gecko's feet	Adhesion
Human brain	Artificial intelligence and computing

1.3.2 Nanomaterials in non-living species

Naturally occurring nanomaterials may originate from one of the following non-living sources:

(i) Natural erosion and volcanic activity

Nanoparticles are part of the mineral world since they are naturally produced from erosion and volcanic explosions.

(ii) Clays

Minerals such as clays are a type of layered nanostructured silicate materials that are characterized by a fine 2D crystal structure. Mica, one among them is the most studied. In mica, the large number of silicate sheets is held together by relatively strong bonds. On the other hand, montmorillonite-a smectic type of clay, have relatively weak bonds between layers. Each layer consists of two sheets are silica held together by cations such as Li^+, Na^+, K^+ and Ca^{2+}. The presence of the cations is necessary for compensating the overall negative charge of the single layers. The layers are 20-200 nm in diameter laterally and come into aggregates called tactoids which can be about 1 nm or more thick. The fine nanostructure of clays determines their properties. As an example, a nanostructured clay swells several times of the original volume when water is added to it, due the opening of the layered structure by the water molecules that replaces the cations. Clay swelling is a significant factor in soil stability and is taken into account in constructing roads etc.

(iii) Natural colloids

Naturally occurring liquid colloids, such as milk, blood, aerosols, *e.g.*, fog; gels *e.g.*, gelatin are some of the examples of natural colloids. In these materials the nanoparticles are dispersed in the medium (liquid or gas) but do not form a solution. All these materials have the characteristic of scattering light and often their color (such in the case of milk and blood) are due to the scattering of light by the nanoparticles that makes them up.

(iv) Minerals

Many of the natural materials such as shells, corals and bones are formed by the self-assembly of calcium carbonate crystals with other natural materials, such as polymers, to form fascinating three-dimensional architectures. For instance, a shell is grown layer by layer coating of protein supported by chitin, a polysaccharide polymer. The proteins act like a nano-assembly mechanism to control the growth of calcium carbonate crystals. Around each crystal remains a honeycomb like matrix of protein and chitin. This relatively "flexible envelop" is fundamental for the mechanical properties of the shell and mitigate cracking. The size of each crystal is around 100 nm. As a result the mollusc's shells have extraordinary physical properties viz. strength and resistance to compression.

1.4 SIGNIFICANT ASPECTS OF NANOSCIENCE

We have learned now that nanoparticles are actually fairly common in many different environments that remained undetected for reasons of non -availability of detection tools as well as ignorance about their significance. In fact, many physical, chemical, and biological processes (both human activities and natural processes) produce nanoparticles. Naturally occurring nanoparticles can be found in volcanic ash, ocean spray, fine sand and dust, and even biological matter (*e.g.* viruses). Synthetic nanoparticles are equally, diverse if not more than their naturally occurring counterparts. Nanoscience being multidisciplinary in nature is claimed by physics, chemistry and biology, all the three important branches of basic sciences, for its ownership.

1.4.1 Nanoscience in physics, chemistry and biology

i. **Physics** is the mother of the natural sciences. In principle, physics can be used to explain everything that goes on at the nanoscale. It has come up to explain the unique optical, electronic, electrical, magnetic properties of nanomaterials based on the concepts such as quantum confinement, density of states, etc. There is active physics research going on in nano-mechanics, quantum computation, quantum teleportation, and artificial atoms. While physics can explain everything, sometimes it is more convenient to think of nanostructures in terms of chemistry where the molecular interactions are described in terms of bonds and electron affinities.

ii. **Chemistry** is the study of molecules and their reactions with each other. Since molecules typically have dimensions of a few nanometers, almost all of nanoscience can be reduced to chemistry. Chemistry research in nanotechnology is concerned with almost all types of materials, self-assembly, and supramolecular chemistry. Chemists have been making polymers, which are large molecules made up of nanoscale subunits, for many decades. The extraordinary catalytic and other chemical reactivity of nano materials find its explanation in availability of larger fraction of atoms at the surface characterized by higher surface energy and unsaturated bonding in nanomaterials as compared to their bulk counterparts. Sometimes the chemical description of a nanostructure is insufficient to describe its function. For instance, a virus can be described best in terms of biology.

iii. **Biology** is sometimes described as nanotechnology that works. Biological systems contain small and efficient motors. There are more than 50 kinds of motors found in cells. Biological systems produce impressive control systems. The brain of a bee is tiny and consumes little power yet regulates complex flying a behavioural patterns. A cell one micron in size can store 1 Gbit of information in DNA. They self- reproduce. They construct tough and strong material. Biology is an important source for inspiration in nanotechnology. Copying engineering principles from biology and applying it to create new materials and technologies is called biomimetic which presently is a very active research area.

1.4.2 Nanomaterials

Materials are some thing what the world is made of. Materials are the most wonderful gift of mother nature to the mankind. They are so important in the development of civilization that we associate Ages with them *i.e.* Stone Age, Bronze Age (3000BC-1200BC), Iron Age (1200BC-1800AD), Polymer ge (1800AD…), etc. The combination of physics, chemistry, and the focus on the relationship between the properties of a material and its microstructure is the domain of Materials Science which has allowed designing of customized materials and provided a knowledge base for the engineering applications. Materials, in general comprises large numbers of atoms and molecules and have properties determined by complex, heterogeneous structures. Based on the interatomic bonding, materials are classified into the categories of metals, semicondctors, ceramics and polymers. A composites is made of different materials in intimate contact (example: fiberglass, concrete, wood) to achieve specific properties. Biomaterials can be any type of material that is biocompatible and used, for instance, to replace human body parts.

The bulk properties of any material are merely the average of all the quantum forces affecting all the atoms. As a material is broken into smaller and smaller particles brought down in nanosize a point closer to the sizes of an atom or molecule, known as nanomaterial, the averaging no longer works. The bulk properties of material change drastically. Composites made from particles of nano-size ceramics or metals smaller than 100 nanometers can suddenly become much stronger than predicted by existing materials-science models. For example, metals with a so-called grain size of around 10 nanometers are as much as seven times harder and tougher than their ordinary counterparts with grain sizes in the hundreds of nanometers. The causes of these drastic changes stem from the weird world of quantum physics.

Nanomaterials are very important and interesting to generate understanding about the origin of their unique properties as well to reailze their applications in diverse areas.They are also intrinsically complicated. Nanomaterials can be nanoscale in zero dimension (fullerene, QDs), one dimension (*e.g.*, surface films), two dimensions (*e.g.*, strands or fibres), or three dimensions (*e.g.*, precipitates, colloids). They can exist in single, fused, aggregated or agglomerated forms with spherical, tubular, and irregular shapes.

If one dimension of the three dimensional nanostructure is at nano scale, then it is called a Quantum Well. If two dimensions of the three dimensional nanostructure is at nano scale, then it is called a Quantum Wire. If all the three dimensions of the nanostructure is at nano scale, then it is called a Quantum Dot. The idea behind the new nano materials focuses on four key questions:

i. Why are nanostructures interesting?

ii. How can they be synthesized?

iii. How can materials of different nature be made in nano size?

iv. How can the relationships between their structures and compositions, matrices and interfaces control the properties of the materials that incorporate them (in composites)?

The last question related to "materials by design" has been a goal of material science since its inception. All conventional materials like metals, semiconductors, glass, ceramics or polymers can in principle be obtained with a nanoscale dimension. The spectrum of nanomaterials ranges from inorganic to organic, crystalline to amorphous particles, which can be found as single particles, aggregates, powders or dispersed in a matrix, over colloids, suspensions and emulsion, nanolayers and nanofilms. Supramolecular structures, such as dendrimers, micelles, DNA, or liposomes also belong to the category of nanomaterials.

Nanoparticles are solid particles, consist of few tens of atoms or molecules which is closer to the size of an individual atoms/molecules rather than the macroscopic objects. Nanoparticles can be generated in a variety of sizes and morphologies. Some important morphologies of nanomaterials include:

Nanowires and Nanotubes. Linear nanostructures such as nanowires, nanotubes or nanorods can be generated from different material classes *e.g.* metals, semiconductors or carbon by means of several production techniques. Carbon nanotubes are one of the most promising linear nanostructures which can occur in a variety of modifications (*e.g.* single-, or multiwalled, filled or surface modified). At present carbon nanotubes can be produced by CVD methods on a several tons per year scale. Nanowires are ultrafine wires or linear arrays of dots, formed by self-assembly. They can be made from a wide range of materials. Semiconductor nanowires made of silicon, gallium nitride and indium phosphide have demonstrated remarkable optical, electronic and magnetic characteristics.

Nanolayers/Nanocoatings. Nanolayers are one of the most important topics within the range of nanotechnology. Through nanoscale engineering of surfaces and layers a vast range of functionalities and new physical effects (*e.g.* magneto electronic or optical) can be achieved. Furthermore, a nanoscale design of surfaces and layers is often necessary to optimize the interfaces between different material classes (*e.g.* semiconductor compound on silicon wafers) and to obtain the desired special properties.

Nanoporous materials. Materials with defined pore sizes in the nanometer range are of special interest for a broad range of commercial applications because of their outstanding properties with regard to thermal insulation, controllable material separation and release and their applicability as templates or fillers for chemistry and catalysis. One example of nanoporous material is aerogel, which is produced by sol-gel chemistry. A broad range of potential application of these materials include catalysis, thermal insulation, electrode materials, environmental filters and membranes as well as controlled release drug carriers.

1.4.3 Synthetic and Characterization Strategies of Nanomaterials

Today's scientists and engineers are finding a wide variety of ways to deliberately make materials at nanoscale to take advantage of their enhanced properties such as higher strength, lighter weight, increased control of light spectrum, and greater chemical reactivity than their larger-scale counterparts. Different approaches used in the synthesis of nanomaterials and nanodevices can accommodate solid, liquid, and/or gaseous precursor materials. In general, most of these techniques can be classified as top-down and bottom-up approaches and strategies that have elements of both. Top-down methods to produce nanoparticles from bulk materials include high-energy ball milling, mechanochemical processing (MCP), electro-explosion, sputtering, and laser ablation. The popular techniques of bottom-up approaches include chemical vapour deposition (CVD), chemical vapour condensation (CVC), plasma or flame spraying synthesis, laser pyrolysis, precipitation, reduction, sol-gel, solvothermal, matrix-mediated (template-assisted) processing, self-assembly, and so on. These chemical processes rely on the availability of metal-organic molecules as precursors. Sol-gel processing differs from other chemical processes because of its relatively low processing temperature. The main focus in both the approaches is the ability to control: particle size, particle shape, size distribution, particle composition and degree of particle agglomeration. The top-down approach to nanofabrication has the advantage that almost any pre-determined structure can be produced.

Much attention however is now focused on processes that involve some degree of molecular self-assembly. Nanoparticles of a wide range of materials - including a variety of organic and biological compounds, but also inorganic oxides, metals, and semiconductors - can be processed using chemical self-assembly techniques. These techniques exploit selective attachment of molecules to specific surfaces, biomolecular recognition and self-ordering principles (*e.g.* the preferential docking of DNA strands with complementary base pairs) as well as well-developed chemistry for attaching molecules onto clusters and substrates (*e.g.* thiol (-SH) end groups) and other techniques like reverse micelle, sonochemical, and photochemical synthesis to realise 1-D, 2-D and 3-D self-assembled nanostructures. The molecular building blocks act as parts of a jigsaw puzzle that join together in a perfect order without an obvious driving force present. In this respect biological materials have remarkable advantages over inorganic materials in the diversity of self-assembled structures that they can produce. Evolution in the natural world has produced an astonishing variety of biomolecular devices, and compared with conventional technologies, many natural molecular devices display enormous functionality. Among the most outstanding examples of synthetic structures now being fabricated are DNA based geometrical structures (including artificial crystals) and functioning DNA-based nanomachines.

Once scientists had the right tools such as the scanning tunneling microscope (STM), atomic force microscope (AFM), the age of nanotechnology was born as one could analyze and authenticate the credibility of a material, whether natural, to be nano. Beside these two techniques almost all other spectroscopic, scattering, microscopic, and analytical techniques, summarized in **Table 1.3**, are used now to determine structural, surface (morphological) and functional characteristics of nanomaterials.

Briefly, common techniques/tools utilized to evaluate a particle's morphological properties include transmission electron microscopy (TEM), scanning electron microscopy (SEM), scanning tunnelling microscopy (STM), atomic force microscopy (AFM) and dynamic light scattering (DLS). X-ray photoelectron spectroscopy (XPS), X-ray diffraction (XRD), UV-visible spectroscopy, Fourier transform infrared spectroscopy (FTIR) and Raman spectroscopy are used to analyse particle's bonding, structural and compositional characteristics. The separation and quantification of nanoparticles with different particle sizes could be processed using High Performance Liquid Chromatography (HPLC) with Size Exclusion Chromatography. A combination of tools including X-ray photoelectron spectroscopy (XPS), transmission electron microscopy (TEM), and

X-ray diffraction (XRD) and several other methods are being routinely used to obtain essential information.

Table 1.3. Techniques for Nanoparticles Characterization

Equipment/technique	Characterization parameters
Dynamic Light Scattering (DLS)/ Particle Size Analyser	Particle size and size distribution of nanoparticles in liquid solutions or suspensions
Zeta Potential Analyser	Surface charge of nanoparticles in aqueous solutions or suspensions
Scanning Electron Microscope (SEM)	Shape of nanoparticles and consistency of nanoparticles surface; information on surface structure
Transmission Electron Microscope (TEM)	Surface property; size and shape; morphology of nanoparticles; proteins (enzymes), viruses at molecular level; Information on internal structure by bright field and dark field image
Atomic Force Microscope (AFM)	Shape and surface morphology (including friction and softness) of nanoparticles with high lateral and vertical resolutions
Scanning Tunnelling Microscopy (STM)	Surface images with atomic scale lateral resolution; modification of material at atomic /molecule/nano scale with high precision
Laser Scanning Confocal Microscope (LSCM)	Non-invasive technique to, investigate the migration of nanoparticles into bio-barrier; provide 3D morphology of nanoparticles
Surface Area Analysers and Pore Size Analyser (BET)	Surface-area analysis, porosity and adsorption capability
X-ray diffraction (XRD)	Crystal structure
X-Ray Photoelectron Spectroscope (XPS, ESCA)	chemical composition (both elemental and chemical state) information on nanoparticle's surface
Fourier Transform Infrared Spectroscope (FTIR)	Assisted analytical tool for chemical composition
Differential Scanning Calorimetry (DSC)	Thermal analysis (and component interactions) of nanoparticles and related materials during fabrication process
High Performance Liquid Chromatography (HPLC)	Detection, separation and quantification of nanoparticles/nanomaterials with different particle size

1.5 UNIQUE CHARACTERISTICS OF NANOMATERIALS

Properties are the way the material responds to the environment. For instance, the *mechanical, electrical* and *magnetic* properties are the responses to mechanical, electrical and magnetic forces, respectively. Other important properties are *thermal* (transmission of heat, heat capacity), *optical* (absorption, transmission and scattering of light), chemical reactivity, and the *chemical stability* in contact with the environment (like corrosion resistance).

We have defined nanomaterials as small objects that are sized between 1 and 100 nanometers. It turns out there are some very interesting things about the nanoscale in that substances at this scale behave very differently as compared to their counterparts with larger particle size. The interest in nanostructured materials arises from the fact that due to the small size of the building blocks and the high density of interfaces (surfaces, grain and phase boundaries) and other defects such as pores, new physical and chemical effects are expected or known properties can be improved substantially. The physical and chemical properties of nanostructured materials (such as optical absorption and fluorescence, melting point, catalytic activity, magnetism, electric and thermal conductivity, etc.) differ significantly from the corresponding coarser bulk material.

For example, the metal copper is transparent at the nanoscale while gold, which is normally unreactive, becomes chemically very active. Gold and silicon which are yellow and grey appear red when in nanosize regime. Gold nanoparticles melt at much lower temperatures (~300°C for 2.5 nm size) than the gold slabs (1064°C). Absorption of solar radiation in photovoltaic cells is much higher in nanoparticles than it is in thin films of continuous sheets of bulk material - since the particles are smaller, they absorb greater amount of solar radiation. Carbon, which is quite soft in its normally occurring form (graphite), becomes incredibly hard when it's tightly packed into an arrangement called a nanotube. In other words, materials can have different chemical and physical properties on the nanoscale even though they are still the same materials. On the nanoscale, it's easier for atoms and molecules to move around and between one another, so the chemical properties of materials can also be different. The interesting, and sometimes unexpected, properties of the nanomaterials are considered to arise because of reduced size (few tens of atoms or molecules), not very far from those of single atoms or a molecule. For comparison, bulk materials with particle size more than one micrometre don't exhibit size dependent changes in their chemical and physical properties. Further, the underlying reasons for the occurrence of novel physical, chemical, mechanical and to an extent biological properties in materials at nanoscale are the manifestation of quantum effects and high surface area-to-volume ratio or a combination of the two.

1. Quantum effects

Quantum effects can begin to dominate the behaviour of matter at the nanoscale - particularly at the lower end - affecting the optical, electrical and magnetic behaviour of materials. Materials can be produced that are nanoscale in zero dimension, one dimension (for example, very thin surface coatings), in two dimensions (for example, nanowires and nanotubes) or in all three dimensions (for example, nanoparticles). In quantum size effect where the normal bulk electronic structure is replaced by a series of discrete electronic levels leading to:

- The transition between molecular and metallic states providing specific electronic structure (local density of states LDOS);
- Plasmon excitation;
- Quantum confinement;
- Short- range ordering; etc.

An example of quantum confinement is the change in the optical properties, in the photoemission of many semiconductor materials as they go nano (see Fig. 6.13). Optical properties which may be considered uninteresting, simply by changing its size to the nanoscale show varying bright colors as a function of size within the nanoregime.

The size dependent color of the nanoparticles was utilized, though without any intention, by artists as far back as the 9th century for generating glittering effects on surfaces of quartz or colors in stained glass. The variation in the size, shape, or type of the nano particles have dramatic effect on their colors. For example, spherical gold nano particles with diameter of 25 nanometers look green, and gold nanoparticles with 100 nanometer in diameter look orange.

2. High surface area-to-volume ratio

Nano-materials have far larger surface areas than similar masses of larger-scale materials. As surface area per mass of a material increases, a greater amount of the material can come into contact with surrounding materials. This make materials more chemically reactive (in some cases materials that are inert in their larger form are reactive when produced in their nanoscale form), and affect their mechanical strength or electrical properties and lower the melting points. All such changes are the result of large percentage of atoms at the surface with higher surface energies and unsaturated bonding with neighbouring atoms as compared to their bulk counterparts. A large number of low-coordination sites such as corners and edges, having a large number of "dangling bonds" show specific chemical properties and the ability to store excess electrons.

A simple thought experiment shows why nanoparticles have phenomenally high surface areas (Fig. 1.9). When the size of building blocks gets smaller, the surface area of the material can increase even while the volume remains the same. For example, dissecting a 1 cm^3 of any material into 1 nm particles increases the total combined surface area from 6 cm^2 to 60,000,000 cm^2; approximately 10 million times larger.

The extremely large surface-to-volume ratio relative to bulk materials lead to the presence of a high percentage of atoms/molecules lying at reactive boundary surfaces. For example, in a particle with 10 nm diameter only around 20 per cent of all atoms are forming the surface, whereas in a particle of 1 nm diameter this figure can reach more than 90 per cent. The increase in the surface-to-volume ratio results in the increase of the particle surface energy, which leads to *e.g.* a decrease in melting point or an increased sintering activity. Furthermore large surface area of particles may significantly raise the level of otherwise kinetically and thermodynamically unfavourable reactions. For instance, even gold, which is a very stable material, becomes reactive when the particle size is small enough.

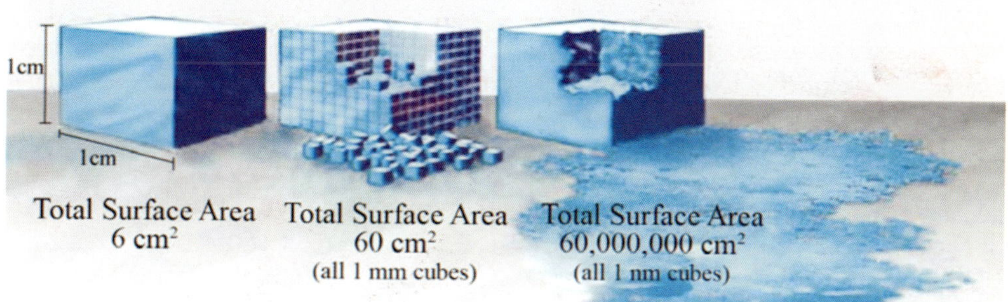

Figure 1.9. Exponential increase in surface area on going towards nano size

Increased surface area and associated improved reactivity in nanostructured materials have helped:

- To create better catalysts.
- To enhance propulsion and explosive characteristics.
- To make nanostructured membranes and materials ideal candidates for water treatment and desalination, among other uses.
- Functionalization of nanoscale material surfaces (adding particles for specific purposes), for applications ranging from drug delivery to clothing insulation.

Table 1.4 Characteristics of nanomaterial and their attributes

Characteristic	Attributes
Size	Key criteria for a nanomaterial
Shape	Carbon nano sheets with a flat geodesic (hexagonal) structure show improved performance in epoxy composites versus carbon fibres.
Surface charge	Surface charge impact adhesion and agglomeration characteristics
Surface Area	High surface energy, high percentage of surface atoms facilitate higher chemical reactivity
Surface porosity	Zeolite-type porous surfaces for maximum absorption/occlusion
Composition	Chemical composition of nanomaterials as catalyst is critical to ensure the correct stoichiometry of product
Structure	Heterogeneity, crystal structure and grain boundaries

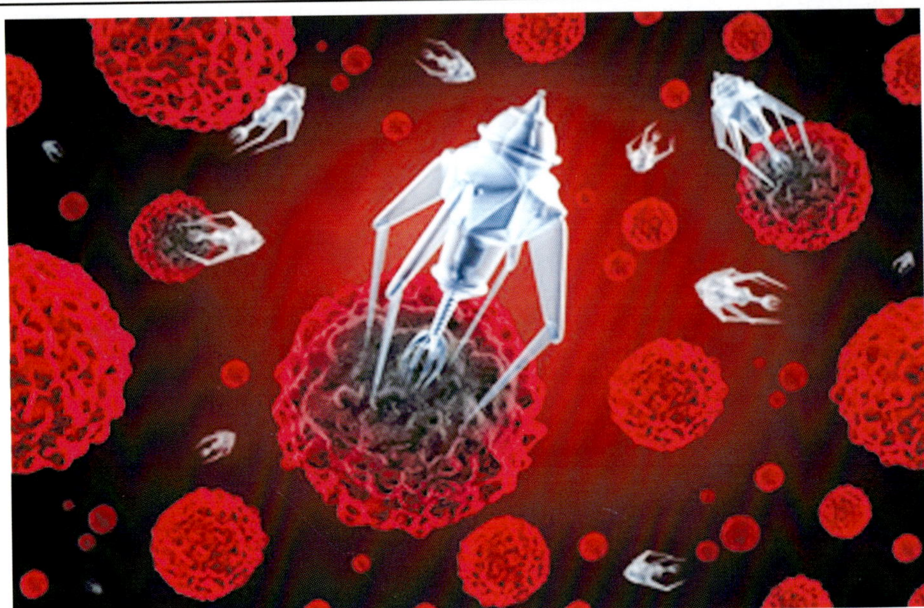

Fig. 1.10. Nanosensors in the bloodstream

While the size effect is mainly considered to describe physical properties, the surface or interface induced effect plays prominent role for chemical processing, in particular in connection with heterogeneous catalysis. Experimental evidence of the quantum size effect in small particles has been provided by different methods, while the surface induced effect is evidenced by measurement of thermodynamic properties like vapour pressure, specific heat, thermal conductivity and melting point of small metallic particles. In general, there are seven key characteristics that contribute to the uniqueness of nanomaterials and these are summarized in **Table 1.4.**

1.6 APPLICATIONS: CURRENT STATUS

Our ability to work on the same scale as nature does to engineer nanomaterial in variable shapes, sizes and morphologies with unique properties has indeed been revolutionary. Nanotechnology is affecting every aspect of our lives, from the medicines we use, to the power of our computers, the energy supplies we require, the food we eat, the cars we drive, the buildings we live in, security we need against both internal and external threats, and the clothes we wear. The word Nanomaterials holds potential to make supercomputers that are small enough to be carried in a shirt pocket or machines that can repair damaged tissues by travelling through the human body. Engineered nanoparticles can be specifically designed and obtained with customized chemical and physical properties in order to fulfill the requirements of specific applications. They can serve as the end product, like sensor for special purposes, pharmaceutical drugs and quantum dots, or they can serve as components in end products, as in the case of carbon black in rubber products. In the following sections we shall discuss the emerging applications of nanomaterials in different sectors. Many of the applications of nanomaterials have already been commercialized with turnover in several billion dollars.

Health Care

Right from the beginning of the National Nano Initiative programme in 2001 by USA, followed by other counties, nanomaterials have been very attractive for medical applications. The simple reason behind it has been properties of nanomaterials, such as size, shape, chemical composition, high surface area and surface energy, aggregation and agglomeration, and solubility. All of such properties greatly influence their interactions with biomolecules and cells. For example, nanoparticles with size-tunable light emission have been employed to produce exceptional images of tumor sites; single-walled carbon nanotubes, having diameters comparable to the width of DNA molecules, have demonstrated an impressive potential as high-efficiency delivery transporters for biomolecules into cells. Application of nanotechnology in medicine currently being developed involves employing nanoparticles to deliver drugs, heat, light or other substances to specific types of cells, such as cancer cells. Particles are engineered so that they are attracted to diseased cells, which allow direct treatment of those cells. This technique reduces damage to healthy cells in the body and allows for earlier detection of disease. Fluorescent nanocrystals have several advantages over organic dye molecules as fluorescent markers in biology. They are incredibly bright and do not photodegrade. Drug-conjugated nanocrystals attach to the protein in an extracellular fashion, enabling protein transport. In addition, nanotechnology in the form of lab-on-a chip also help in early diagnosis of a variety of serious disease. **Table 1.5** briefly summarize use of different nanomaterials in health care.

Table 1.5 Nanomaterials and their specific applications

Type of nanomaterial	Applications in healthcare
Metallic and magnetic nanoparticles such as gold, silver, iron @cobalt, iron (III) oxide, etc.	• Theronaustic agents • Hyperthermia and photodynamic cancer treatment • Selective magnetic bio-separations • Coated with antibodies to cell-specific antigens, for separation from the surrounding matrix • Membrane transport studies • Drug delivery • Magnetic Resonance Imaging contrast agent
Silver nanoparticles	• Anti-microbial agent

Type of nanomaterial	Applications in healthcare
	• Incorporated into a wide range of medical devices, including bone cement, surgical instruments, surgical masks, aprons etc.
Conjugated gold shell nanoparticles	• Improve solubility of drugs • Permit further conjugation
Carbon nanomaterials: Fullerenes and Carbon nanotubes (CNTs)	• Fullerenes in drug delivery systems to support the optimal transport and release of medicines to the right target inside the body • Fullerenes as lubricants for prosthetics and surgical implants • Functionalised CNTs: o therapeutic delivery o vascular stents and neuron growth and regeneration o gene therapy using DNA bonded CNT
Quantum dots	• Disease diagnosis and screening technologies, tag multiple biomolecules to monitor complex cellular changes and events associated with diseases
Dendrimers	• Used as a carrier for a variety of drugs (*e.g.* anti-cancer, anti-viral, anti-bacterial, etc.) with capacity to improve solubility and bioavailability of poorly soluble drugs
Lipid-based nanoparticles	• Can fuse with the cell membrane and deliver molecules inside the cells
Porous and biocompatible ceramic nanoparticles: ZnO, TiO_2; nanophosphate of calcium (mimic structure and composition of natural bone)	• Drug vehicles, UV protection of skin, synthetic bone in treatment of fractures and soft tissue injuries
Nanotubes, nanowires, magnetic nanoparticles	• Disease diagnosis and screening technologies, including 'lab on a chip

Electronics

In some sense, electronic miniaturization has been the true driving force for nanotechnology research and applications. It's a sure thing that silicon era is on the way up. Only nanotechnology can radically change ICT systems in order to continue to follow Moore's law. Nanotechnologies are therefore, expected to enable the production of smaller and cheaper devices with increasing efficiency. The current nanotechnology applications are focused on miniaturized electronic devices with special emphasis on:

- Computer chips with efficacy improvement by a factor of millions,
- Data storage at multi-terabit level,

- Efficient sensors for chemical, biological, and radiological contaminants, etc.

Nanoelectronics hold the capability of electronics devices that can be expanded while reducing their weight and power consumption. Reduction in size of transistors in accordance of Moore's law makes it possible to happen. In another example, Quantum dots (QDs), because of having both photoluminescence and electroluminescence are one of the most promising optoelectronic materials for application in next-generation displays. Compared to organic luminescent materials used in organic light emitting diodes (OLEDs), QD-based materials have purer colors, long lifetime, low manufacturing cost, and lower power consumption. Another key advantage of quantum dot displays is that, because QDs can be deposited on virtually any substrate, it is easy to expect printable and flexible – even rollable – displays of all sizes. Such displays have high application potential in consumer products such as cameras, PDAs, laptops, televisions, and other as yet undreamt of devices.

Energy

Nanomaterials have emerged as a potent source in harvesting renewable energy (solar, wind) storing (hydrogen), and generating energy (fuel cells, batteries). In photovoltaic devices they absorb sunlight more strongly than dye molecules or bulk semiconductor material; therefore high optical densities can be achieved while maintaining the requirement of thin films. Perfectly crystalline CdSe nanocrystals are also an artificial reaction center, separating the electron hole pair on a femtosecond timescale. Silicon nanowires arrays embedded in a polymer results in low-cost but high-efficiency solar cells.

Nanotechnologies can improve capacity and safety of lithium-ion batteries decisively, as for example, using nano ceramic, heat-resistant and still flexible separators and high-performance electrode materials. The lithium-ion technology is regarded as the most promising variant of electrical energy storage. The efficient storage of hydrogen is regarded as one of the critical factors of success on the way to a possible hydrogen management. Nanomaterials, particularly based on nanoporous metal-organic compounds, provide development potentials in this regard. Epoxy containing carbon nanotubes is being used to make windmill blades. The resulting blades are stronger and lower weight and therefore the amount of electricity generated by each windmill is greater.

Food

Nanotechnology also has applications in the food sector. Many vitamins and their precursors, such as carotinoids, are insoluble in water. However, when skillfully produced and formulated as nanoparticles, these substances can easily be mixed with cold water, and their bioavailability in the human body also increases. Many lemonades and fruit juices contain these specially formulated additives, which often also provide an attractive color.

Cosmetics

The applications of nanotechnology and nanomaterials can be found in many cosmetic products including moisturisers, hair care products, make up and sunscreen. Nano zinc oxide and titanium dioxides have been commercialized as UV absorbers and incorporated in sun creams; the small particles filter the high-energy radiation out of sunlight. Because of their tiny size, they remain invisible to the naked eye and so the cream is transparent on the skin.

Automobiles

Nanocrystals of various metals have been shown to be 100 percent, 200 percent and even as much as 300 percent harder than the same materials in bulk form. Because of high wear resistance metal nanocrystals might be incorporated into car bumpers, making the parts stronger, or into aluminum, making it more wear resistant. A plastic nanocomposite which is scratch-resistant, light-weight, and rust-proof, and generates improvements in strength and reductions in weight, which lead to fuel savings and increased longevity have been developed and used by leading auto makers in car bumpers that makes it 60% lighter and twice as resistant to denting and scratching.

Nano Catalysts

A variety of chemical reactions take place on the surface of the catalyst, and the larger the surface area, the more active the catalyst. Nanoscale catalysts thus open the way for numerous process innovations to make many chemical processes more efficient and resource-saving – in other words more competitive. A very promising application of gel-based nanoscale catalyst includes liquification of coal and subsequent conversion into gas. This technology is expected to economically transform coal into diesel and gasoline thus reducing the dependency of coal rich countries on oil import.

Pollution Control

Nanotechnology is being used in several applications to improve the environment. This includes cleaning up existing pollution, improving manufacturing methods to reduce the generation of new pollution. Potential applications include:

- Cleaning up of organic chemicals polluting groundwater. The iron nanoparticles disperse throughout the body of water decompose the organic impurities.
- Generating less pollution during the manufacture of materials. For example, the use of silver nanoclusters as catalyst can significantly

reduce the polluting byproducts generated in the manufacture of propylene oxide used in plastics, paint, detergents and brake fluid.

- Filters based on nanoalumina fibers are used to removal of viruses in flowing water. It is also useful for sterilization of biological, pharmaceutical and medical serums, protein separation, collector/concentrator for biological warfare detectors, and several other applications such as sterilizing drinking water. Graphene membranes are at the threshold of commercialization for desalination and purification of water.

Paints and Textile

Clothes are also receiving the nano-treatment. Again, nano-silver is playing a lead role due to its anti-microbial properties, with the risk being that the nanoparticles will be in direct contact with skin over prolonged periods of time. Clothes can also be treated with nano-films to make them stain, water and static charge resistant. These films, which are only a few atoms thick, could be in contact with skin over prolonged periods.

Dispersions of polymer nanoparticles, ranging from ten to several hundred nanometers in size, find applications in exterior paints, coatings and adhesives, or in the finishing of paper, textiles and leather.

Construction Industry

Energy consumption in commercial buildings and residential houses for heating, lighting, and air conditioning is increasing day-by day. The use of nano-material/ nanotechnology in the construction industry is being considered not only for enhancing material properties and functions but also in the context of energy conservation. It is reported that nanotechnology has a significant impact in the construction sector, safety of the buildings, facilitating the ease of maintenance and to provide increased living comfort. Nanoparticles of TiO_2, Al_2O_3 or ZnO are applied as a final coating on construction ceramics to bring desired effects. For instance, TiO_2 is being used for its ability to break down dirt or pollution when exposed to UV light and then allow it to be washed off. ZnO is used for UV resistance in coatings and paints. Nanosized Al_2O_3 particles are used to make surfaces scratch resistant.

Sports Goods

Sports goods have been improved significantly by nanotechnology. Strength of tennis racquets is being increasing by adding nanotubes to the frames which increases control and power when the player hist the ball. Filling any imperfections in golf club shaft materials with nanoparticles; this improves the uniformity of the material that makes up the shaft and thereby improving your swing. Reducing the rate at which air leaks from tennis balls so they keep their bounce longer. Metal nanocrystals can be used to produce bearings that last longer than their conventional counterparts, new types of sensors and components for computers and electronic hardware.

Defence and Security

Nanotechnology has brought qualitative technological changes for a wide range of applications in defence and security sector through providing light weight platforms; enhancing range, efficiency and lethality of weapons; stealthy structure; aerodynamics stability; higher mobility; sensing and mitigation of chemical, biological, radiological or nuclear (CBRN) threats; energy generation and management; smart structures and materials; resilience and robustness etc. Areas currently under active investigation include nanoparticles to make 'smart' coating' on vehicles, nanosensors in light-weight uniforms, sensors to detect life signs, and advanced computing power for code breaking and encryption.

Agro Forestry

The essential nanotechnology based inputs in agriculture are in the forms of nanosensors for monitoring quality of soil, to detect soil contamination, viruses and harmful pathogens; nanomaterials for soil decontamination (zero valent nano iron is one such material); Use of nano encapsulated pesticides and insecticides to mitigate user's health hazards; use of nanoporous materials for retention and sustained release of water and nutrients. Nanotechnology offers great potential to make agriculture smart, more efficient, increase yield and product with enhanced nutritional benefits.

1.7 RISKS ASSESSMENT AND MANAGEMENT

1.7.1 Risks Assessment

Although nanomaterials in may offer a large number of benefits different sectors but at the same time they may also expose users and occupational workers to new risks. There is still a knowledge gap in the information available regarding the toxicity of manufactured nanomaterials, which makes performing risk assessments difficult. The main challenge is to understand the possible hazards workers may face when working with manufactured nanomaterials or nanodevices. An additional issue is that, while a nanomaterial itself may not be toxic, it could act as a Trojan horse, meaning that a more toxic material may attach itself to the nanomaterial and gain entry to the body, organs or cells. The most adverse effects of nanomaterials have been found in the lungs. There is a variety of pathways through which nanomaterials can enter the human body that may pose an occupational health hazard:

- Inhalation is the most common route of exposure to airborne nanoparticles in the workplace. Inhaled nanoparticles can deposit in the respiratory tract and the lungs depending on their shape and size.

- Ingestion can occur from unintentional hand-to-mouth transfer from contaminated surfaces, or by ingestion of contaminated food or water. Ingestion may also occur as a consequence of inhalation of a nanomaterial, as inhaled particles that are cleared from the respiratory tract via the mucociliary escalator may be swallowed.
- Dermal penetration is another possibility. Though intact skin seems to be a good barrier against the uptake of nanomaterials however, damaged skin seems to be less effective, but the level of uptake is likely to be lower than that associated with inhalation.
- Nanomaterials can also be introduced into the human body via the parenteral route; accidentally through needle stick injury, cuts and other damage to the skin.

It is important to properly assess and manage the possible occupational risks of nanomaterials in order to protect workers' safety and health adequately.

1.7.2 Risks Management

According to EU Directive 89/391/EEC, employers must conduct regular workplace risk assessments and put in place adequate prevention measures, and this also applies to the potential risks of nanomaterials in the workplace. In addition, Directive 98/24/EC on chemical agents at work imposes more stringent provisions on the management of risks from substances at work, which also apply to nanomaterials as these fall within the definition of 'substances'. There is guidance and tools available to help manage the risks of nanomaterials in the context of current limitations and of the current state-of-the-art research. The measures recommended in sectors (*e.g.* for research laboratories) are partially applicable to occupational worker working in industrial and health sectors as well. The main guiding preventive measure include:

1. Elimination and substitution

As with all other dangerous substances, elimination and substitution should be given priority over other prevention measures (*i.e.* the aim is to prevent all workers from exposure to nanomaterials). The consideration of the balance between the desired properties and effects, on the one hand, and health risks, on the other hand, should always be borne in mind and elimination and substitution given thorough consideration. In addition, it may be possible to:

- avoid the presence of nanomaterials that could become airborne (such as powders or dusts) by using a less hazardous form, *e.g.* by solubilizing powder forms of nanomaterials into liquids, pastes, granulates or compounds or by binding them to solids; and
- reduce the hazardous behaviour by modifying the surface of the nanomaterial, for example by coating it in order to adjust the dustiness, solubility and other properties.

2. Engineering controls

It may not be possible to have technical systems everywhere for reducing or preventing exposure to nanomaterials at source, for example closed systems that create a physical barrier between a person and the nanomaterial. However, engineering controls at source are feasible in other operations, such as the preparation of drugs containing nanomaterials, *e.g.* tablets or ointments in glove boxes. Clean benches with high-efficiency particulate air (HEPA) filters are another effective risk prevention measure for many activities such as the preparation of nanoparticles. Exposure to spills, dusts or vapours of nanomaterials from samples or sample preparation products should be controlled by the use of a high-throughput ventilation system in combination with PPE, particularly gloves and masks. Local exhaust ventilation systems are normally in place in laboratory workplaces, surgery rooms or areas with high safety standards (*e.g.* because of the risk of infection) and in storage areas. Such systems also capture nanomaterials. However, in the case of nanomaterials, the use of normal multistage filters, with high-efficiency particulate air filters (HEPA H14) or ultra-low penetration air filters, as the final filter before re-introduction of the extracted air is recommended. In any case, the suitability of the filtration systems in place should be evaluated.

3. Organizational Measures

Risk prevention measures in workplaces where hazardous nanomaterials may exist includes:

- areas or workplaces specifically dedicated for handling nanomaterials should be separated from other workplaces and clearly indicated with appropriate signs;
- minimize the number of workers be exposed to nanomaterials;
- minimize the duration of workers' exposure to nanomaterials;
- prohibit access to unauthorized personnel;
- regular cleaning (wet wiping) of work areas should be done where nanomaterials are used or handled;
- monitoring of air concentration levels, *e.g.* in comparison with background levels when no handling of nanomaterials occurs;
- priorities in risk management should be given not only to known risks, but also to the assessment and management of nanomaterials at workplaces in which hazard and exposure information are missing, incomplete or uncertain;
- time pressure should be avoided;

- sufficient training should be provided to ensure that workers have the skills and knowledge to perform the work safely and to protect themselves from exposure to any nanomaterial releases;
- instructions and information should always be provided to all workers, particularly when workers are contracted for only one task and/or are not familiar with chemical risks in general and risks from nanomaterials specifically (*e.g.* cleaning personnel, student assistants). Information can be documented in workplace instructions;
- taking a precautionary approach to risk prevention for nanomaterials: all measures available should be implemented, to reduce the release of nanomaterials.

4. Personal protective equipment (PPE)

PPE should be used as a last resort when exposure cannot be reduced effectively enough with the above-mentioned measures. Workers in the healthcare sector are likely to use PPE in their activities because of other health risks (*e.g.* biological agents). However, the PPE used needs to be evaluated with regard to its suitability for nanomaterials.

Respiratory protection: For activities involving airborne nanomaterials, for example grinding, sanding or milling of bridges or implants containing nanomaterials, respiratory protection must be used. High efficiency particular air (HEPA) filters, respiratory cartridges and masks with fibrous filtering materials are effective against airborne nanomaterials.

In cases where the respiratory devices do not cover the eyes, tight-fitting safety goggles should be used for eye protection.

Gloves: In the case of handling nanomaterials, gloves from synthetic polymers, such as latex, nitrile or neoprene, have been found to be effective. The thickness of the glove material is a major factor in determining the diffusion rate of the nanomaterial. Therefore, the use of two pairs of gloves at the same time is recommended.

Protective clothing: Non-woven textiles (air-tight materials), such as high-density polyethylene (low dust retention and low dust release), should be preferred. Protective clothing made with cotton fabrics woven textiles should be avoided.

Prevention of explosion and fire: As a result of their small size, nanomaterials in powder form may present risks of explosion that their respective coarse materials do not. Care should be taken when nanopowders are generated (*e.g.* grinding, sanding or polishing of implants and bridges containing nano-materials) or handled (*e.g.* by mixing, cleaning or disposing of such powders). The preventive measures for nanomaterials in powder form are essentially the same as those for any other explosive and flammable coarse material and explosive dust clouds, and should meet the following requirements:

- Materials should be solubilized by wetting the workplace (prevention of dusts);
- Low-spark equipment and other ignition sources, or conditions facilitating electrostatic charging, should be removed from the work-place; instead, intrinsically safe equipment (signal and control circuits operating with low currents and voltages) should be used, when possible;
- Dust layers should be removed by wet mopping up; and
- Storage of explosive or flammable materials at workplaces should be minimized. Anti-static bags may be used.

1.8. FUTURISTIC OUTLOOK

One of the greatest challenges facing society in the 21st century is providing better living standards to all people while minimizing the impact of human activities on the global environment and climate as the world population reaches 8–10 billion by 2050. Currently, the world is facing great challenges to meet rising demands for basic commodities (*e.g.*, food, water and energy), affordable means for transportation and communication, shelters to live, healthcare and employment. At the same time a pollution free environment is required to live a happy life. Soon after the inception of the National Nanotechnology Initiative (NNI), in 2001, it was envisioned that nanotechnology could provide more sustainable solutions to the global challenges facing society in terms of improving agricultural yields, provide more efficient and cost-effective water treatment and desalination technologies, and enable the development of clean and renewable energy sources including highly efficient solar photovoltaic cells.

Nanotechnology has a vast future ahead of it as breakthroughs are constantly being made in this area every day. It is envisaged that nanotechnology will play a big role in our lives, in the very near future, in everything ranging from health to food to clothes to energy to water to clean environment to transportation to security etc. Below are a few areas where the impact of wondrous possibilities of nanotechnology will be felt further.

Health

Nanotechnology will affect the health industry far greater than it will anything else. There will be a lot of things that nanotechnology will be able to accomplish in the health will include accurate and early diagnosis and treatment of life threatening disease such as cancer, Alzheimers, Parkinsons, Aids, etc. together with the replacement of damaged organ at the cellular level. It has been discovered that shining X -rays on a special type of nanoclusters will produce their own electrons that can be controlled in order to target specific cancer cells and destroy them without harming much of the surrounding area. This involves no radiation (other than X-rays) and absolutely no chemotherapy. Nanotechnology is the cure-all medicine that we've been searching for years and it turns out that it's not actually a "medicine" at all. In all likelihood it is feasible to introduce tiny nanosensors inside the body of a patient enabling doctors to personalise their treatment by getting much more detailed information with less hassle to the patient as represented in Figure 1.10.

The possibilities are endless, ranging from monitoring inflammation and post-surgery recovery to more exotic applications whereby electronic devices actually interfere with our body's signals for controlling organ function.

Security

Nanotechnology is expected to bring a whole new meaning to the word "security". Well now you can have nanocameras that are too small to even know they're there. The nanocameras could even be programmed to follow movement and alert the police if something (specifically, someone) gets too close. It is anticipated that with the application of nanomaterials (*i.e.* metamaterials) on soldiers' uniforms or surfaces of combat weapons it may be possible to make them invisible to enemy's surveillance. Further, the use of nanomaterials in propellants and explosives will enhance the speed of missiles and lethality of the weapons at much lower pay loads. Nanotechnology will also allow us to make body armor that is a hundred times stronger than steel but light as a feather. That means that soldiers and police officers will be much safer while performing the dangerous duties that they take on in order to protect us.

Food Preservation

Nanotechnology can help to preserve food in at least two different ways: The first way that nanotechnology can improve food is by sealing the bags that the food is in. The way nanotechnology can help this is by forming a criss-cross net over the container that takes air a very long time to get through. The other way that nanotechnology can help preserve food is mostly limited to bread, specifically. Nanotechnology can help this by reinforcing the proteins to make them stronger which means that the fat cells take much longer to take over. With these two preservation techniques just from adding nanotechnology to the food, we may keep food fresh for longer time and without consuming energy.

High Strength Low Density Materials

Nanotechnology will be able to reinforce materials to make them not only stronger, but also lighter. There are structures called nanotubes (or carbon nanotubes) that are just as stiff as diamonds and one hundred times stronger than steel, yet they're six times lighter than steel. Carbon nanotubes are great because athletes (such as golf players or tennis ball players) are constantly looking for equipment that is both strong and light. With the help of these carbon nanotubes, those athletes will be able to get exactly what they're looking for.

Self-healing Structures

In case if cracks appear in a given structure such as that of a building or an aircraft then nanotechnology could play the role of healing. For example, coating the buildings with water repelling or dispersing nanoparticles throughout a material so that they can migrate to fill in any cracks and act as self-healing materials for everything from aircraft cockpits to microelectronics, preventing small fractures from turning into large, more problematic cracks.

Energy

Nanotechnology is already playing a role to harvest, convert and store energy by way of creating batteries that can store more energy for electric cars and has enabled solar panels to convert more sunlight into electricity. The common trick in both applications is to (for example nanowires or carbon nanotubes) turn a flat surface into a three-dimensional one with a much greater surface area. This means that there is more space for the reactions that enable energy storage or generation to take place, so the devices operate more efficiently. It is believed that with further research, nanotechnology could also enable objects to harvest energy from their environment. New nano-materials and concepts are currently being evolved that show potential for producing energy from movement, light, variations in temperature, glucose and other sources with high conversion efficiency.

Water Purification

Water is the most essential element for life. It constitutes 70% of the total mass of our body and at the same time most of the diseases are caused due to intake of water contaminated with biological and organic impurities in turn introduced by agricultural and industrial activities. Improving water purification really is a global issue. It is estimated that many developing and under- developed countries are withdrawing more than 25% of their renewable freshwater resources for agricultural, domestic, and industrial uses annually. By the year 2025 it is predicted that over 2.8 billion people in 48 countries will be "water stressed". To combat this growing issue, countries are turning to unconventional options for wastewater treatment and desalination. The currently used technologies for water desalination and purification such as osmosis, reverse osmosis, electro dialysis, though effective but costly and large energy consuming. There is therefore urgent requirement for newer, faster, cheaper, and overall better way to provide fresh water on the global scale. Nanotechnology has proven to be a wonderful alternative to previously used methods in energy and medicine, for example, and we may be close to seeing its deployment for a similar improvement in the case of water purification.

Membranes based on carbon nanotubes and graphene oxide are fast emerging as the new resources for purification of water contaminated from a variety of biological and organic impurities and removal of harmful inorganic salts soluble in water. These tiny, flexible, and surprisingly resilient materials may very well be our new direction in sustainable, large-scale water purification. The major advantage of CNTs is that water passes through them in a nearly frictionless manner due to something called hydrophobicity (*i.e.* preference to be away from water). However, their limitation lie by low salt rejection rates and the difficulty of producing highly aligned and high-density CNT arrays. Because flux across a membrane scales inversely with the membrane's thickness, new types of ultrathin membranes offer the promise of greatly increased water permeability. Graphene oxide, which consists of a 2D sheet of sp2-bonded carbon atoms in a hexagonal honeycomb lattice, is the ultimate thin membrane emerging for both water purification and desalination.

Graphene oxide (GO) membranes consist of many layers of two-dimensional graphene oxide layers stacked on top of each other to give a laminate of two- dimensional GO sheets. These stacked sheets have interconnected channels running through them which act as the membrane pores. The pores are uniform in size and only 0.9 nm in width. This small width means the membrane is highly selective as only ions and molecules smaller than 0.9 nm can permeate through. Any ions or molecules larger than 0.9 nm is prevented from passing through the membrane by a process called physical extrusion.

The channels inside the GO membrane can be modified with different methods based on what type of membrane is required. Modifying the membrane pores allow membrane permeability to be fine-tuned, meaning certain compounds and ions can be selected in such a way that the membrane is highly permeable to them. Additionally, unlike polymer membranes, GO membranes are more chemically inert, which means they have a longer service lifetime.

In summary it is to be stated that nanotechnology and synthesis of new nanomaterials with desired properties will open new frontiers in the design of catalysts and catalyst technology for the petroleum, chemical, automotive, pharmaceutical, and food industries. The design of catalyst supports commensurate with biological structures will be an important bridge between conventional and enzymatic catalysis. In fact, oxidation catalysis can be performed today more efficiently in a zeolite, "ship-in-a-bottle" catalytic complex than with natural enzymes. This is but one example of an entire array of anticipated future developments. It is anticipated that in the near future interactions between nanoscale scientists and system designers will enable to do nanomanipulation in three dimensions to guide nanoassemblies in bulk as well as on surfaces. An important element of this interaction will be prototyping methods, an intermediate level of implementation between lab-scale demonstration and mass production. The new nanomaterials obtained in high purity and at large scale will impact not only the performance of the most advanced computational and electronic devices, objects of daily use familiar to every consumer, such as cars, appliances, films, containers, and cosmetics paving the way for their commercialization.

Suggested further readings

1. R.P. Feynman, "There's plenty of room at the bottom", presentation at the *American Physical Society Conference*, Caltech, California, 1959.
2. K. E. Drexler, Engines of Creation, "The Coming Era of Nanotechnology", Anchor Press, Doubleday, 1986, New York; "Nanosystems: Molecular Machinery, Manufacturing and Computation", 1992, New York: John Wiley & Sons.
3. C. Neinhuis and W. Barthlott, "Characterization and distribution of water-repellent, self-cleaning plant surfaces", *Annals of Botany*, **79** (1997) 667-677.
4. R. Blossey, Self-cleaning surfaces virtual realities, *Nature Materials*, **2** (2003) 301-306.
5. A.R. Parker and H. E. Townley, Biomimetics of photonic nanostructures, Nature Nanotechnology 2 (2007) 347 – 353.
6. J. C. Grossman, "A Little Bit About Nanotechnology", *Outreach Program*, 2007, MIT.
7. R. O. Ritchie, "Armoured Oyster Shells", *Nature Materials*, **13**(2014) 435-437.
8. P. Go′mez-Romeroa and C. Sanchezb, "Hybrid materials. Functional properties. From Maya Blue to 21st century materials", *New J. Chem.*, 29 (2005) 57–58.
9. M. R. Gwinn and V.Vallyathan, "Nanoparticles: Health Effects—Pros and Cons", *Environ. Health Perspect.* **114** (2006): 1818–1825.
10. N Kumar and S Kumbhat, *Essentials in Nanoscience and Nanotechnology,* J Wiley, NY, 2016
11. A. Nel, T. Xia, L. Medler and N. Li, "Toxic Potential of Materials at the Nano Level", *Science*, 311 (2006) 622–627.

Review Questions

1. Define nanoscience and nanotechnology.
2. What is so unique about materials at nanoscale?
3. What do you understand by natural, incidental and engineered nano materials? Give some examples.
4. Give some examples of existence of nanoscience in nature.
5. Who is considered as the 'Father of Nanoscience'? And why is he so known ?

6. Which technological discovery in the eighties accelerated the growth of nanoscience and nanotechology?
7. Give the name and contribution of scientists in the area of carbon nano materials.
8. Write a note on 'Lotus Effect' and outline its technological impact.
9. What is unique about Gecko? How do you correlate Gecko effect with nano science?
10. Classify the nanomaterials on the basis of their dimension. Give some examples.
11. Which are the sectors likely to be benefitted in future from the advancements in the nanotechnology?
12. In what way nanotechnology will help in pollution control?
13. What is the health risk factors associated with the increasing use of nanoparticles?
14. Discuss various pathways through which nanomaterials can enter human body.
15. Discuss various measures to manage the risk associated with use of nanoparticles.

■■■■

Accounts of Nanomaterials 2

2.1 Introduction

2.1.1 Dimensionality of nanomaterials

2.1.2 Morphology of nanostructured materials

2.2 Inorganic nanomaterials

2.2.1 Nanometals and alloys

2.2.2 Colloidal metal nanoparticles

2.2.3 Nano-oxides of transition and non-transition elements

2.2.4 Non-oxide inorganic nanomaterials

2.2.5 Zeolites

2.2.6 Quantum dots

2.2.7 Nanoclusters

2.3 Carbon- based Nanomaterials

2.3.1 Fullerene

2.3.2 Carbon nanotubes (CNTs)

2.3.3 Graphene

2.3.4 Carbon nanofibers

2.3.5 Carbon nano-onions

2.3.5 Nano diamond

2.4 Organic Nanomaterials

2.4.1 Polymeric nanoparticles

2.4.2 Dendrimers

2.4.3 Cellulose

2.5 Biological nanomaterials

2.6 Self -Assembled and Supramolecular Nanomaterials

2.6.1 Liquid Crystalline structures

2.6.2 Supramolecular organic frameworks (SOF)

2.7 Nanocomposites

2.7.1 Ceramic matrix nanocomposites (CMNCs)

2.7.2 Metal matrix nanocomposites (MMNCs)

2.7.3 Polymer matrix nanocomposites (PMNCs)

2.8 Nano coatings

2.8.1 Functional nano coatings

2.8.2 Protective nano coatings

2.8.3 Smart nano coatings

2.1 INTRODUCTION

When the size of a material is reduced to a smaller size, initially its properties remain the same. However, when size drops below 100 nm, dramatic changes in properties can be observed, and the material is termed as nanomaterials. In general, a nanomaterial may be of a natural, incidental or man-made origin with at least one external dimension in the nanoscale (~1 nm to 100 nm) and or having internal structure or surface structures in the nanoscale. Nanomaterials have a unique and more pronounced physico-chemical characterristics compared to those of the same material in bulk. Nanomaterials are the building blocks of nanotechnology and can be manipulated for specific application. Much of the nanoscience and nanotechnologies are oriented with the production of new nanomaterials with enhanced properties. Nanomaterials are generally classified based on their dimensionality, morphology, and composition. The majority of the nanostructures are synthesized using organic substances, necessary for stabilizing and functionalization of inorganic nanoparticles, therefore, nanomaterials are usually hybrid in nature. Many of such classification schemes are simplistic, overlooking the complexity of nanometerials and the multivariable dependency of a property on specific features. Nanostructures are the ordered system of nanometer scale in a certain pattern.

2.1.1 Dimensionality of Nanomaterials

Size and shape-dependent variation in properties is a significant aspect of nanoscience. Many unique properties of nanomaterial arise due to spatial confinement of electrons, phonons, and electric fields around the particles as well as due to their large surface to volume ratio which can influence their chemical reactivity, as well as magnetic, opto-electronic and mechanical properties. Changes in the properties arise due to the quantum-mechanical nature of physics being followed in the domain of the nanoscale. Restricted motion of electrons (*i.e.* quantum confinement), holes, excitons, phonons, and plasmon with respect to the physical shape of an object is the reason for the change in properties of typical nanostructure. The most important change that is manifested is in color, due to the confinement of electrons and consequent changes in electronic energy levels.

Isotropic and anisotropic nanomaterials are categorized on the basis of dimensionality as shown schematically in **Fig. 2.1**. In the case of isotropic or 0D particles (*e.g.* sphere), because of the confinement of electrons to the same extent in all the three dimensions, properties will be more or less the same regardless of directions. Anisotropic nanomaterials can be categorized as 1-, 2-, and 3D nanostructures wherein the electron motions are possible in these many dimensions.

0D (Nanoclusters)

Nanoclusters
&
Nanoshperes

1D (Nanotubes, fibers rods)

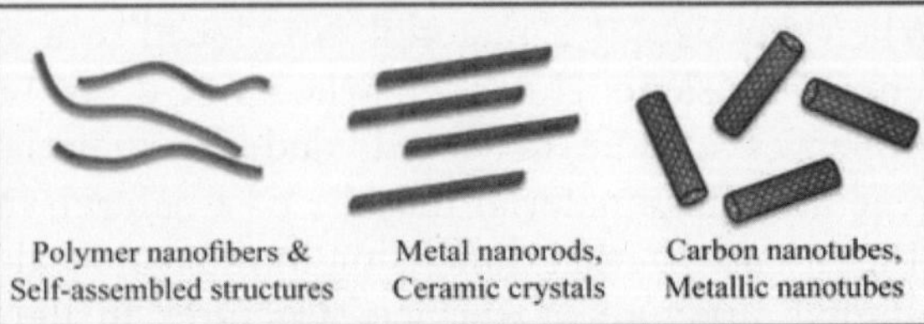

2D (Films, sheets, layered structures)

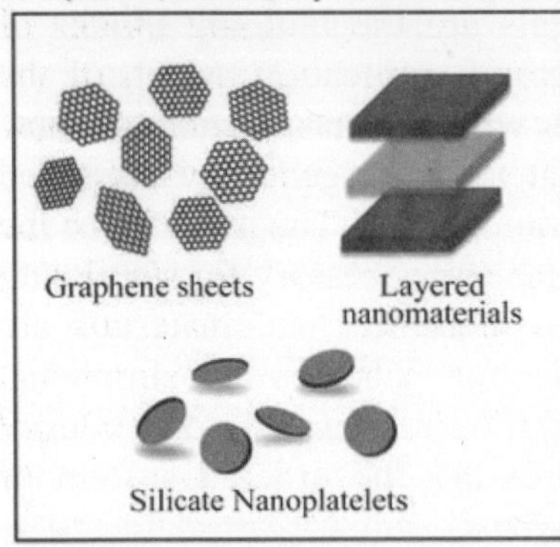

3D (Polycrystals, spherical)

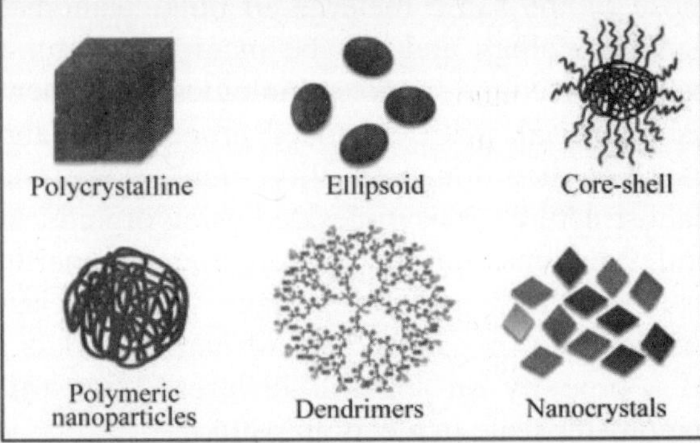

Fig. 2.1. Dimensionality in nanomaterials: (a) 0D-spheres and clusters; (b) 1D-nanofibers, wires, and rods; (c) 2D-films, plates, and networks; (d) 3D-nanocrystalline particles

0-D Nanomaterials. In 0-dimensional materials all the dimensions are measured within the nanoscale. The most common representative of 0-dimensional nanomaterials are nanoparticles. Nanoparticles can be amorphous, crystalline, polycrystalline, composed of single or multi elements and may exhibit various shapes and forms. These particles can be metallic, ceramic, or polymeric in nature and may exist individually or incorporated in a matrix. Quantum dots, nano spheres, nano lenses, nano onions are examples of 0-D nanoparticles.

1-D Nanomaterials. In these materials, one of the dimension is not confined at nanoscale while two other dimensions are at nanoscale which lead to needle-shaped materials. Examples include nanotubes, nanofibers, nano wires and nano rods etc. 1-D nanomaterials can be amorphous or crystalline; single crystalline or polycrystalline; stand alone materials or embedded within another medium; metallic, ceramic, or polymeric.

2-D Nanomaterials. Anisotropic geometry, wherein, two of the dimensions are not confined to the nanoscale while one dimension is at nanoscale. Such materials exhibit a plate-like structure and include nanofilms, nanolayers, and nanocoatings, ribbons, plates, sheets, triangles, etc. 2-D nanomaterials can be amorphous or crystalline and made up of various chemical compositions. They are used as a single layer or as multilayer structures and can also be deposited on a substrate or integrated in a matrix material.

3-D Nanomaterials. Anisotropic geometry, wherein bulk nanomaterials that are not confined to the nanoscale in any dimension. These materials are thus characterized by having three arbitrary dimensions above 100 nm. These materials possess a nanocrystalline structure with features at the nanoscale. 3-D nanomaterials may contain dispersions of nanoparticles, bundles of nanowires, and nanotubes as well as multi-nanolayers, colloids, and free nanoparticles with various morphologies such as nanocubes, nanocages, nanoballs, etc. Examples of 3-D nanomaterials are thin films deposited under conditions that generate atomic-scale porosity, nanocoils, nanoflowers, and nanoporous materials. Owing to the large specific surface area, porosity, and other superior properties arising from quantum size effect over their bulk counterparts, 3-D nanomaterials assume great significance in many of their applications.

2.1.2 Morphology of Nanostructured Materials

Nanoparticles possess a variety of shapes and morphologies. Amorphous particles usually adopt a spherical shape whereas anisotropic nano materials may acquire a variety of shapes such as wire, rods, prism etc. Particle anisotropy offers features and functions that are difficult to obtain simply by size-tuning of spherical nanoparticles. For example, a slight change in particle geometry can produce great changes in the surface plasmon resonance (SPR) peak position of a metallic nanoparticle, which can hardly be achieved in the spherical nanosystems through a similar change in diameter. In addition, the optical properties of anisotropic gold or silver nanorods (NRs) or prisms are tunable throughout the visible, near infrared (NIR), and infrared regions of the spectrum, as a function of their aspect ratios. When anisotropy is added to the nanoparticles, such as NRs or prisms, the SPR is not only enhanced but also becomes strongly tunable as a function of aspect ratio. Introducing anisotropy into the nanoparticles can also make a substantial change in their magnetic properties.

The morphological characteristics of nanomaterials to be taken into account are flatness, sphericity and aspect ratio. High aspect ratio nanoparticles include nanotubes and nanowires, with various shapes, such as helices, zigzags, belts, or perhaps nanowires with diameter that varies with length. Small-aspect ratio nanoparticles include spherical, oval, cubic, prism, helical, or pillar morphologies. Collections of many particles exist as powders, suspension, or colloids. Controlling the morphology of nanoparticles is of key importance for exploiting their properties in several emerging technologies. **Table 2.1** summarzes various nano terms and definitions as approved by ISO TC 229 on Nanotechnology.

Owing to their small size effect, large surface-to-volume ratio effect and quantum tunnel effect, the nanoparticles demonstrate special chemical, biological and physical properties and can be widely used in a variety of applications. With further processing steps, nanostructured powders and dispersions can be used to fabricate coatings, components or devices that may or may not retain the nanostructure of the particulate raw materials. Industrial scale production of nanoparticulate materials like carbon black, polymer dispersions or micronized drugs has been established for a long time.

Table 2.1. Terms used for nanomaterials and their ISO approved definitions

Terms	Definition
Nanoparticle	Nano object with three external dimension in nanoscale
Nanoplate	Nano object with one external dimension in nanoscale and two other external dimensions significantly larger
Nanofiber	Nano object with two similar external dimension in nanoscale and the third dimension significantly larger
Nanotube	Hollow nanofiber
Nanorod	Solid nanofiber
Nanowire	Electrically conducing or semiconducting nanofiber
Quantum dot	Crystalline nanoparticle that exhibit size- dependent properties due to quantum confinement effects on the electronic states
Nanostructured materials	One of the five types: nanostructured powder; nanocomposite; solid nanofoam; nanoporous materials, fluid nanodispersion
Nano-onion	Spherical nanoparticle with a concentric multiple shell structure

Nanowires, Nanorods and Nanotubes are linear nanostructures and distinguish themselves from other nanomaterials by their unique one-dimensionality (1D). The peculiar electronic structure exhibited by 1D systems such as density of states singularities, molecule-like states over large distances, and spin-charge separation associated with a Luttinger liquid make these morphology elements attractive for new nanoscale devices. Nanowires, nanotubes or nanorods can be generated from different material classes, *e.g.* metals, semiconductors or carbon by means of several production techniques. Carbon nanotubes are one of the most promising linear nanostructures which can occur in a variety of modifications (*e.g.* single-, or multiwalled, filled or surface modified) and has been discussed separately.

1D is the smallest dimension for efficient transport of electrons and optical excitations, which at least qualifies nanowires as interconnections for nanoscale devices. Vertical nanowire arrays display many desirable characteristics for device integrations, such as enhanced capacity and rate capability in lithium ion batteries and improved light coupling and electron lifetime in dye-sensitized solar cells. In recent years, a variety of inorganic nanowires of elements, oxides, nitrides, carbides and chalcogenides have been generated by employing various physical or wet chemical methods and characterized. One of the crucial factors in the synthesis of nanowires is the control of composition, size and crystallinity. The unique setting of nanowires aligned on a conducting substrate have facilitated the studies on properties and applications of these novel materials in field emission, gas sensors, dye-sensitized solar cells/photo electrochemical devices, field effect transistors, etc.

Nanowires are similar to nanofibers albeit considerably smaller in diameter, of the order of 4 nm. Nanotubes constitute an exciting class of 1D nanomaterials of which carbon nanotubes are recognized widely as materials of importance. Tungsten disulphide (WS_2) nanotubes were synthesized in 1992 just after the discovery of carbon nanotubes [39]. Since then nanotubes of several inorganic materials, *i.e.* metals, metal oxides, metal chalcogenides, nitrides, etc. have been reported. Typical inorganic nanotube materials are 2D-layered solids such as tungsten(IV) sulphide (WS_2), molybdenum disulphide (MoS_2) and tin(IV) sulphide (SnS_2)WS_2 and SnS_2/tin(II) sulphide (SnS). Traditional ceramics like titanium dioxide (TiO_2) and zinc oxide (ZnO) also form inorganic nanotubes. More recent nanotube and nanowire materials are transition metal/chalcogen/halogenide (TMCH), described by the formula $TM_6C_yH_z$, where TM is transition metal (molybdenum, tungsten, tantalum, niobium), C is chalcogen (sulphur, selenium, tellurium), H is halogen (iodine), and the composition is given by $8.2<(y+z)<10$. TMCH tubes can have a sub nanometer-diameter, lengths tunable from hundreds of nano-meters to tens of microns and show excellent depressiveness owing to extremely weak mechanical coupling between the tubes. **Figure 2.2a** depicts the SEM image of TiO_2 nanotubes obtained by calcination of poly(vinyl alcohol) titanium tetra isopropoxide (PVA-TTIP) hybrid nanofibers. Inorganic nanotubes can be synthesized by a variety of methods such as template inside, template outside (membrane and inverse micelle technique) and non-template methods.

Metal-oxide nanotubes have properties quite different from those of carbon nanotubes. They can be made more hydrophilic than carbon and can load nearly 50% of their weight with water. Controlling dimensions of nanostructures is critical because properties such as electronic band-gap largely depend on dimensions. Dimension control has proven difficult in carbon nanotubes, but control is possible with different chemical-process conditions across a broad range of metal-oxide materials, according to researchers. Inorganic nanotubes are heavier than carbon nanotubes and not as strong under tensile stress, but they are particularly strong under compression, leading to potential applications in impact resistant applications such as bulletproof vests. Metal oxides nanotubes show superior catalytic properties as compared to respective nanoparticles. For example, the photocatalysis of the TiO_2 nanotubes is reported to be better than that of commercially available anatase type TiO_2 nanoparticles. Inorganic nanotubes are most likely the best conductor of electricity on a nanoscale level.

Their thermal conductivity is comparable to that of the diamond along the tube axis. Mechanically they are the stiffest, strongest, and toughest fiber that can ever exist. Strong van der Waals attraction in nanotubes leads to spontaneous roping of many nanotubes, leading to self-assembled structure which may prove to be important in many applications.

A high-speed photo detector for visible and near-infrared light based on individual WS_2 nanotubes has been prepared in the laboratory. Since inorganic nanotubes are hollow and can be filled with another material, they can be used to preserve or guide it to a desired location or generate new properties in the confined filler material. To this goal, inorganic nanotube hybrids were made by filling WS_2 nanotubes with molten lead, antimony or bismuth iodide salt by a capillary wetting process, resulting in PbI_2@WS_2, SbI_3@WS_2 or BiI_3@WS_2 core-shell nanotubes

Nanorods are solid nanostructures morphologically similar to nanowires but with a low aspect ratios of approximately 3 : 5. They are formed from a variety of materials including metals, semiconducting oxides, diamonds and organic materials. **Figure 2.2d** shows an image of ZnO nanorods grown from aqueous solution. Aggregated diamond nanorods produced from fullerenes are the hardest material so far discovered. Nanorods show characteristic size dependent properties with the greatest effects observed in the 1–10nm size range. This is due to the large surface area-to-volume ratio of nanoparticles, which increases surface free energy to a point that is comparable to their lattice energy. Nanorods have the capacity for large variations in composition. In recent years, there has been an escalation in the development of techniques for synthesis of multicomponent nanorods and subsequent surface functionalization. Multifunctional nanoparticles exhibit characteristic electronic, optical, and catalytic properties significantly different from those of their individual constituent metals. Nanorods are under considerable investigation for applications in the area of gas sensors; video displays; computer components; nanoelectronics and nano-optoelectronic components; MEMS devices; and solar energy conversion. Nanorods are produced by a number of techniques, including: a vapor-liquid-solid approach; mechanical alloying; direct chemical synthesis using ligands for shape control; plasma arc discharge; laser ablation; and catalytic decomposition. A novel nanostructure fabrication technique is introduced in the production of nanorods called glancing angle deposition (GLAD). GLAD is a physical vapour deposition technique in which the substrate is rotated in the polar and azimuthal directions by two stepper motors programmed by a computer.

Nano belts are a class of nanostructure often viewed as a type of nanowire, usually made from semiconducting metal oxides (such as SnO_2, ZnO, In_2O_3 or CdO, or selenides such as CdSe). Nano belts form ribbon-like structures with widths of 30-300 nm, thicknesses of 10-30 nm, and lengths in the millimetre range. They are chemically-pure and structurally uniform single crystals, possessing rectangular cross sections, clean edges, and smooth surfaces. By virtue of their unique geometry (**Figure 2.2c**), nano belts have enormous surface areas per unit mass and most of the material is at the surface thus making them highly attractive element of gas sensors for detection of nerve gas, ethanol and hydrogen sensing. Nanobelts represent a unique way to manipulate light at the microscopic scale. Nano belts are also unique because the plasmonic waves occur across their width, not along their length. Nanobelts scatter light at a particular wavelength (or color), depending on the aspect ratio of their cross sections (width divided by height) which makes them highly tunable. Nanobelts possess essentially similar properties to nanowires; they can exhibit field effect transistors (FET) behaviour and have functionalizable surfaces.

Fig. 2.2. SEM images of nanomaterials of different morphologies: (a) TiO_2 nanotubes; (b) Nylon-6 and polystyrene nanofiber; (c) SnO_2 nanobelts; (d) ZnO nanorods

Nanolayers are one of the most important topics within the range of nanotech-nology. Through nanoscale engineering of surfaces and layers a vast range of functionalities and new physical effects (*e.g.* magnetic, electronic or optical) can be achieved. Furthermore, a nanoscale design of surfaces and layers is often necessary to optimize the interfaces between different types of materials (*e.g.* semiconducting material on silicon wafers) and to obtain the desired special properties.

A Nanocrystallite is generally understood to possess crystalline order in addition to nanoscale size. If one of the three-dimensional nanostructure is at nanoscale, then it is called a Quantum Well. If two dimensions of the three-dimensional nanostructure is at nanoscale, then it is called a Quantum Wire. If all the three dimensions of the nanostructure is at nanoscale, then it is called a Quantum Dot.

2.2 INORGANIC NANOMATERIALS

The spectrum of nanomaterials ranges from inorganic to organic, crystalline to amorphous particles, which can be obtained as single particles, aggregates, and powders or dispersed in a matrix, over colloids, suspension and emulsion, nanolayers and nanofilms. Nanomaterials of varied chemical compositions and their characteristics are described in the following sections.

Inorganic nanomaterials include metals, alloys, metal oxides and metalloid oxides, semiconductor quantum dots and carbon based nanomaterials (carbon nanotubes, graphene, fullerenes etc.).

2.2.1 Metals and Alloys

Nano-metals have an enormous potential for applications in electronics, construction, power transformation, energy storage, telecommunications, information technology, medicine, catalysis and environment protection. The most exploited properties of nanopowder are based on common phenomena at nanoscale such as high specific surface area, soft magnetic properties and new deformation mechanisms. Metallic nanomaterials included in **Table 2.2** are most promising for a variety of applications.

Table 2.2 Important metallic nanomaterials

Metallic nanomaterials	Examples
Metals and alloys	Ti, Ti-Al, Ti-transition metals (Fe, Ni,Cu) alloy; Fe-transition metal (Co, Ni, Cr, Cu, Zr) alloy; Fe-Cu-Nb-Si-B alloy; Al, Al-transition metal (Fe, Ni, Ti, Zr) alloy; Mg, Mg-Ni alloy; Mg-Al alloy
Nobel nanometals	Ag Au Ru Pt

Wet chemical reduction procedures have been used practically for all transition metals with different types of stabilizers and chemical reducing agents. Some of the commonly used reducing agents for preparing nano metals are diborane, borohydrides, alcohols, CO, formic acid/sodium formate, formaldehyde, or benzaldehyde, silanes, hydrogen, tetra-alkyl ammonium hydro tri-organoborates, tetrakis-(hydroxymethyl)-phosphonium chloride (THPC), hydrazine, hydroxylamine, etc. Among others tetra alkyl ammonium hydro tri-organoborates offers several advantages:

- The method is generally applicable to salts of metals in groups 4–11 of the periodic table.
- The method yields extraordinarily stable metal colloids that are easy to isolate as dry powders.
- The particle size distribution is nearly monodisperse.
- Bimetallic colloids are easily accessible by co-reduction of different metal salts.
- The synthesis is suitable for bulk preparations and easy to scale up.

Gold nanoparticles are mostly used in medical applications, in particular for in-vitro diagnostics. Other applications include catalysts, optics, solar cells, inks for printable electronics, sensors and surface coatings. Moreover, due to high thermal and electrical conductivity, gold NPs also find application in electronics. Gold is also widely used in biocatalysis and catalysis. Nano gold has been widely used to construct biosensors due to its excellent ability to immobilize biomolecules such as proteins, enzymes and antibodies whilst maintaining the biocatalytic activities of those biomolecules. Supported gold catalysts provide a high catalytic activity in numerous reactions such as the reduction of nitrogen oxide and oxidation of CO and the epoxidation of propene at low temperatures. The high catalytic activity of gold is not uniquely strange as bulk gold is quite inert. Reactive molecules such as CO and H_2 do not adsorb on the surface of gold. However, this behaviour dramatically changes when gold is highly dispersed as nanosized particles on certain metal oxides including TiO_2.

Silver nanoparticles were first produced in 1880. It was used for a long time in photographic film applications. Today, nano silver is mostly used in antimicrobial applications where a high release of silver ions is needed to kill the bacteria. Silver nanoparticles are synthesized either by using traditional Creighton methods where the $AgNO_3$ is reduced with $NaBH_4$, or through non-traditional methods such as photo reduction and electrolysis of Ag salt solution. Similar to nano Au, the optical properties of nano Ag can be systematically manipulated by adjusting the size, composition, and shape. **Figure 2.3** iluustrate vivid colors of plasmonic silver nanoplates with spectral properties optimized for red, blue, green, colours. Antimicrobial uses of nano silver include in textiles for hospitals, wound plasters, and anti-odor sportswear, bed mattresses, socks or inner wears. There are also reported uses in toys, household appliances such as refrigerators and washing machines, cosmetics, bathing soaps, containers for contact lenses, non-textile antimicrobial coatings, inks for inkjet printers, printable electronics, catalysts, photovoltaic, displays and fuel cells.

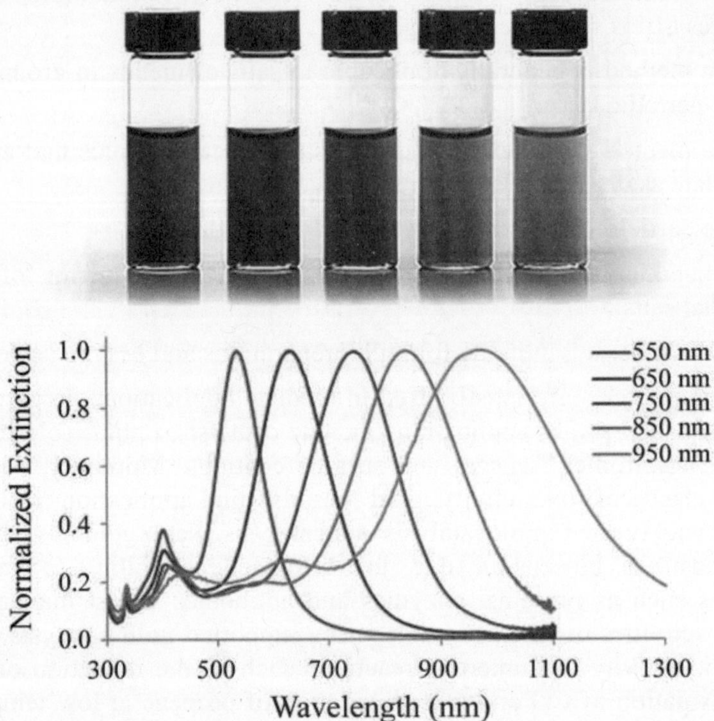

Fig. 2.3 Variation in colour of plasmonic silver nanoplates on changing dimensions

Copper nanopowder finds application in electronics and, to a small degree, in inks. Copper nano-powder is obtained by the simple reduction of metal salts with reducing agents such as $NaBH_4$, H_2, or hydrazine in the presence of a stabilizing agent Eq. (2.1).

$$Cu^{2+} + 2Cl^{-} + 6H_2O + 2NaBH_4 \rightarrow Cu^0 + 2NaCl + 7H_2 + 2B(OH)_3 \quad (2.1)$$

Theoretically any metal with a larger standard reduction potential (E^0) than the reducing agent (*e.g.*, −0.481V for borohydride ion) is a candidate for reduction to its metallic form. This includes most of the first row transition metal ions, and many others from the main group/transition metal series. However, it should be noted that solution pH and side-reactions (*e.g.* metal ions being converted to borides by BH_4^- rather than reduction), often provide a barrier toward successful metal ion reduction.

Iron nanoparticles are mostly used in magnetic recording tapes, waste water and land remediation. Nano zero valent iron (nZVI) can be synthesized in a variety of pathways such as the reduction and precipitation of ZVI from an aqueous iron salt solutions using sodium borohydride or reduction of iron oxides or hydroxides by hydrogen. nZVI is highly reducing and generates reactive oxygen species (ROS) through Fenton chemistry. nZVI may react with oxygen/water and get oxidized to ferrous ions (Fe^{2+}), where the electrons released are available to reduce other compounds (Eqns. 2.2, 2.3).

$$2Fe^0 (s) + O_2(aq) + 4H^+ (aq) \rightarrow 2Fe^{2+} (aq) + 2H_2O(l) \quad (2.2)$$

$$Fe^0 (s) + 2H_2O(aq) \rightarrow Fe^{2+} (aq) + H_2(g) + 2OH^- (aq) \quad (2.3)$$

Nano iron-based magnetic alloys are used for efficient energy transmission due to small size of the grains compared to the magnetic domain size

Nanoparticles of Platinum, Palladium and their alloys are mostly used in electronics (production of multi-layer ceramic capacitors). Other uses include catalysis (including combustion exhaust purification) and energy technologies. Their applications in data storage and healthcare field have also been investigated. Rhodium nanoparticles are reported to be used in catalysis. Titanium nanoparticles are the material of choice for medical implants. Nano titanium and Nano aluminium-based alloys are increasingly used as light weight alloy in aerospace and automotive sector due to their superior mechanical properties for structural applications. Nano magnesium and its alloys are used for hydrogen storage. Other metal nanoparticles *e.g.* nickel, cobalt, aluminium, zinc, manganese, molybdenum, tungsten, lanthanum, lithium are used in electronics. Coatings based on metallic NPs are used for improving tribological properties such as higher wear resistance, less friction, better corrosion resistance and sustainable production process.

2.2.2 Colloidal metal nanoparticles

Nanometal colloids are generally defined as isolable particles in size between 1 and 50 nm that are prevented from agglomeration by capping/protecting shells. Depending on the protection shell used they can be re-dispersed in water ('hydrosols') or organic solvents ('organo sols). The number of potential applications for these colloidal metallic nanoparticles is growing rapidly because of their unique electronic structure and their extremely large surface areas.

Nanostructured metal colloids can be obtained by both the so-called 'top down' and 'bottom up' methods. A typical top down method for example involves the mechanical grinding of bulk metals and subsequent stabilization of the resulting nanosized metal particles by the addition of colloidal protecting agents. The bottom up methods of wet chemical nanoparticles preparation rely on the chemical reduction of metal salts, electrochemical pathways, or controlled decomposition of metastable organometallic compounds. A large variety of stabilizers, *e.g.* donor ligands, polymers, and surfactants, are used to control the growth of the primarily formed nanoclusters and to prevent them from agglomerating. The chemical reduction of transition metal salts in the presence of stabilizing agents to generate zero valent metal colloids in aqueous or organic media was first published in 1857 by Faraday, and this approach has become one of the most common and powerful synthetic methods in this field.

The basic mechanism involved in metal colloid formation is that the metal salts are reduced to zero valent metal atoms in the embryonic stage of nucleation. These can collide in solution with further metal ions, metal atoms, or clusters to form an irreversible 'seed' of stable metal nuclei. Depending on the difference of the redox potentials between the metal salt and the reducing agent applied, and the strength of the metal-metal bonds, the diameter of the 'seed' nuclei can be well below 1 nm. Nanostructured colloidal metals require protective agents for stabilization and to prevent agglomeration. The two basic modes of stabilization which have been distinguished are electrostatic and steric (**Figure 2.4**). Electrostatic stabilization involves the columbic repulsion between the particles caused by the electrical double layer formed by ions adsorbed at the particle surface (*e.g.*, sodium citrate) and the corresponding counter ions. By coordinating sterically demanding organic molecules that act as protective shields on the metallic surface, steric stabilization is achieved. In this way nanometallic cores are separated from each other, and agglomeration is prevented.

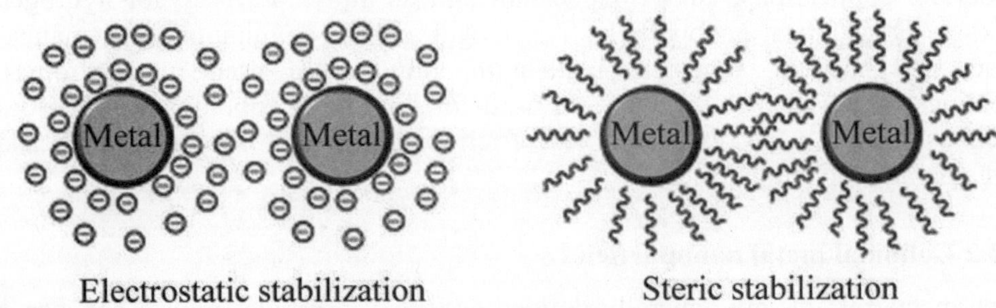

Fig. 2.4. Electrostatic and steric stabilization of nanostructure metal colloids

The main classes of protective groups are polymers and block copolymers; P,N,S donors (*e.g.*, phosphines, amines, and thioethers); solvents such as THF, THF/MeOH, or propylene carbonate; long chain alcohols; surfactants; and organometallics. In general, lipophilic protective agents make metal colloids soluble in organic media ('organosols') while hydrophilic agents yield water-soluble colloids ('hydrosols'). In Pd organosols stabilized by tetraalkylammonium halides the metal core is protected by a monolayer of the surfactant coat. Metal hydrosols, in contrast, are stabilized by zwitter ionic surfactants which are able to self-aggregate, and are enclosed in organic double layers.

Due to their nanosized dimensions, colloidal gold has been increasingly receiving considerable attention in nanotechnological advancement and development. Gold sols are prepared by the reduction of $[AuCl^{4-}]$ with sodium citrate or sodium borohydride or silver metal.

$$3Ag\,(s) + HAuCl_4\,(aq) \rightarrow Au(s) + 3AgCl\,(aq) + HCl\,(aq) \qquad (2.4)$$

Depending on the particle size, shape, and agglomeration gold colloids can be red, violet or blue, exhibiting interesting optical properties. Colloidal gold nanoparticles are mostly used in medical applications, in particular for in-vitro diagnostics. Other applications include catalysts, optics, solar cells, inks for printable electronics, sensors and surface coatings. Gold colloids also find applications in areas of Histochemistry and Cytochemistry where they are used as electron-dense labeling agents. Moreover, due to their high thermal and electrical conductivity, gold nanoparticles also find application in electronics.

2.2.3 Nano- oxides of transition and non-transition elements

Nanoparticles of transition and non-transition metal oxides are among the very important class of nanomaterials from the point of view of both of unique properties and application in many diverse areas.

Silicon dioxide (SiO_2). Various forms of synthetic amorphous silica are precipitated silica, silica gels, colloidal silica (or silica sols) and fumed (or pyrogenic) silica. Most forms are either used as stable dispersions of non-agglomerated SiO_2 particles (colloidal silica) or as agglomerated or aggregated particles (other forms of silica). Colloidal silica is stabilized dispersions of non-agglomerated, mostly spherical SiO_2 particles. Their main uses are in the paper industry (*e.g.* providing anti-slip properties; retention aids and in coatings of ink jet paper; better handling of recycled paper); chemical-mechanical planarization (CMP) slurries (*e.g.* polishing agent for Si wafers used to produce computer chips); coatings, paints, inks and adhesives (increasing strength, scratch and abrasion resistance); precision metal casting and refractory (*e.g.* molds for casting around wax originals); food industry (*e.g.* as an aid for clarifying wine, beer, fruit juices, etc.); bulk plastics and composites; photography; metal surface treatment; catalysis; textile; leather; and building industry (*e.g.* thermal and acoustic insulation).

Titanium dioxide (TiO_2). Nano titanium dioxide exists in various crystalline modifications, including rutile and anatase. Titanium dioxide powder has been used extensively as white pigment showing maximum reflectivity at a particle size of ~300 nm and also acts as an effective UV filter. The nano titanium oxide (~50 nm) is transparent, which provides an aesthetic advantage for uses in sunscreens (mostly rutile). NanoTiO_2 in the anatase modification has specific electrical and photocatalytic as well as antimicrobial properties. The nanoTiO_2 in anatase form is reported to be more reactive than the bulk form. Apart from sunscreens, the UV filter property of TiO_2 is also used in coatings for plastics and metals, varnishes for wood preservation, in textile fibers and in packaging films. Another main use is as catalysts (*e.g.* decomposing NO_x into nitrates or N_2). The photocatalytic and antimicrobial properties are used in "self-cleaning" products (*e.g.* windows, cement, tiles and textiles for use in hospitals) and air purification systems. Use in tribological coatings prevent deposits in engines and enhances fuel efficiency. TiO_2 nanoparticles are also used to increase scratch-resistance of coatings and in the production of electronic components and dental impressions. TiO_2 can also be used in dye-sensitized solar cells to produce electricity, though efficiency is currently lower than traditional silicon solar cells. Sol-gel method has been widely used for the production of nanoTiO_2. However, the sol-gel derived material is amorphous in nature and often requires thermal treatment (>350°C) to induce crystallization.

Zinc oxide (ZnO). Like titanium dioxide, zinc oxide in nanoform is colorless and an effective UV-filter with a different spectrum than titanium dioxide. It also has antimicrobial properties (though less strong than TiO_2) and can be used as an active agent in self-cleaning products. Major uses are as a UV-filter in cosmetics (where it competes with bulk zinc oxide but has the advantage of being transparent), in varnishes (as a UV-filter and self-cleaning agent), ceramics and electronics. Nanoform zinc oxide is also used in rubber, improving toughness, increasing abrasion resistance (*e.g.* reducing wear loss in tyres) and preventing UV and bacterial degradation. An emerging use is zinc oxide nanowires for UV nano lasers. Uses are also reported in liquid crystal displays and solar cells.

Several methods of synthesizing ZnO nanocrystals have been developed including hydrothermal, condensation template and surfactant assisted solution methods. Out of these, hydrothermal synthesis has proven to be a facile method for the synthesis of nano ZnO. Microwave assisted hydrothermal is also an alternative option that offers many advantages such as low cost, short reaction time, high purity and large scale production.

Aluminum oxide (Al_2O_3). Aluminum oxide nanoparticles are widely used as fillers in polymers and tyres and to increase scratch- and abrasion-resistance in coatings. Nano Al_2O_3 powders and dispersions are used inter alia in scratch- and abrasion-resistant coatings (*e.g.* for cutting and grinding tools, automobile exteriors, safety glasses and scratch-resistant windows for bar code scanners, flooring), as abrasive particles in slurries for polishing semiconductor and precision optical components, in the coating of light bulbs and fluorescent tubes, as a flame retardant, as fillers for polymers and tyres, in coatings of high-quality inkjet papers, in catalysts including the support structure in automobile catalytic converters, in refractory materials and as ceramic filtration membranes. Nano alumina can also be used for the manufacture of transparent ceramic bodies for high-pressure lamps.

Aluminum hydroxides and aluminum-oxo hydroxides. Different forms of aluminium hydroxide (*e.g.* bayerite and gibbsite) and aluminium-oxo hydroxide (*e.g.* boehmite and diaspore) particles in nanoform are available. Aluminium hydroxide $Al(OH)_3$ in powder form is used as a flame retardant and as filler in carpets, rubbers, plastics, and foamed plastics. Moreover, it is used in toothpaste and cosmetics. Aluminium-oxo hydroxides are often used in the dye and plastics industries as thickeners and fillers and as agents that reduce adhesiveness and increase scratch resistance. Besides, they also serve to enhance the color saturation of paints and varnishes.

Iron oxides. There are several types of nano iron oxides. The most common ones are hematite (ferric oxide, Fe_2O_3) and magnetite (ferro-ferric oxide, Fe_3O_4). Nano-sized ferric oxide are commonly used in pigment applications (*e.g.* in automotive industry or in cosmetics) offering clean shades of a number of colours but with high transparency while still offering protection from UV light. Magnetite particles have been used for a long time for data storage in magnetic tapes, hard drives, etc. Another use of magnetite nanoparticles is in ferro fluids, which are stable colloidal suspensions of magnetic nanoparticles in a liquid carrier. Ferro fluids are used in electronic components like loudspeakers and hard disks (preventing dirt particles from entering the hard drive), or in shock absorbers in the automotive industry. Other emerging applications of nano iron oxides are in medicine. After selectively attaching nanoparticles to tumour cells, those cells can selectively be destroyed by applying electromagnetic energy. Drugs and diagnostic agents attached to magnetic nanoparticles could selectively be transported to targets in the body. Other uses of iron oxide nanoparticles include polishing media, catalysts, components in fuel cells, oxygen sensors, ceramics and optoelectronic devices, and soil and groundwater remediation and water treatment.

Cerium dioxide (CeO_2). Ceria is a rare-earth oxide with specific optical properties. Nanostructured non-stoichiometric CeO_{2-x} films are used for applications in optical, electro-optical, microelectronic and optoelectronic devices. Nano ceria is used inter alia as a polishing material for glass surfaces and silicon wafers, to finish photo masks and disk drives, as an anticorrosion material, *e.g.* in exterior architectural paint, on steel and other metal plates, and in fuel cells. Another major application is as a catalytic diesel fuel additive, decreasing toxic diesel emissions and increasing fuel efficiency.

Zirconium dioxide (ZrO_2). Ceramic materials made by sintering nano zirconia (ZrO_2) powder have a number of unique properties, including some forms with very high fracture toughness. The most significant application of ZrO_2 is in optical connectors, followed by fuel cells, lithium-ion batteries, catalysts and ceramic membranes. Other developing applications are in structural and electronic ceramics, dental fillings, prostheses, fluorescent lighting, polishing agent etc.

Mixed metal oxide nanomaterials. Indium tin oxide (ITO) is a semi-conducting material used as thin-film material for the production of transparent electrodes in liquid crystal displays, touch screens, organic LEDs, thin-film solar cells, semiconducting sensors etc. Due to its IR reflectivity it is often used as thermal insulation coating on window glass. Its anti-static properties make it additionally suitable for packing and storage of sensitive electronic components. The high price and increased global demand of ITO has intensified the research for its alternatives. Antimony tin oxides (ATO) have similar IR-radiation reflecting properties and considered as alternative to ITO.

Other oxygen containing inorganic nanomaterials include barium titanate, barium sulphate, strontium titanate, strontium carbonate. Barium titanate powders are the dominating raw material for the production of ceramic dielectric layers in low-temperature multilayer ceramic capacitors (MLCC). Nano calcium carbonate also exists, although most of the fine-ground calcium carbonate is generally in a particle size above 100 nm. Ultra-fine calcium carbonate is used as advanced filler in sealants and in plastic for window frames. Fine-ground calcium carbonates are widely used as fillers in paper, plastics, paints, coatings, adhesive and sealants. They are also used as a food additive.

2.2.4 Non-oxide inorganic nanomaterials

Non-oxide inorganic nanomaterials include aluminium nitride, silicon nitride, titanium nitride, titanium carbon nitride, tungsten carbide, tungsten sulphide. Aluminium nitride is used in the electronic industry in various particle sizes, including nano size regime. Titanium nitride powders with a particle size from nano- to micrometres are used as additive in the production of wear-resistant sintered materials. Furthermore it is added to plastics, particularly to poly ethyl terephthalate (PET). TiN nanoparticles improve the thermal properties of the material and allow increasing the production output of PET bottles. Tungsten carbide nanoparticles are used mainly for hardening the surfaces of cutting tools to improve wear and temperature resistance. Tungsten carbide nanoparticles are at the threshold of large-scale production. Molybdenum and tungsten sulphide nanoparticles seem to be a promising lubricant for harsh conditions.

2.2.5 Zeolites

Zeolites are one of the best known example of nanoporous materials in the world today. They represent the largest group of micro porous (pore size <2 nm) crystalline inorganic polymers based on a three-dimensional tetrahedral arrangement of SiO_4 and AlO_4 (aluminosilicate) connected through their oxygen atoms. **Figure 2.5(a-c)** shows the structure of four commonly used zeolites: zeolite A, zeolite Y and zeolite L and ZSM-5. Due to their ion-exchange properties, as well as adsorption and reactions of molecules within its cages, zeolites have found use in numerous applications in catalysis and separations.

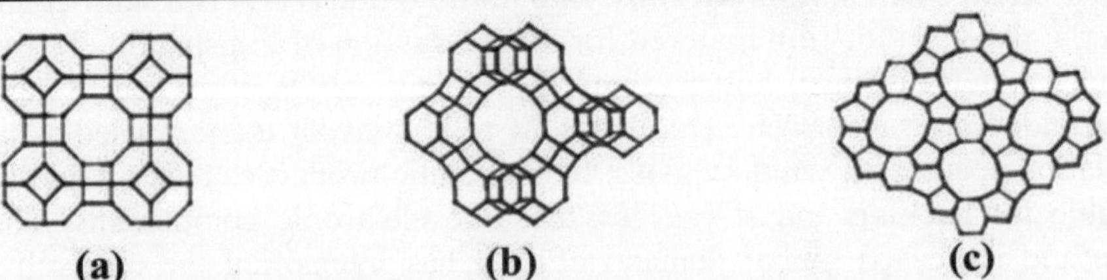

Fig. 2.5. Representative open pore zeolite frameworks: (a) zeolite A:3D, (b) zeolite Y:3D, (c) Zeolite L:1D

2.2.6 Quantum dots

Quantum dots, discovered in 1980, are tiny particles or nanocrystals of a semiconducting material with diameters in the range of 2-10 nanometres (10-50 atoms). Quantum dots display unique electronic properties, intermediate between those of bulk semiconductors and discrete molecules that are partly the result of the unusually high surface-to-volume ratios for these particles. The most apparent result of this is fluorescence, wherein the quantum dots can produce distinctive colours determined by the size of the particles **(Fig. 2.6)**.

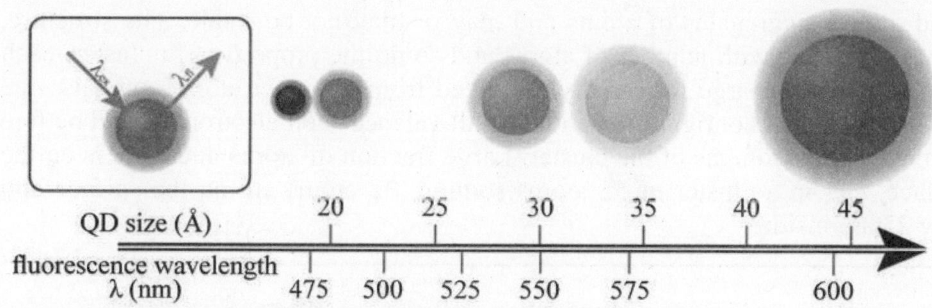

Fig. 2.6. Fluorescent colour and emission wavelength properties of CdSe QDs

Quantum dots can be classified as core-, core shell- and alloyed-quantum dot.

Core-type quantum dots are single-component materials with uniform internal compositions, such as chalcogenides (selenides or sulphides) of metals like cadmium or zinc; example, ZnS, CdS or CdSe. The photo- and electrolumine-scence properties of core-type nanocrystals can be fine-tuned by simply changing the crystallite size.

Core-shell quantum dots (CSQDs) are nanomaterials with small regions of one material embedded in another with a wider band gap, for example CdSe in the core and ZnS as the shell. Coating quantum dots with shells improves the quantum yield by passivating non-radiative recombination sites and also makes them more robust to processing conditions for various applications. Whereas, further coating of core-shell QDs with hydrophilic polymers make them water-soluble and bio-compatible for various medical applications.

Alloyed quantum dots are formed by alloying together two semiconductors with different band gap energies and exhibit interesting properties distinct not only from the properties of their bulk counterparts but also from those of their parent semiconductors. Thus, alloyed nanocrystals possess novel and additional composition-tunable properties aside from the properties that emerge due to quantum confinement effects.

2.2.7 Nanoclusters

Nanoclusters are agglomerates of a finite number of atoms or molecules, from a few to thousands, forming a bridge between small molecules and crystalline materials (**Fig. 2.7**), representing an intermediate stage between single atoms/ molecules and bulk. Usually clusters size is below 1 nm whereas size of nanoparticles is defined in 1-100 nm range. At such a small size, the shape of the cluster and interactions between clusters decide their properties. This opens up a huge possibility of tailor making the materials with different properties just due to their size, shape and/or assembly. Molecules have fixed properties due to their composition, structure, bond lengths, and bond angles. The clusters on the other hand are the aggregates of atoms and may or may not be stable. The structure, bonds, etc. vary with number of atoms and so do the properties. In cluster, each atom's positive charge is spatially separated from the other atoms and each one has its ionic shell configuration and thus all valence shell electrons would be free to form an electron gas of the cluster. Large fraction of atoms in cluster is on the surface, *e.g.* in a cluster of 55 atoms sodium, 32 atoms are on the surface and only 23 are inside.

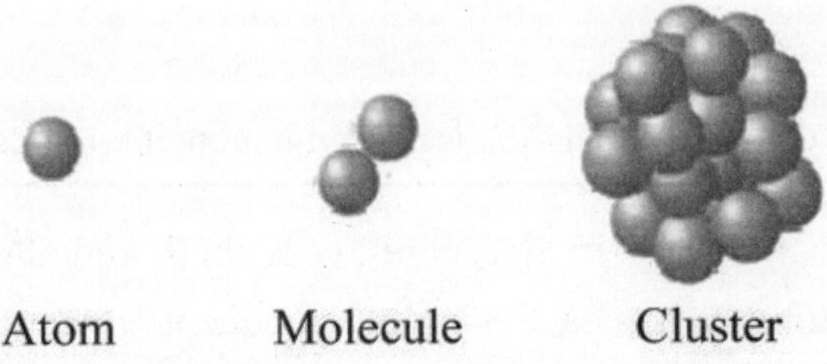

Fig. 2.7. Nanocluster from atom/molecule

Depending upon the number of atoms *(n),* clusters can be micro clusters *(n* ~ 2 to 10); small clusters (*n* ~10 to 100); large clusters *(n* ~ 100 to 1000) and *n* more than 10^3 and below 10^6 atoms, it is a nano-particle. Clusters can also be classified on the basis of their composition:

- Homogeneous clusters containing only single type of elements, *e.g.* Na, K, Ar, Au, Pb, C, Si
- Heterogeneous clusters containing combination of atoms of several elements in subunits with equal stoichiometry, *e.g.* 3Na+3K, 2Na+3K , (NaCl)n, $(Na_{n+1}Cl_n)^+$, $(Na_nCl_{n+1})^-$, $(SF_6)_n$, $(H_2O)_n$, $(Cu_3Au)_n$, $(ClCH_3C_6H_3CO_2H)_n$, $(TiO_2)_n$.

Clusters of nonmetals, *e.g.* carbon, silicon are held together by covalent bonds. In a silicon cluster, the stable configurations arise having 3, 4, 5, 6, 13 atoms with energetically favoured structures **(Fig. 2.8)** and not the fragments of the bulk lattice.

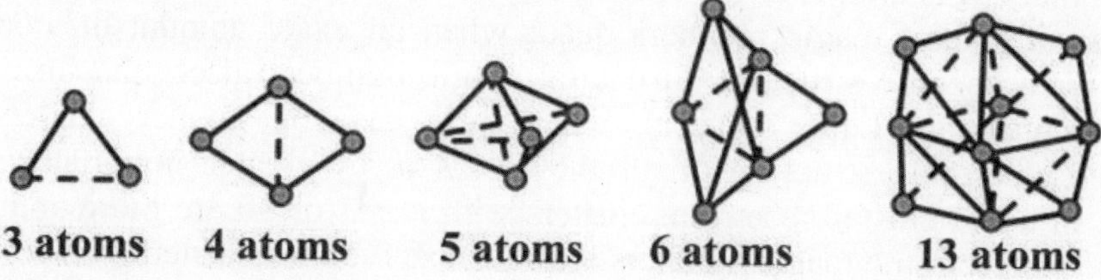

Figure 2.8 Clusters of silicon

Clusters of rare gases, *e.g.* Ar, Kr (close shell atom), are held together by weak van der Waals forces and are formed at low temperatures. Clusters with hydrogen bonds, for example a $(H_2O)n$ cluster is a strongly held molecular cluster, stronger than a cluster of rare gases but much weaker than metallic, covalent or ionic clusters. In case of metals, some critical number of atoms characteristic of the material assumes local stable sizes, depending on the ionization energy. Higher the ionization energy more stable is the cluster. When the ionization potentials of sodium clusters are plotted as a function of the number of atoms in the clusters, peaks are observed at clusters with n=2 and 8. This can be compared to the high ionization potentials observed in rare-gas atoms, where electrons are removed from the filled orbital. One can get more insight about metal clusters by using a Jellium model. The jellium model has successfully been used in the theoretical modeling of nanoclusters. It envisions a cluster of atoms as a single large atom, where the distribution of the ionic cores is replaced by a constant positive background, the so-called 'jellium density', and only valence electrons are treated explicitly. It can be seen from the jellium model that the electronic structure of a cluster varies from that of a single atom, but it also varies greatly from that of a bulk material. When a bulk lattice is formed, the discrete energy levels of the atoms are smudged out into energy bands. *E.g.* valence bands and conduction bands exist in semiconductors and metals. As particles shrink in size there is a dramatic increase in the energy band as well as discrete energy levels as the continuous density of states in bulk is replaced with a set of discrete energy levels.

Nanoclusters are obtained by initial vaporization of atoms from a bulk target using high energy sources. The evolution of their properties and structures can be traced as a function of size, and quantum effects unique to finite systems. A cluster may be considered as being formed shell by shell around a central atom. As the number of shells increases, total number of atoms goes on increasing and the percentage of surface atoms compared to interior atoms goes on decreasing. The magic number (n) of atoms in a cluster can be obtained by considering basic structures like cube, octahedron, icosahedrons, tetrahedron and dodecahedron. One of the common metal cluster is the 13-atom icosahedrons. It has one central atom and 12 surrounding atoms, 20 faces, 12 vertices and 30 edges. Number of atoms in the nth shell of an icosahedron can be obtained from $10\ n^2 + 2$. The magic number, n is unique to the cluster defined in terms of their structure or the electrons. Structural magic numbers occur when an exact amount of atoms is needed for a specific structure. Structural magic numbers affect the formation and abundances of larger clusters, whereas in case of smaller clusters the combined electronic structure of all the atoms is of greater importance. For instance in a Pb_n cluster spectrum, clusters with $n = 7$ or 10 are more abundant than others. Electronic magic numbers are also observed for clusters.

2.3 CARBON NANOMATERIALS

Carbon is the most fascinating element in the periodic table having the ability to bind itself to nearly all other elements. Carbon is important for many technological applications, ranging from drugs to synthetic materials. This role is a consequence of carbon's ability to bind to itself to nearly all elements in almost limitless variety. The resulting structural diversity of organic compounds and molecules is accompanied by a broad range of chemical and physical properties. Carbon is the base material for DNA and thus provides the basis for life on earth. Carbon can exist in several different allotropic forms, possibly by altering the periodic binding motif in networks consisting of sp^3, sp^2, and sp^1, hybridized carbon atoms. Diamond (sp^3) and graphite (sp^2) represented the only known allotropes of carbon for a long time. Serendipitous discovery of fullerenes by HW Kroto, RE Smalley and co-workers in 1985 at Rice University, marked the beginning of an era of synthetic carbon allotropes (SCb As) or better known as carbon nanomaterials. The new members added to this family include the carbon nanotubes (CNTs) in 1991 by S Iijima and the rediscovery of graphene in 2004 by KS Novoselov and AK Geim. All of these new carbon allotropes are bonded by sp^2 carbon atoms. Synthetic carbon allotropes consist of fully conjugated π-electron systems and can be considered as topologically confined objects in 0-, 1-, or 2-dimensions, respectively (**Fig. 2.9**).

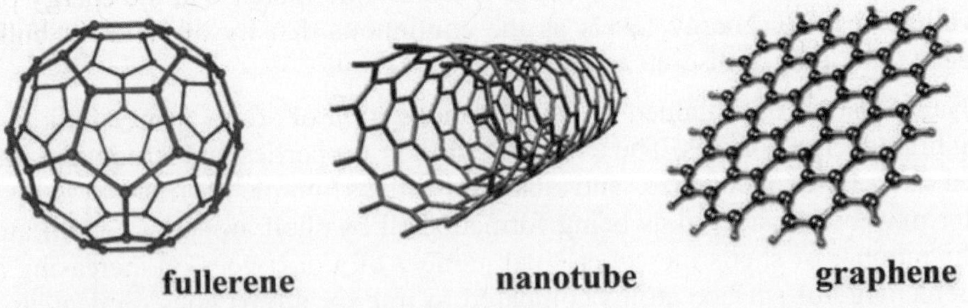

Fig. 2.9. Allotropes of carbon nanomaterials

Carbon nanomaterials are characterized by stability under ambient conditions currently represent one of the most promising material families with enormous potential for high-performance applications in the fields of nanoelectronics, optoelectronics, hydrogen storage, sensors, and reinforce-ments of polymers based on their unprecedented electronic, optical, mechanical, and chemical properties.

2.3.1 Fullerene

The fullerenes, of which C_{60} is the smallest stable and most prominent form, is a spherical network of 60 structurally equivalent sp^2 hybridized carbon atoms in the shape of a football (more technically, an I_h-symmetrical truncated icosahedron) composed of 12 pentagons and 20 hexagons with approximately 0.7 nm in diameter. In the fullerene structure, all C sites are equivalent and the bond lengths are 0.14nm for the double bond and 0.146 nm for the single bond. The C_{60} fullerene is also known Buckminster fullerene or Bucky ball. The name is a reference to Buckminster Fuller, as C_{60} resembles his trademark geodesic domes. The first successful preparation of 0-dimensional fullerene in macro scale quantities involved the evaporation and re-condensation of graphite using an arc-discharge method.

Many representatives of higher fullerenes have been isolated and characterized as D_{5h}-C_{70}, chiral D_2-C_{76}, D_{6h}-C_{84} and D_{5h}-C_{90}. Furthermore, a lot of chemistry has been performed on fullerenes to create fullerene salts, open-cage fullerenes and quasi-fullerenes. The latter are spherical closed structures composed of pentagons and hexagons combined with rings of other sizes.

Properties of fullerene: Fullerenes are large, closed-cage, carbon clusters and have several special properties that were not found in any other compound before. Fullerenes are nontoxic and soluble in many organic solvents. Functionalization of fullerenes further enhances its solubility. Spherical curvature in fullerene molecules, however, produces angle strain that allows fullerenes to undergo characteristic reactions of addition to double bonds (hybridization turns from sp^2 to sp^3), thus forming a large number of derivatives known as fullerides. C_{60} fullerene in organic solvents exhibit 5 stages of reversible oxidation/reduction thereby making it either as electrophiles or nucleophiles. This characteristic conjugation leads to the pronounced redox activity and imparts outstanding electronic properties. Further, chemical functionalization enables a dramatic increase in the solubility of fullerenes, in any solvent, which is required for their development as new materials or as bioactive redox drugs. Further, chemical functionalization also provides the possibility of combining the unique properties of fullerenes with those of other compounds. Exohedral addition reactions to the conjugated π- system represent the most important functionalization methods realized so far. The driving force for exohedral additions is the reduction of strain energy stored in the spherical carbon framework. Additions usually occur regioselectively at fusions of two six-membered rings, because in contrast to the neighbouring [5, 6]-bonds, these [6, 6]-bonds are shorter and have more double-bond character. Cluster-opened fullerenes have been prepared that can be filled with small molecules, such as hydrogen, and subsequently reclosed to form

$H_2@C_{60}$. This opens the door to a whole world of new endohedral derivatives. Endohedral fullerenes, incorporates guest atoms or molecules inside the carbon framework. The first endohedral fullerene containing an Ih-symmetrical C_{60} cage is $Li@C_{60}$. The recent examples include $La@C_{82}$ and $Sc_{3N}@C_{80}$. Other types of fullerene derivative for example, salts, such as K_3C_{60} and Tetrakis-dimethyl-amino-ethylene(TDAE)-C_{60} shows unprecedented properties such as superconductivity and ferromagnetism respectively. The substitution of carbon atoms with heteroatoms such as nitrogen, realized in the hetero fullerene $(C_{59}N)_2$ allows for further variation of the electronic properties of the fullerene core. Hydrogenation of fullerene, catalyzed with metals at low pressure, produces fullerene hydrides.

Fullerenes packed, in fcc structure, can be intercalated with alkali and alkaline earth metal atoms which provide electrons to the conduction band (from semiconducting to metallic behaviour). K-doped fullerene (K_3C_{60}) revealed superconductivity behaviour at 18K. Physical and chemical properties may be tuned via the addition of element and molecular species into the fullerene lattice ($C_{59}N$), within the cage (N at center of C_{60}), or coating the surface of fullerene with transition metals, thereby forming a variety of *new* fullerenes. Fullerene and its derivatives are considered potential candidates for a number of applications, such as:

i. To catalyse the conversion of methane into hydrocarbons and prevents coking reaction.
ii. Fullerene- based capacitors as sensors to detect H_2S in N_2 and water in isopropanol.
iii. Fullerene can be transformed to diamond at high pressure at room temperature.
iv. Inclusion of fullerenes imparts improved mechanical and electrical of metallic alloys.
v. Fullerene derivatives find several application in biomedical fields such as inhibition of human HIV replication; biological antioxidant (radical sponge); binding specific antibiotics to the structure of resistant bacteria and even target certain types of cancer cells such as melanoma; and light-activated antimicrobial agents.

2.3.2 Carbon Nanotubes (CNTs)

CNTs are extended tubes of rolled graphene sheets and have assumed an important role in the context of nanomaterials, because of their novel chemical and physical properties. The electrical conductivity and the outstanding mechanical properties of carbon nanotubes can be mainly explained in terms of the graphite-like structure of the walls of these materials. The sp2-hybridized carbon atoms in conjugated C=C bonds, in analogy to those described for graphite, enable the electron delocalization. All of these remarkable properties give CNTs a range of potential applications in the field of composites, sensors, nanoelectronics and display devices. CNTs are now available commercially. They can be grown by several techniques such as arc discharge, CVD and laser ablation. However, the selective and uniform production of CNTs with specific dimensions and physical properties is yet to be achieved.

There are two types of CNTs, one tubular, i.e single-walled carbon nano tube (SWCNTs) and several concentric tubes, *i.e.* multi-walled carbon nano tubes (MWCNTs). Both of these are typically a few nanometres in diameter and several micrometres to centimetres in length. The SWCNTs may bundle into a rope of several SWCNTs. Most single-walled nanotubes (SWCNTs) have a diameter of close to 1 nanometer, with a tube length that can be many millions of times longer. Carbon nanotubes can also be classified on the basis of structures such as: armchair, zigzag, and chiral nanotubes, depending on how the two-dimensional graphene sheet is 'rolled up' (**Fig. 2.10**). Multiwalled carbon nanotubes (MWCNTs) have more than one graphene cylinder nested one into another. Typical MWNTs have an inner diameter of 1-3 nm and an outer diameter of approximately 10 nm with the inter layer separation, of ~0.34 nm.

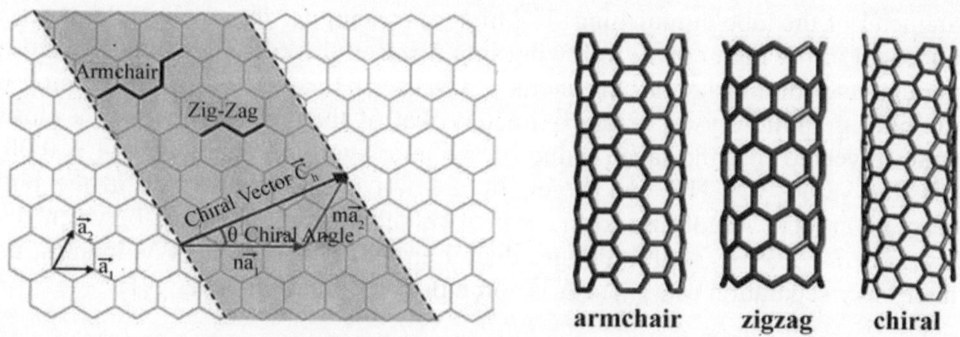

Fig. 2.10. Different structural forms of CNTs

The way the graphene sheet is wrapped is represented by a pair of indices (*n,m*). The integers *n* and *m* denote the number of unit vectors along two directions in the honeycomb crystal lattice of graphene. The diameter (*d*) of an ideal nanotube can be calculated from its (*n,m*) indices as follows (5.1), where $a = 0.246$ nm.

$$d = \frac{\alpha}{\pi}\sqrt{\left(n^2 + nm + m^2\right)} \qquad (2.4)$$

In zigzag SWCNTs, the zigzag pattern is around the circumference and $m = 0$; in chiral SWCNTs the twisting of hexagons is around tubulular body; and in armchair SWCNTs the chair-like pattern is around the circumference and $n = m$. To discriminate between different types of carbon nanotubes, the chiral angle and vector play an important role in determining the important properties of nanotubes. Armchair nanotubes are formed when $n = m$ and the chiral angle is 30°. Zigzag nanotubes are formed when either n or m is zero and the chiral angle is 0°. All other nanotubes, with chiral angles intermediate between 0° and 30°, are known as chiral nanotubes. The name 'chiral' means handedness and indicates that the tubes may twist in either direction. The geometry of the chiral SWCNT lies between that of the armchair and zigzag SWCNTs. The chiral form is believed to be the most commonly occurring SWNT.

The properties of nanotubes are also determined by their diameter, which depends on n and m. A nanotube is usually characterized by its diameter d and the chiral angle θ ($0 \leq |\theta| \leq 30°$). The integers (n, m) determine d and θ. Further, n and m also determines the conducting behaviour of nanotubes; if $(n–m) \neq 3\times$ integer; it is semiconducting behaviour, whereas when $(n–m) = 3\times$ integer; a metallic conducting behaviour. Multiwalled carbon nanotubes (MWCNT) are nanotubes with more than one graphene cylinder nested one into another. Typical sizes for MWNTs are an inner diameter of 1-3 nm and an outer diameter of approximately 10 nm. The spacing of inter shell by using image of TEM with high resolution; spacing found to vary from 0.34 nm, augmenting with the diameter of the tube diminishing. The biggest spacing for the smallest diameter is allocated in the high curve, following in a distasteful force augmented, linked to the diameter diminished by the shells of CNTs. The value of 0.34 nm spacing in the bulk graphite crystal is approximately that of the CNTs. However, a closer study revealed that the mean value of the interlayer spacing is 0.3444 ± 0.001 nm; the values in CNTs are larger, by a few per cent, than those in the bulk graphite crystal. According to theoretical calculations the distance between two layers is $d = 3.39$ Å, slightly bigger than in graphite. Based on TEM images, the inter layer separation of $d = 3.4$ Å is commonly reported for MWCNT.

Functionalization of CNTs

Functionalization is a surface treatment which brings functional groups at the surface of nanoparticles. This is particularly important for CNTs having a strong tendency towards agglomeration. The functionalization might also accomplish changes in the chemical structure and physical properties of CNT. The chemical reactions involving the cleavage of C=C can lead to decreased electrical conductivity.

The functionalization methods can be classified in five main groups:

i. ***Strong oxidation:*** The main goal of this procedure is introduction of carboxylic acids and other carbonyl groups at the surface of the carbon nanotubes. This method is mostly suitable when an optimal reinforcement effect in a matrix of the composite is primarily required.

ii. ***Cycloaddition reactions:*** In this method, the C=C bonds of the CNTs react with dienes or tetrazines. In this process the C=C bonds are broken leading to changes in the electrical conductivity.
iii. ***Radical-based reactions:*** This method is based on the cleavage of C=C bonds by free radicals, generated in-situ during the functionalization process.
iv. ***Mild redox reactions***: In these procedures, the C=C bonds are broken through reduction with metals followed by another functionalization step. This method has high selectivity and the functionalized CNTs exhibit a remarkable dispersability.
v. ***Non-covalent functionalization***: This method facilitates the CNT dispersion in polymeric matrices and solvents by making use of surfactants, aromatic hydrocarbons (pyrenes) conjugated polymers aromatic polyimide bio macromolecules, cytosine, thiophene, as modifiers of carbon nanotubes. The main advantage of this method is that no cleavage of C=C is necessary, so the electrical, thermal and mechanical properties of the CNTs are not affected.

Purification of CNTs

Purification of carbon nanotubes is an important step in realizing their application. Normally CNTs are associated with impurities such as graphite sheets, amorphous carbon, metal catalyst and the smaller fullerenes irrespective of their mode of synthesis. These impurities will interfere with most of the desired properties of the CNTs. The common purification techniques can be divided into two main categories, *i.e.* structure selective and size selective separations. The first one will separate the SWNTs from the impurities; the second one will give a more homogeneous diameter or size distribution.

The purification techniques include oxidation, acid treatment, annealing, ultra sonication, micro filtration, ferromagnetic separation and cutting, functionalization and chromatography techniques. Oxidative treatment of the SWNTs is a good way to remove carbonaceous impurities and other impurities attached to the metal surface. Acid treatment removes the metal catalyst.

Annealing at high temperatures (873–1873 K) leads to the removal of graphitic carbon and the short fullerenes through their pyrolysis. High temperature treatment under vacuum also facilitates the removal of metal through melting. Technique of ultra-sonication facilitates de-agglomeration of CNTs particles in presence of suitable surfactant and solvent. Magnetic Purification helps in removal of ferromagnetic particles (metallic catalyst) from the graphitic shells.

Unique properties of carbon nanotubes

Carbon nanotubes show extraordinary mechanical, electrical, and thermal properties because of their C=C covalent bonding and seamless hexagonal network architecture. Electronic, molecular and structural properties of carbon nanotubes are determined to a large extent by their nearly one dimensional structure.

Chemical reactivity: The chemical reactivity of CNT arises due to the presence of C=C bonds and functional group. The chemical reactivity of a CNT is more as compared to the graphene sheet as a result of the curvature of the CNT surface.

Electrical conductivity: Depending on their chiral vector, carbon nanotubes with a small diameter are either semi-conducting or metallic. The differences in conducting properties are caused by the molecular structure that results in a different band structure and thus a different band gap. The differences in conductivity can easily be derived from the graphene sheet properties. It is shown in **Figure 2.10(d)** that a (n, m) nanotube is semiconducting if $n - m \neq 3 \times$ integer whereas shows metallic conductivity when, $n = m$ or $n–m = 3 \times$ integer, where n and m are defining coordinates of the nanotube. The resistance to conduction is determined by quantum mechanical aspects and was proved to be independent of the nanotube length.

Optical activity: Theoretical studies have revealed that the optical activity of chiral nanotubes disappears if the nanotubes become larger. Therefore, it is expected that other physical properties are influenced by these parameters too.

Mechanical strength: Carbon nanotubes have a very large Young modulus of over 1 Tera pascal in their axial direction, making CNTs as stiff as diamond. The nanotube as a whole is very flexible because of the great length. Therefore, these compounds are potentially suitable for applications in composite materials that need anisotropic properties.

2.3.3 Graphene

Graphene, the youngest allotrope of elemental carbon, is a one-atom-thick sheet of sp^2-bonded carbon atoms in a hexagonal arrangement **Fig. 2.9**. Graphene is at the cutting-edge of materials science and condensed matter physics research. It is the thinnest known material in the world and conceptually a basic building block for constructing other nanocarbon materials. It can be rolled into one-dimensional carbon nanotubes (CNTs), and stacked into three-dimensional (3D) graphite. With the addition of pentagons it can be wrapped into a spherical fullerene. Chemically graphene is obtained in the form of graphene oxide by oxidation of graphite to graphite oxide followed by exfoliation. Myriads of applications are being developed for graphene sheets as electrodes in ultra-capacitors for fast recharging in minutes, replacing indium in flat screen TVs, DNA attached graphene as sensors for rapid disease diagnostics.

Unique properties of graphene

Graphene is expected to out-perform CNTs, graphite, metals and semiconductors as an individual material or as a component in a hybrid/ composite material for its unique properties such as zero band gap, the fastest electron mobility of ~15,000 $cm^2\ V^{-1}\ cm^{-1}$ or $10^6\ \Omega$ cm (lower than Ag), a super high mobility of temperature-independent charge carriers of 200,000 $cm^2\ V^{-1}\ s^{-1}$ (200 times higher than Si), and an effective Fermi velocity of 10^6 m s^{-1} at room temperature, similar to the speed of light.

Charge carriers in graphene are described by a Dirac-like equation, instead of the usual Schrödinger equation. Graphene is assumed to be a perfect crystal structure having low-energy quasiparticles which obey a linear dispersion relation, similar to massless relativistic particles. The essential characteristic of a gap less semiconductor has led to many observations of peculiar electronic properties including ballistic transport, pseudo spin chirality based on the "Berry phase", a room-temperature half-integer "chiral" quantum Hall effect, and conductivity without charge carriers, that make it a promising choice for future electronic materials, both as a device and as an interconnect. Graphene has not only unique electronic properties, but also excellent mechanical, optical, thermal and electrochemical properties that are superior to other carbon allotropes such as graphite, diamond, fullerene and CNTs.

As yet, graphene appears to be one of the strongest materials ever tested. Single-layer graphene has excellent mechanical properties with a Young's modulus of 1.0 TPa and a stiffness of 130 GPa (19,000,000 psi). Graphene has a breaking strength 200 times greater than that of steel.

Graphene is almost transparent with optical transmittance of ~97.7% (absorbing 2.3% of white light), and superior thermal conductivity of 5000 W $m^{-1}\ K^{-1}$ (about 100 times that of Cu). It also has a high theoretical specific surface area of 2620 $m^2\ g^{-1}$, extreme electrical conductivity and good flexibility. Graphene has temperature independent high charge carrier mobility (10,000 - 200,000 VS/cm^2) in suspended state with high carrier concentration ($n=2x10^{11}$) leading to ballistic transport in micrometer range. Graphene is thus a candidate for high frequency electronic devices and is a potential candidate for high tech applications such as ultra-sensitive gas detectors, Spintronics, quantum computing, etc. Graphene is a very low density material, density being 0.77 mg/ m^2, calculated on the basis of the fact that the unit hexagonal cell of graphene contains two carbon atoms and has an area of 0.052 nm^2.

Electronic structure of graphene

The electronic structure of graphene is rather different from usual three-dimensional materials. Its Fermi surface is characterized by six double cones, as shown in **Fig. 2.11**. In intrinsic (undoped) graphene the Fermi level is situated at the connection points of these cones. Since the density of states of the material is zero at that point, the electrical conductivity of intrinsic Graphene is quite low and is of the order of the conductance quantum σ~e2/h; the exact prefactor is still debated. The Fermi level can however be changed by an electric field so that the material becomes either n-doped (with electrons) or p-doped (with holes) depending on the polarity of the applied field. Graphene can also be doped by adsorbing, for example, water or ammonia on its surface. The electrical conductivity for doped graphene is quite high, may even be higher than that of copper at even room temperature. The black line in **Figure 2.11** represents the Fermi energy for an undoped graphene crystal. The energy, *E*, for the excitations in graphene as a function of the wave numbers, kx and ky, in the x and y directions. The black line represents the Fermi energy for an undoped graphene crystal. Close to this Fermi level the energy spectrum is characterized by six double cones where the dispersion relation. Close to the Fermi level the energy spectrum is characterized by six double cones where dispersion relation for electrons and holes is linear. Since the effective masses are given by the curvature of the energy bands, this corresponds to zero effective mass. The equation describing the excitations in graphene is formally identical to the Dirac equation for mass less fermions which travel at a constant speed. The connection points of the cones are therefore, called Dirac points. This gives rise to interesting analogies between graphene and particle physics, which are valid for energies up to approximately 1 eV, where the dispersion relation starts to be nonlinear. One result of this special dispersion relation is that the Quantum Hall effect becomes unusual in graphene.

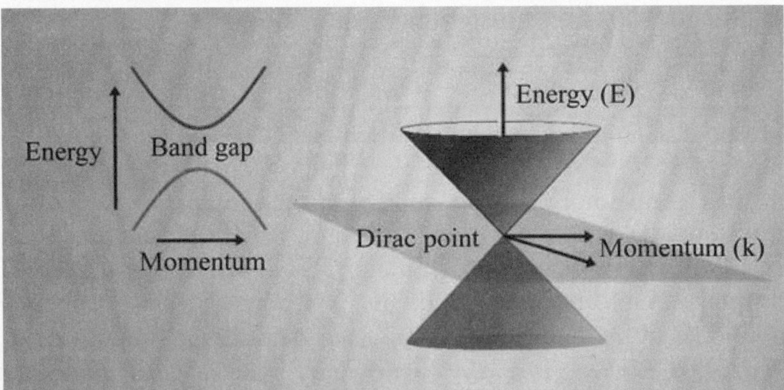

Fig. 2.11. Band structure of graphene

Band structure

Band structure of graphene shows the π-bands which are responsible for charge carrier transport. In contrast to semiconductors, which possess a parabolic dispersion relation, graphene exhibits a linear dependence of the electron energy on the wave vector. The first Brillouin zone of graphene is illustrated in the horizontal plane and labelled with some points of interest. The conduction and valence bands cross at points *K* and *K'* are the two non-equivalent corners of the zone, also known as the Dirac points. Graphene differs extremely from conventional 2D electron gas systems created in semiconductor heterostructures. The reason is the linear dispersion relation $E(\boldsymbol{k})$ of the charge carriers in the vicinity of the *K*-point of the hexagonal Brillion zone, where the bonding π-band meets the anti-bonding π*-band. The band structure thus has the form of a double cone, which formally is equivalent to the dispersion relation of rest-mass-less particles. The symmetry of the lattice requires a two-component wave function, similar to particles known from relativistic quantum mechanics. The charge carriers and their behavior are described by Dirac equation for mass-less fermions. This has several interesting consequences. An example is the unusual Landau level spectrum when the system is subject to a magnetic field, and the resulting new half-integer quantum Hall effect. Other interesting properties of charge carriers in graphene are their scattering and interference phenomena. Graphene thus provides new and interesting physics to study both experimentally and theoretically.

Graphene oxide

As graphene is expensive and relatively difficult to produce in large quantity, great efforts are made to find effective yet inexpensive ways to make and use graphene derivatives or related materials such as graphene oxide (GO) is a single-atomic layered material, made by the oxidation of graphite which is cheap and abundant. Graphene oxide is an oxidized form of graphene, laced with oxygen-containing groups such as hydroxyl, carbonyls and epoxy. It is considered easy to process since it is dispersible in water (and other solvents), and it can even be used to make graphene. Graphene oxide is obtained by the oxidation of graphite to graphite oxide using strong oxidizing agents, followed by exfoliation. Graphene oxide can further be reduced to graphene oxide having many properties similar to pristine graphene. Intermediate graphite oxide is synthesized using four basic methods: Staudenmaier, Hofmann, Brodie and Hummers. One of the advantages of graphite oxide is its easy dispersability in water and other organic solvents, as well as in different matrices, due to the presence of the oxygen functionalities (**Fig. 2.12**). This remains as a very important property when mixing the material with ceramic or polymer matrices while trying to improve their electrical and mechanical properties.

GO is heavily oxygenated, bearing hydroxyl and epoxy groups on sp^3 hybridized carbon on the basal plane, in addition to carbonyl and carboxyl groups located at the sheet edges on sp^2 hybridized carbon. Hence, GO is highly hydrophilic and readily exfoliated in water, yielding stable dispersion consisting mostly of single layered sheets (graphene oxide). GO consists of graphene sheets decorated mostly with epoxide and hydroxyl groups. Rapid heating of GO results in its expansion and delamination caused by rapid evaporation of the intercalated water and evolution of gases from pyrolysis of the oxygen containing functional groups. Such thermal treatment has recently been suggested to be capable of producing individual functionalized graphene sheets.

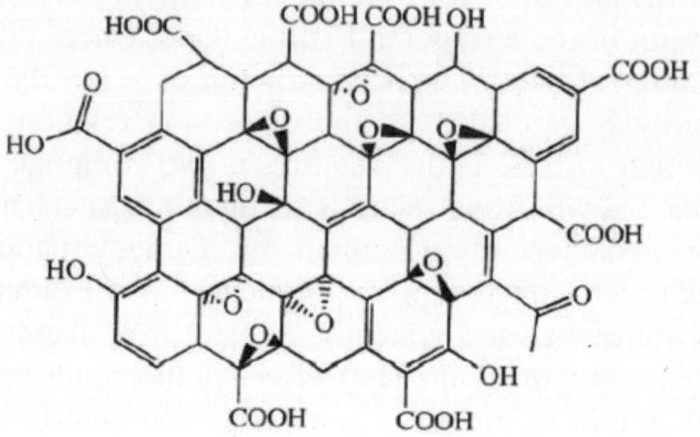

Fig. 2.12. Graphite oxide

Graphene oxide is not a good conductor, but processes exist to augment its properties including the conductivity. In terms of electrical conductivity, graphene oxide is often described as an electrical insulator, due to the disruption of its sp^2 bonding networks. In order to recover the honeycomb hexagonal lattice, and with it the electrical conductivity, the reduction of graphene oxide has to be achieved.

Functionalization of graphene oxide can fundamentally change its properties. The resulting chemically modified graphene could then potentially become much more adaptable for a lot of applications. There are many ways in which graphene oxide can be functionalized, depending on the desired application. For optoelectronics, bio-devices or as a drug-delivery material. For example, it is possible to substitute amines for the organic covalent functionalization of graphene to increase the dispersability of chemically modified graphene in organic solvents. It has also been proved that porphyrin-functionalized primary amines and fullerene-functionalized secondary amines could be attached to graphene oxide platelets, ultimately increasing nonlinear optical performance.

The surface modifications to functionalize graphene oxide can be achieved by two approaches *i.e.* covalent- and non-covalent bonding. In covalent functionnalization, oxygen functional groups on graphene oxide surfaces, including carboxylic acid groups at the edge and epoxy/hydroxyl groups on the basal plane can be utilized to change the surface functionality of graphene oxide. Graphene oxide had been treated with organic isocyanates to give a number of chemically modified GO. Treatment of isocyanates reduces the hydrophilicity of graphene oxide by forming amide and carbamate esters from the carboxyl and hydroxyl groups of graphene oxide respectively. The subsequent addition of nucleophilic species, such as amines or alcohols, produced covalently attached functional groups on graphene oxide via the formation of amides or esters. **Figure 2.13,** shows isocyanate treatment of GO where organic isocyanate reacts with the hydroxyl and carboxyl groups of the graphene oxide sheets. The resultant amine functionalized graphene oxide has demonstrated various applications in optoelectronics, drug-delivery materials, biodevices, and polymer composites. The attachment of hydrophobic long, aliphatic amine groups on hydrophilic graphene oxide improved the dispensability of modified graphene oxide in organic solvents, while porphyrin-functionalized primary amines and fullerene-functionalized secondary amines introduced interesting nonlinear optical properties. The amine groups and hydroxyl groups on the basal plane of graphene oxide have also been used to attach polymers through either grafting-onto or grafting-from approaches. To grow a polymer from graphene oxide, an atom transfer radical polymerization (ATRP) initiator was attached to graphene surfaces. Besides the carboxylic acid groups, the epoxy groups on graphene oxide can be used to attach different functional groups through a ring-opening reaction.

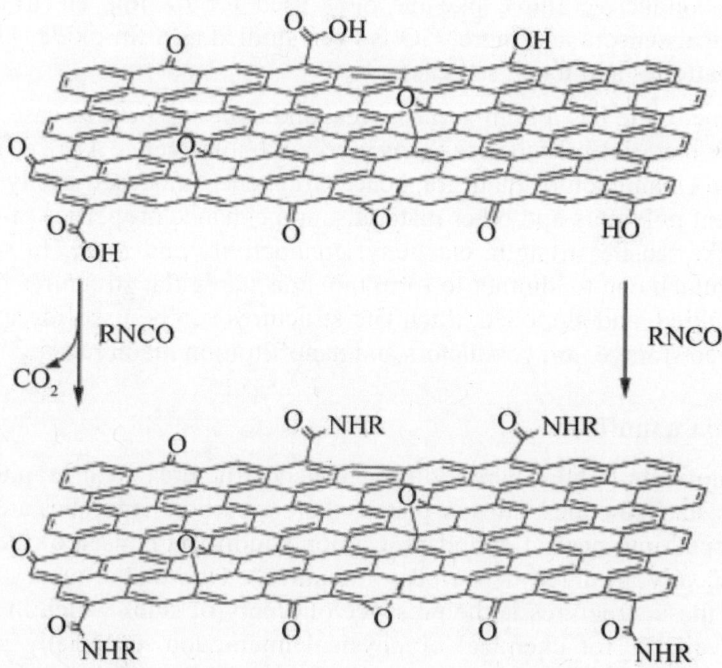

Figure 2.13 Isocyanate treatment of GO [Stankovich *et al.* ©2006, Elsevier] Applications of graphene oxide

The **non-covalent functionalization** of graphene oxide utilizes the weak interactions (*i.e.* p–p interaction, Van der Waals interactions and electrostatic interaction) between the graphene oxide and target molecules. The sp^2 network on graphene oxide provides p–p interactions with conjugated polymers and aromatic compounds that can stabilize reduced graphene oxide resulted from chemical reduction and produce functional composite materials. During the chemical reduction of graphene oxide, reduced graphene oxide nanosheets are stabilized via the π–π interaction between aromatic molecules and reduced graphene oxide nanosheets. Aromatic molecules have a large aromatic plane and can anchor onto the reduced graphene oxide surface without disturbing its electronic conjugation, providing stability for reduced graphene oxide. Dye-labelled DNA has also been used to functionalize graphene oxide to detect proteins and DNA. In the presence of a target, the binding between the dye-labelled DNA and target molecule will alter the conformation of dye labelled DNA, and disturb the interaction between the dye-labelled DNA and graphene oxide. Such interactions will release the dye-labelled DNA from the graphene oxide, restoring dye fluorescence.

Graphene oxide (GO) films can be deposited on essentially any substrate, and later converted into a conductor. It is especially fit for use in the production of transparent conductive films, like the ones used for flexible electronics, solar cells, chemical sensors and more. GO is even studied as a tin-oxide (ITO) replacement in batteries and touch screens.

Graphene oxide has a high surface area, and so it can be a fit case for its use as electrode material of batteries, capacitors and solar cells. Also, it is cheaper and easier to manufacture than graphene. Graphene oxide can easily be mixed with different polymers and other materials, and enhance properties of composite materials like tensile strength, elasticity, conductivity and more. In solid form, GO flakes attach one to another to form thin and stable flat structures that can be folded, wrinkled, and stretched. Such GO structures can be used for applications like hydrogen storage, ion conductors and nanofiltration membranes.

2.3.4 Carbon nanofibers

Carbon nanofibers (CNFs), are cylindrical nanostructures with graphene layers arranged as stacked cones, cups or plates whereas carbon nanotubes are graphene layers wrapped into perfect cylinders. Carbon nanofibers consist of the graphene sheet completely arranged in various orientations. One of the most outstanding features of these structures is the presence of plenty of sides which in turn make the sites available for chemical or physical interaction, especially adsorption. Carbon nanofibers can be obtained in sizes ranging from 5 to several hundred microns on length and between 100 and 300 nm in diameter. From electronic microscopy studies, it is possible to define sequence of the events leading to the formation of carbon nanofibers. **Figure 2.14** shows HRTEM image of plasma assisted grown carbon nanofibers. Carbon nanofibers have high electrical conductivity (5×10^5 $S.cm^{-1}$) and thermal conductivity (2000 W/m.K). Its mechanical strengths may be as high as 8.7 GPa, which approaches the strength of graphite microfibers. These properties makes CNFs as the ideal reinforcing material in polymer matrices to achieve highly strong and electrically and thermally conducting nano composites.

2.3.5 Carbon nano-onion

Carbon nano-onions (CNOs), a fullerene-related material having concentric graphitic shells, represent another new allotropic nanophase of carbon materials. Essentially a multiwall fullerene since center shell is a C_{60} molecule. CNOs have already been shown to offer a variety of potential applications such as solid lubrication, electromagnetic shielding, fuel cells, heterogeneous catalysis, gas and energy storage, and electro-optical devices owing to their outstanding chemical and physical properties. According to a recent study, CNOs can also be used to produce ultra-high power micrometre-sized super capacitors due to their accessible external surface area for ion adsorption.

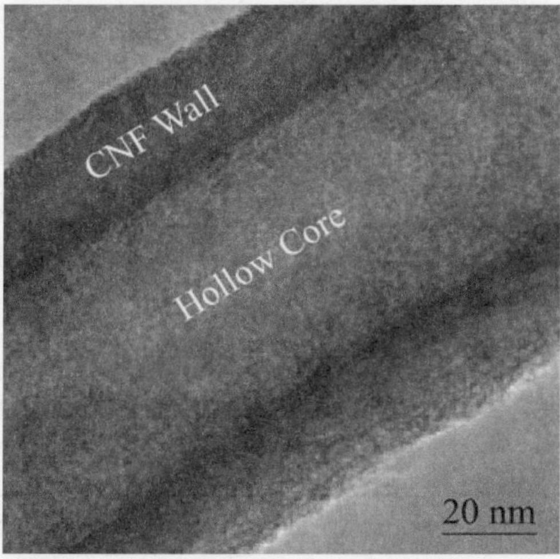

Fig. 2.14. HRTEM micrographs of PR-25 carbon nanofiber

Since the first observation of the onion-like structure by Ugarte in 1992, who obtained them by intense electron irradiation of carbon soot, several methods have been reported to prepare CNOs. For example, by irradiating carbon material consisting of different carbon nanostructures (fullerenes, nanotubes, amorphous carbon); high-temperature annealing of diamond nanoparticles under vacuum; high-dose carbon ion implantation into copper and silver; arc discharge between two graphite rods immersed in water. In addition, chemical vapor deposition was also considered as a viable method to synthesize CNOs. Other methods to synthesize CNOs includes, laser irradiation, shock compression, and high-energy ball-milling. However, most of the above-mentioned methods require high energy input, and CNOs are sometimes of low yield or just as a by-product. **Figure 2.15** illustrates the mechanism proposed for the formation of carbon onions via spiroid intermediates. At any time there is a subtle equilibrium between concentric, closed species and spiroid, the open forms.

Fig. 2.15. Mechanism proposed for the formation of carbon onions via spiroid intermediates in a subtle equilibrium with concentric, closed species

2.3.6 Nano diamond

Diamond is a true gift of nature and is well known as the hardest natural material. The properties of diamond and its practical and technological applications have been extensively investigated for centuries.

In the last two decades, a 'new' form of nanometer-sized diamond particles has attracted much attention. Such particles were formed even before the formation of the solar system and have been found in primitive meteorites, proto-planetary nebulae, interplanetary dust, and in several earth sediments dated around 13,000 BC. The study of the relative stability of diamondoid hydrocarbon molecules with the diamond structure, terminated by hydrogen atoms shown in **Fig. 2.16**, is helpful to assess the ultimate depth of crude-oil preservation on earth. The first diamondoid (adamantane) was identified in crude oil in 1933 but it is only a few years ago that larger molecules (with 22 C atoms or more) were recovered from oil.

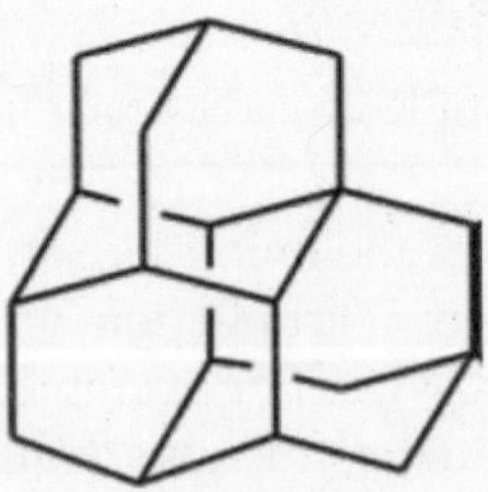

Fig. 2.16. Nanodiamond, Triamantane, a diamondoid molecule

There are several methods for the synthesis of nanodiamonds (ND) such as detonation technique, laser ablation, high energy ball milling at high pressure high and temperature (HPHT), chemical vapor deposition (CVD), autoclave synthesis from supercritical fluids, and ion irradiation of graphite and ultrasound cavitations. Out these, the first three methods are used commercially for the synthesis of NDs.

2.4 ORGANIC NANOMATERIALS

The organic nano world is predominated by polymeric nanomaterials available in the form of particles, tubes, fibers, films and branched structures (*i.e.* dendrimers). A brief description of their characteristics and potential applications are described here.

2.4.1 Polymeric nanoparticles

Polymer nanoparticles are nanoscale polymeric units such as polyacylbenzene-polydiene (PAB-PD) nanoparticles. They are used in drug delivery systems or as filler material in polymer/metal/ceramic nanocomposites. Polymer nanotubes, nanowires and nanorods have potential applications in electronic, magnetic, optical, optoelectronic, and micromechanical devices. One of the promising polymeric nanotubes is polyaniline (PANI) nanotubes which show good conductivity and may be used in conductive fabrics. Polyglycidylmethacrylate (PGMA) fibres, utilized in making smart fabrics are called as "smart fibres", as they can change their properties depending on the environmental conditions. Textiles based on PGMA fibers may switch between contradictory properties such as, hydrophobic and hydrophilic; conductive and non-conductive; acidic and basic or may switch to different colours.

Polymeric nanofilms

Polyalkylthiophene, polystyrene-polyethylene oxide (PS-PEO), poly (methyl methacrylate) (PMMA, the acrylic glass) are reported to form polymeric nano-scale thin films. They are mainly used as coatings in the bio-medical applications. There are also other nanofilms, *e.g.* based on styrene-ethylene-butylene-styrene (SEBS), polyacrylonitrile nanostructures (PAN) are found in semiconductors, solar cells, sensors and membranes in filters. Their electrical properties are based on a variable and controllable band gap for semiconductor use.

Polymeric nanofibers

Nanofibers, in particular, are of great interest because of their long length scales, resulting in very high aspect ratios, and the high degree of possible orientation. Nanofibers are generally prepared by electro spinning since the technique is potentially scalable and requires little equipment. Polymers in the form of solution or melt are used in electrospinning of nanofiber. A wide variety of synthetic polymers such as polystyrene (PS), polycaprolactone (PCL), poly(methyl methacrylate) (PMMA) and poly(vinyl alcohol) (PVA) have been electro-spun from solution.

2.4.2 Dendrimers

Dendrimers, first reported in 1978 by Fritz Vögtle, are three- dimensional globular monodisperse highly branched polymers nanometer-scale dimensions. Dendrimers consist of three components: a central core, an interior dendritic structure (the branches), and an exterior surface with functional groups. The varied combination of these components yields products of different shapes and sizes with shielded interior cores that are ideal candidates for applications in both biological and materials sciences. Dendrons are monodisperse wedge-shaped sections of dedrimer is with a single focal point reactive function (**Figure 2.17**). Dendrimers are prepared in a series of repetitive reactions from simple branched monomer units emitted from a central core with an exterior corrugated surfacing whose size and shape can be precisely controlled using either convergent or divergent step-growth polymerization.

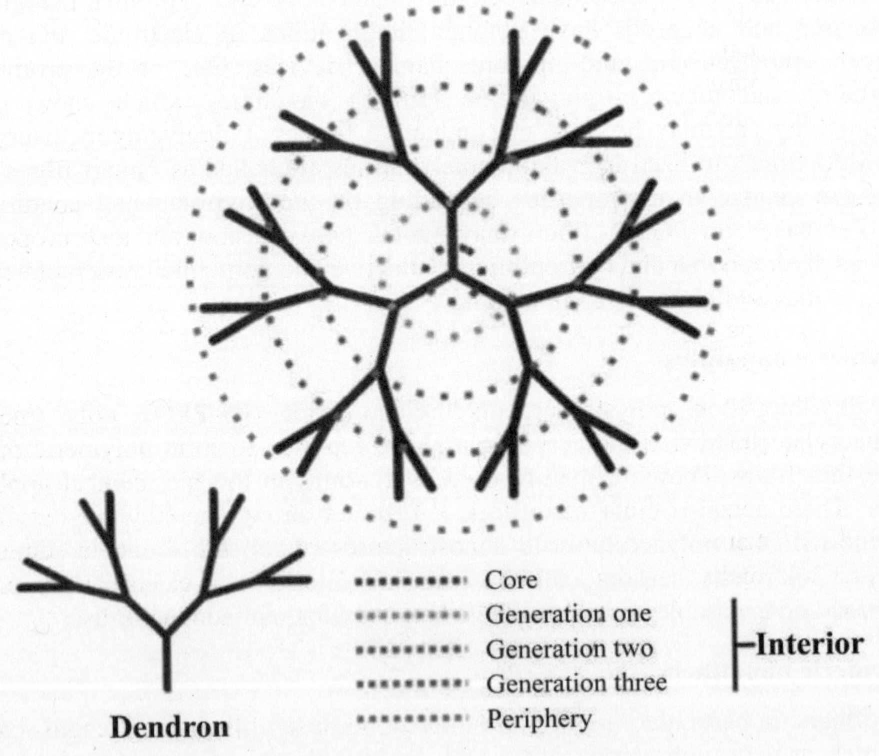

Fig. 2.17. Typical structure of a dendrimer

They are characterized by a high specific surface area and, when dispersed, by a non-linear mass-viscosity relation. The well-defined structure, monodispersity of size, surface functionalization capability, and stability are properties of dendrimers that make them attractive drug carrier candidates. Drug molecules can be incorporated into dendrimers via either complexation or encapsulation. Dendrimers are being investigated for both drug and gene delivery, for instance, as carrier for penicillin, and in anticancer therapy. Other dendrimers applications include catalysis and molecular encapsulation for light harvesting. Thus numerous applications of dendrimers include pharmaceuticals, light-emitting diodes and lasers, catalyst carrier crystals, cross-linking agents in radiation-curable surface coating resin, semi-permeable membranes, polymer additives and biotechno-logical applications.

2.4.3 Nano cellulose

Cellulose is the most abundant natural biopolymer on earth, which is renewable, biodegradable, as well as non-toxic. It is a carbohydrate polymer made up of repeating D-glucopyranose units and have three hydroxyl groups per anhydro-glucose unit (AGU) giving the cellulose molecule a high degree of functionality. The knowledge of the molecular structure of cellulose is of prime importance as it explains the characteristic properties of cellulose, such as hydrophilicity, chirality, biodegradability and high functionality. As a renewable material, cellulose and its derivatives have been widely studied, focusing on their biological, chemical, as well as mechanical properties. The materials based on cellulose and its derivatives have been used for more than 150 years in a wide variety of applications, such as food, paper production, biomaterials and pharmaceuticals. Nano crystalline cellulose (NCC) obtained from acid hydrolysis of cellulose fibres has been realized as a new class of nanomaterials. Compared to cellulose fibres, NCC possesses many advantages, such as nanoscale dimension, high specific strength and modulus, high surface area, unique optical properties, etc. These amazing physicochemical properties and wide application prospects have attracted significant interest from both fundamental research and industrial applications. Nano cellulosic materials can be used as a reinforcement material in composites and for medical implants. Nanocrystalline cellulose derived from acid hydrolysis of native cellulose possesses different morphologies depending on the origin and hydrolysis conditions.

NCCs are rigid rod-like crystals with a diameter in the range of 10–20 nm and lengths of a few hundred nanometers. Nanocrystalline cellulose materials from tunicates and green algae have lengths in the range of a few micrometres and crystallites from wood and cotton have lengths of the order of a few hundred nanometers, while some spherical shape NCCs were also produced during the acid treatment. An important characteristic of NCCs, when prepared in sulphuric acid, is that they possess negative charges on their surface due to the formation of sulphate ester groups during acid treatment, which enhances their stability in aqueous solutions. Nanocrystalline cellulose derived from acid hydrolysis using various forest product sources can disperse in water due to their negative charged surfaces. At low concentration, NCC particles are randomly oriented in aqueous suspension as an isotropic phase, and when the concentration reaches a critical value, they form a chiral nematic ordering, where NCC suspensions transform from an isotropic to an anisotropic chiral nematic liquid crystalline phase. As the concentration increased further, aqueous NCC suspensions showed a shear birefringence phenomenon. NCC suspensions exhibited negative diamagnetic susceptibility anisotropy as they dried under the influence of a magnetic field.

Chemical functionalization of NCC improves its dispensability in organic solvents thus expanding its potential applications in various sectors ranging from iridescent pigments to bimolecular NMR contrast agents. In addition, NCC, based on their solidified liquid property has the potential of use in security paper. It is believed that NCC may be used in lithium battery as a mechanical reinforcing agent for low-thickness polymer electrolytes.

Cellulose nanofibers derived from plant biomass are a particularly desirable group of nano-products due to the unlimited availability of the raw material, its biodegradability and biocompatibility Some of the areas in which the use of cellulose nanofiber is expected are: medicine (drug carriers, surgical materials, prostheses, dressings), cosmetics (creams and nutritional ingredients, tampons, masks), the environment (sensors, filters, nanofilters, absorbers), energy (electric cells, hydrogen storage), chemistry (catalysts with high efficiency, ultra-light materials and composites), electronics (computers, shields for electromagnetic radiation, electronic equipment), textiles (clothing and functional products), defense (special-purpose clothing, face masks).

2.5 BIOLOGICAL NANOMATERIALS

Supramolecular structures of biological origin such as DNA, RNA, liposomes etc. are biological nanomaterials. The biological nanomaterials are materials of biological origin that are used for nanotechnological applications, whereas the term biomaterials, has often been used to describe the materials of non-biological origin used in medical applications as bio-compatible materials. In biology, practically all materials can be considered as nano-materials in one way or another. For example enzymes have well defined structures at the nanometer-level and work as nano-machines. Inorganic nanomaterials, such as alumina, can be used as implants; this may be called biomaterials. Several examples of biological nanomaterials exist in the groups of proteins and nucleic acids as well as lipids which constitute parts of nanomaterials. Bacterial or animal cells are typically in the micrometre range and are not included as nanomaterials, however in some cases it may be convenient to add categories such as 'viruses' that can contain components of all three previously mentioned groups. Representative Biomolecules are shown in **Fig. 2.18**.

In biological nanomaterials there are two common and prominent features which are exploited for nanotechnological applications: self-assembly and specific molecular recognition properties. Biological nanomaterials can broadly be classified into following groups:

(a) *Protein-based nanomaterials*: Most of the examples in this category are dealing with 2D crystallization and 3D self-assembly of proteins and peptides making into lattices. Such 3D constructions can be used as structural or functional elements such as molecular sieves or to be used for drug delivery. The self-assembling protein or peptide can be inserted by a functionality *e.g.* binding activity for a semi-conducting material or an antigenic determinant for immunization purposes. Advantages of protein- and peptide-based nanomaterials is their homogeneity over organic and inorganic materials; producing structures with predictable geometry and stoichiometry.

(b) *Nucleic acids*: Nucleic acid molecules can be used to form nanoscale devices *e.g.* resistors, capacitors, inducers, transistors, wires, switches, memory devices and nanoscale containers for drug or other materials. Also nanoscale filters and molecular sieves can be constructed. DNA molecules may be used in various forms, such as 2D DNA lattices, 3D cages and networks, hybrid structures of DNA and protein, DNA as templates, nanomechanical DNA device.

(c) *Nano carbohydrates*: Nano carbohydrate of animal origin *e.g.* olectin is used for molecular recognition

(d) *Small molecules*: Virus particles, structured materials with virus as the core component.

(e) Lipids: 3D assemblies of lipids can be used to deliver biomolecules across cell membranes. Liposomes consisting of lipid bilayers act as template for self-assembly.

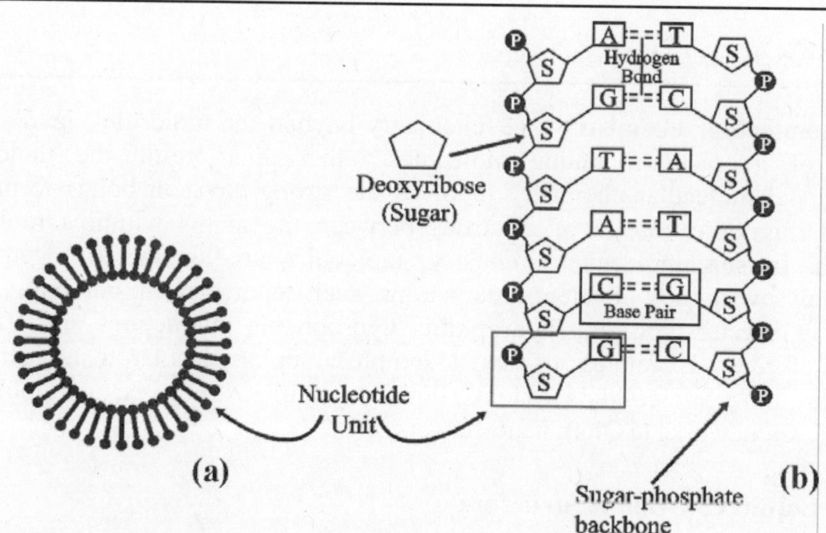

Fig. 2.18. Representative biomolecules: (a) liposome; (b) DNA

2.6 SELF-ASSEMBLED AND SUPRAMOLECULAR NANOSTRUCTURES

Self- assembly is defined as the spontaneous organization of individual components into an ordered structure, without human intervention, involving weak non-covalent interactions (**Fig. 2.19**). Self-assembly is one of the general biomimicking strategies for generating nanostructures. The biological structures, from RNA molecule, to functional organelles such as ribosomes or flagellar nanomotors to viruses and complete cells, are all, self-assembled structures. Organisms themselves can be considered as a collection of self-assembled materials: lipid bilayers, the extracellular matrix, tendon and connective tissue, skin, spider silk, cotton fiber, wood, and bone are all self-assembled biological materials, with an internal hierarchically ordered structure from the molecular to the macroscopic scale. Technologically, man-made, self-assembled nanostructure offers endless opportunities in the diverse areas of electronics, catalysis, energy, magnetic storage, biotechnology and smart materials which adapt according to their environment.

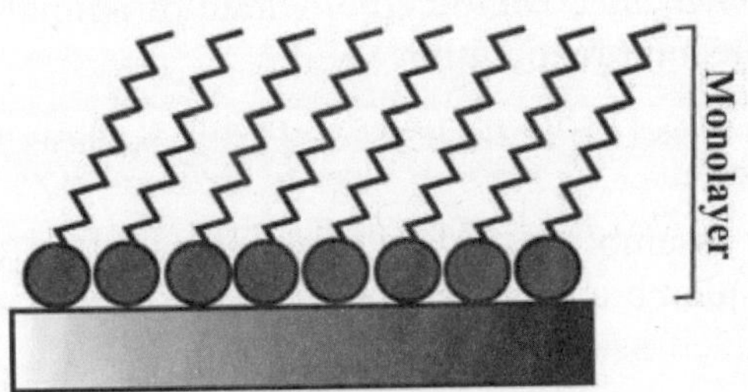

Fig. 2.19. Self assembled monolayers

Supramolecular chemistry *i.e.*, 'chemistry beyond the molecule', involves the study of interactions among molecules rather than within the molecules. Traditional molecular chemistry involves the strong covalent bonds formed by the sharing or exchange of electrons between the atoms within a molecular system. In supramolecular chemistry, molecules are assembled together by relatively weak non-covalent interactions such as hydrogen bonding, π— π stacking, electrostatic, and hydrophilic, hydrophobic interactions and Van der Waals forces. Examples of self-assembled supramolecular nano structure includes, inorganic colloids systems, liquid crystalline structures, supramolecular organic frameworks (SOFs), nano flowers, etc.

2.6.1. Liquid Crystalline Structure

A **liquid crystal** is a fourth state of matter: it has properties between those of a conventional liquid and those of a solid crystal (**Fig. 2.20**). Liquid crystals are partly ordered materials, somewhere between their solid and liquid phases. As the physical properties of LCs fall in between those of the crystalline solid and amorphous liquid states, they are referred to as intermediate phases or mesophases. Some mesophases are very sensitive to small external forces, such as temperature, pH, electric and magnetic fields, which is the basis for their application in information display. Many bio-molecules, such as carbohydrates, lipids, nucleic acids, have been found to exist in various liquid crystalline phases]. The formation of LC phases by nano DNAs (small fragment of DNA) indicates the significant role of LCs in the evolution of life in the prebiotic world.

Fig. 2.20. Illustration of different phases of matter

Liquid crystals (LCs) are divided into Thermotropic and Lyotropic groups. Thermotropic liquid crystals consist of organic molecules, typically having coupled double bonds, and exhibit a phase transition as temperature is changed. Lyotropic liquid crystals consist of organic molecules, typically hydrophilic (water-loving) and exhibit a phase transition as a function of both temperature and concentration of the liquid crystal molecules in a solvent (typically water). The various liquid-crystal phases can be characterized by the type of ordering. Among them, there are mainly nematic, smectic, cholesteric, and discotic phases (**Fig. 2.21**). We can introduce chirality into a liquid crystal molecule to form chiral nematic and chiral smectic phases.

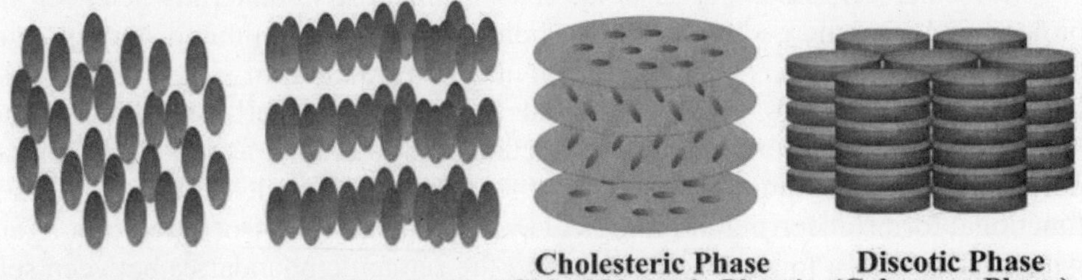

Fig. 2.21. Different phases of liquid- crystalline materials

Novel supramolecular liquid-crystalline materials have been obtained by the hydrogen bond formation between different and independent molecules. The self-assembly of carboxylic acid and pyridine moieties that function as H-bond donors and acceptors, respectively, results in the formation of mesogenic complex structures. A wide variety of liquid-crystalline low molecular weight complexes have been prepared using this approach. This concept has been extended to the polymeric systems. Hydrogen-bonded liquid-crystalline polymers that exhibit mesophases over wide temperature ranges, ferroelectricity or photo-responsibility have been prepared by the molecular association. Moreover, liquid-crystalline polymer networks that show reversible smectic-isotropic phase transitions have been formed by the hydrogen bonds.

2.6.2. Supramolecular organic frameworks (SOFs) are the two-dimensional self-assembled nanostructures which have implications for sensing and separation technologies, energy sciences and biomimetics. Solution-based processing of SOFs offers several advantages for transferring materials to a dry state without losing their structural integrity. It also allows capturing of guest molecules and the amplification of chemical signals in biomimetic sensing applications.

The self-assembly of well-defined 2D polymeric supramolecular systems in solution has been a challenge because such polymers tend to precipitate out of solution, making them difficult to manipulate and characterize. To meet this challenge, a combination of self-assembling tripods and macrocycle rings is used to form a porous framework with honeycomb periodicity, similar to that of a metal organic frame work (MOF) materials, but which remains rigid in solution, equipping the tripods with bulky hydrophilic groups that resist stacking preserved the solubility and single-layer 2D architecture of the framework. **Figure 2.20** shows the self-assembly of a water-soluble supramolecular organic framework (SOF) from a rigid triangular molecule, which has three "arms" made of 4,4′-bipyridin-1-ium (BP), and cucurbit[8]uril (CB[8]). One CB[8] macrocycle encapsulates two BP units in water. As a result, mixing the two molecules in a 2:3 molar ratio leads to the formation of a single-layer ordered honeycomb-styled network in water.

Molecular self-assembly is ubiquitous in chemistry, materials science, and biology and has emerged as a discrete field of study as a synthetic strategy even before self-assembly has been talked about. The formation of molecular crystals, colloids, lipid bilayers, phase-separated polymers, and self-assembled monolayers (SAMs) are all examples of molecular self-assembly, as are the folding of polypeptide chains into proteins and the folding of nucleic acids into their functional forms occurring in biological systems. Even the association of a ligand with a receptor is a form of self-assembly; the semantic boundaries between self-assembly, molecular recognition, complexation, and other processes such as supramolecular that form more ordered from less ordered assemblies of molecules.

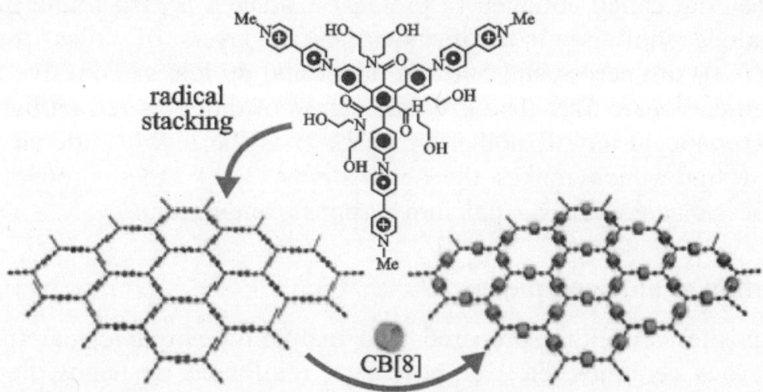

Fig. 2.22 Rigid triangular structure self-assembled in combination with macrocycle rings in solution to create a supramolecular organic framework (SOF)

2.7 NANOCOMPOSITES

Interest in nanocomposite materials is mainly due to the vast range of properties that can arise from the combination of the peculiar characteristics of each component, nanoparticles (NPs) and host matrix such as polymers, ceramics or metals. In general, the composites are solids made up of more than one material. One of the components is a continuous phase called 'matrix' while the second component is a discontinuous phase, known as 'reinforcement'. The two materials work together to give composite, different and superior properties than those of individual materials. To achieve the desired properties, a particular reinforcement is chosen for a particular matrix.

The old perception, "materials by design" has been a goal of material scientist. Fibre glass reinforced plastic (FRP) is considered to be the first modern composite material, where matrix is a plastic and the reinforcement is glass fibre. Some advanced composites are now made using carbon fibres instead of glass which are used in aircraft structures and expensive sports equipment, being lighter and stronger than fiberglass. A nanocomposite is a composite material, in which one of the components has at least one dimension in nanometer scale.

Nature has mastered the remarkable ability to synthesize the nanocomposites using natural bioorganic and inorganic nanomaterials as reinforcement dispersed in polymers matrix such as carbohydrates, lipids, and proteins, to create strong composites. Some remarkable examples includes, wood, bone, nacre (mother of pearl) (aragonite/protein), Abalone shell (calcium carbonate/protein), granite (quartz/feldspar) and so on. Wood is a composite, made up of long cellulose fibers (a polymer) held together by a much weaker substance called lignin. Cellulose and lignin, both weak substances, when bind together form a much stronger one. Bones in our body is also a composite. It is made of a hard but brittle material called hydroxyapatite (mainly calcium phosphate) and a soft and flexible material called collagen (a protein). Collagen is also found in hair and nails. Abalone shell is made of thousands of layers of "tiles" of calcium carbonate (~10 μm across and 500 nm thick) and protein as adhesive that binds the calcium carbonate tiles. The irregular stacks of thin tiles refract light to yield the characteristic luster of mother of pearl. It is the highly ordered brick-like structure design which makes the shell almost 3000 times stronger than the materials it is made out of, *i.e.* calcium carbonate and protein.

Classification of nanocomposites

Nanocomposites are either prepared in a matrix of inorganic materials (glass, metals. porous ceramics, etc.) or polymers, reinforced by nanoscale particles/nanostructures, dispersed throughout the host matrix. The mechanical, electrical, thermal, optical, electrochemical, catalytic properties of the nanocomposites differ markedly from that of the component materials. In mechanical terms, nanocomposites differ from conventional composite material due to the exceptionally high surface- to- volume ratio of the reinforcing phase and/or its exceptionally high aspect ratio. A uniform dispersion of nanoparticles leads to a very large matrix/filler interfacial area, which changes the molecular mobility, the relaxation behavior and the consequent thermal and mechanical properties of the material. Fillers with a high ratio of the largest to the smallest dimension (*i.e.*, aspect ratio) are particularly interesting because of their high specific surface area, providing better reinforcing effects.

The matrix contributes greatly to the properties of the composites. The ability of composites to withstand heat, or to conduct heat or electricity depends primarily on the matrix properties since this is the continuous phase. Therefore, the matrix selection depends on the desired properties of the composite being constructed. Matrices hold and protect the reinforcing material from environmental and physical damage. The strength, stiffness, and density of the composite material are dependent on the reinforcing material. The ultimate tensile strength of a composite is a result of the synergy between the reinforcement and the matrix. There are three main categories of matrices: polymer, metal, and ceramic and the types of reinforcements includes continuous fibers, discontinuous fibers (short fibers), whiskers, and particulates. The elements selected as reinforcements are usually the lower atomic numbered elements of groups 2 (Be, Mg), 13 (B, Al), 14(C, Si), 15(N), and 16(C) because of the desirable engineering traits of light weight, high strength, and stiffness. The stable covalent bonding contributes to the strength and stiffness of the composites.

On the basis of nature of matrix, nanocomposites can be divided into three groups:

1. **Ceramic-matrix nanocomposites** (**CMNC**): Inorganic matrix, mainly ceramics, reinforced by nanoscale particles of inorganic (*e.g.* metal) or organic nature (*e.g.* carbon-based).

2. **Metal-matrix nanocomposites (MMNC)**: A ductile metal or alloy matrix reinforced by nanoscale rigid ceramics. The reinforcement can be either in the form of particles (silicon carbide, aluminum oxide), fibers (silicon oxide, carbon) or a mixture of both, *i.e.* hybrid reinforcement.
3. **Polymer-matrix nanocomposites (PMNC)**: Organic matrix mainly polymers reinforced by inorganic (*e.g.* clay) or organic nanoscale particles.

Each of the above groups may further be subdivided on the basis of type of reinforcement. Nanocomposites are generally represented by molecular formula of matrix/reinforcement.

Typical examples of these groups are included in **Table 2.3.**

Table 2.3. Classification of nanocomposites and typical examples

Class	Examples
I. Ceramic Matrix Nanocomposites (CMNCs)	
a. Nanoceramic-ceramic matrix composites	a. Al2O3/SiC, Al2O3/Si3N4, Al2O3/ TiO2, Al2O3/SiO2
b. Nanometal-ceramic matrix composites	b. Al2O3/W, MgO/Fe, SiO2/Ni, Fe/ Al2O3, Fe/SiO2
c. CNTs- ceramic matrix composites	c. Al2O/CNT, MWCNT-Si3N4,
d. Graphene-ceramic matrix composites	d. GPL/Si3N4
II. Metal-matrix nanocomposites (MMNCs)	
a. Nanoceramic-metal matrix composites	a. Ni/Al2O3, Cu/Al2O3, Fe-Cr/Al2O3, Al/SiC, Mg/SiC, Al/Al2O3, Mg/ Al2O3, Al/AlN, Fe/MgO
b. CNT-metal matrix composites	b. Al/CNT, Mg/CNT
III. Polymer-matrix nanocomposites (PMNCs)	
a. Polymer-inorganic nanocomposites	a. Polyester/TiO2, PMMA/γ-Al2O3, PANI/V2O5, PMMA/Fe, PMMA/ CoPt, PPF/WS2, PPy-PTS/Fe2O3
b. Polymer-clay nanocomposites	b. PET/exfoliated clay, PS/clay, PMMA/ Na-MMT
c. Polymer-carbon nanocomposites	c. PVA/ SWCNT, PPF/GO, PEO/GO
d. Polymer- polysaccharide nanocomposites	d. PVOH/cellulose whiskers, HPMC/MCC
e. Starch-polymer nanocomposites	e. PVOH/SNC
f. Chitin-polymer nanocomposites	f. HPMC/ CS–TPP

PMMA: polymethylmethacrylate, PANI: polyaniline, PTS: para-toluenesulfonate, PPF: polypropylene fumarate, PPy: polypyrrole, PTS: para-toluenesulfonate, PET: poly(ethylene terephthalate), PS: polystyrene, MMT: montmorillonite, PVA: poly(vinyl alcohol), PEO: poly-(ethylene oxide),GO: graphene oxide, MCC: microcrystalline cellulose, PVOH: polyvinyl alcohol, HPMC: Hydroxy propyl methyl cellulose, MCC: microcrystalline cellulose, SNC: starch nanocrystal, CS–TPP: chitosan- tripolyphosphate.

2.7.1 Ceramic matrix nanocomposites (CMNCs)

Inorganic nanocomposites are mainly based on ceramic matrices. Ceramics, the non-metallic high-performance chemical compounds derived from oxides, nitrides, borides, carbides and silicides of metal sought in many applications, viz. highly efficient gas turbines, aerospace materials, automobiles etc. Ceramics based nanocomposites with improved mechanical, tribological, corrosion resistant etc are referred to as structural nanocomposites whereas functional nanocomposites demonstrate improved optical, electrical and magnetic properties.

The incorporation of fine SiC and Si_3N_4, nanoparticles in an alumina matrix (Al_2O_3), a structural ceramic material have shown to improve the fracture toughness from 3 to 4.8 MPa $m^{1/2}$ and the strength from 350 to 1050 MPa at only 5 vol. % additions of SiC.

Composites made from nanocrystalline ceramic materials and nanotubes fillers are much less brittle than monolithic ceramic materials. For example, a material made with nanophase alumina particles as the matrix and dispersions of both single and double walled carbon nanotubes as fillers exhibited a 25% increase in fracture toughness, without compromising the strength of the ceramic matrix.

Graphene with its combination of large specific surface area, two dimensional high aspect ratio sheet geometry, and outstanding mechanical properties shows great promise as nanofiller in composite materials. The majority of work in graphene nanocomposites has focused on polymer matrices

Granular films made up of a ceramic phase embedded with nanosized metal granules (such as $Fe/Al_2,O_3$, Fe/SiO_2) represents an example of functional nanocomposite. Such films show enhanced transport, optical and magnetic properties. The inclusion of nanosized metals in a thin ceramic film can transform it from an insulator to a conductive film.

Nanostructuring of bulk magnetic materials leads to soft as well as hard magnets (permanent magnets) with improved properties. "Finmet" (Fe-Si-B/Nb-Cu), is an excellent example of nanocrystalline soft magnetic alloys, consist of nano size melt-spun Fe-Si-B alloys containing small amounts of Nb and Cu.

2.7.2 Metal matrix nanocomposites (MMNCs)

Incorporation of nanoparticles and nanotubes such as nanofibers, nano-wires or nano-platelets of SiC, TiC, WC, TaC, TiB_2, AlN, and Al_2O_3 into a ductile metal or alloy matrix result into metal matrix nanocomposites (MMNCs). These materials exhibit much improvement in the physical, mechanical and tribological properties compared to MMC with micron-size reinforcements hence are strong candidates for structural applications. Carbon nanotubes reinforced metal matrix composites (CNTs-MMC) are emerging as new materials having advantage of the high tensile strength and electrical conductivity of carbon nanotubes. The inorganic nanotubes such as tungsten disulphide nanotubes have also proved to be good reinforcing agent.

2.7.3 Polymer matrix nanocomposites (PMNCs)

Polymer nanocomposites (PNCs) form a new generation of materials showing upgraded and diversified properties of polymers. Based on the nature of nanoparticle reinforcement PNCs can also be classified:

(i) Polymer- inorganic (metal and semiconductor) nano composites (PINC)

A polymer–Au nanocomposite is the first reported example of PINC. Frequently employed inorganic nano-fillers include *metals* and *metal alloys* (*e.g.* Au, Ag, Cu, Ge, Pt, Fe, CoPt), *semiconductors* (*e.g.* PbS, CdS, CdSe, CdTe, ZnO), $CaCO_3$, other oxides (*e.g.* TiO_2, SiO_2, ferric oxide or mixed metal oxides of iron, cobalt, zinc, nickel). The choice of polymer matrix depending on the applications includes, industrial plastics (*e.g.* nylon 6; nylon MXD6; polyimide; polypropylene; conducting polymers (polypyrrole, polyaniline); coplymer of aniline-formaldehyde (**Figure 2.23**); transparent polymers (polymethylmethacrylate, polystyrene).

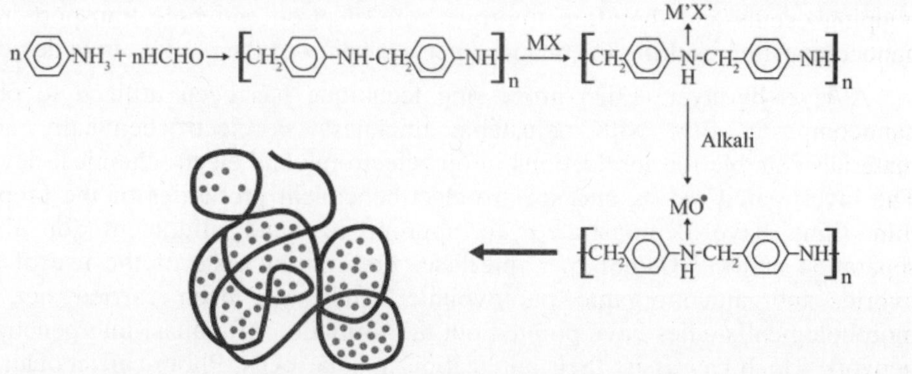

Fig. 2.23. Illustration of a simple route for synthesis of copolymer-metal oxide Nanocomposite

Applications of PINCs include optical applications, magnetic applications, mechanical applications, catalysis, electrochemical applications, electrical and thermal applications, and biomedical applications. Catalytic and electroche-mical applications, such as sensors, are other applications of PINCs where metals including Ge, Cu, Pt, and oxides including SiO_2, TiO_2 are selected as inorganic nano-fillers. Very recently, PINCs are also used in electrical and thermal applications, where nano-fillers such as CNTs and carbon nanofibers are used to enhance the electrical/thermal conductivity of the conductive polymer matrix. While Au and iron oxide nanoparticles were frequently used in biomedical applications, their polymeric nanocomposites have also recently been used as an intelligent therapeutic system and bioimaging agents. Optical and magnetic applications are two of the most important application aspects of PINCs. For optical applications of PINCs, inorganic nanoparticles of metals such as Au or Ag and semiconductors such as TiO_2, ZnO, or PbS are commonly used as 'optically effective additives', while transparent polymers such as PMMA are used as polymer matrix. For magnetic applications, metals and metal alloys such as Fe or CoPt; oxides such as ferric oxide; and ferrite are always used as inorganic nano-fillers.

Energy conversion: Systems based on semiconductor NCs such as CdSe, ZnO, PbS or TiO_2 incorporated in conjugated polymers are of interest for their potential application in the field of energy conversion. Semiconductor NCs can be coupled with conducting polymers for the fabrication of heterojunctions in which the high surface-to-volume ratio and high conductivity of NPs combine with the flexibility and versatility of the polymeric medium. The resulting hybrid heterojunctions can exhibit high electron mobility or improved spectral coverage and can be therefore exploited for photovoltaic applications. A recently developed hybrid heterojunctions by blending nanocrystalline TiO_2 nanorods and high hole mobility polymer {poly(3-hexylthiophene), P3HT}. Anisotropic shaped NCs combined with the polymer properties can reduce the number of interparticle hopping events and therefore improve both electron and hole transport in the nanocomposite, resulting in enhanced performances of the photovoltaic device.

A layer-by-layer (LbL) processing technique has been utilized to obtain nanocomposite film with adjustable thickness as electrochemically active materials suitable for applications in optoelectronic and electrochemical devices. The investigated red-ox and spectro-electrochemical properties of the prepared thin films have demonstrated an appropriate band alignment for charge separation at the NCs/polymer interface. Preliminary tests of the use of such hybrids inorganic/inorganic photovoltaic cells have been carried out, and morphological studies have pointed out the formation of a quasi-interpenetrating network which can assist their application in solar cells. Photo carriers can thus be efficiently separated and transported in such hybrid systems, as required for the fabrication of detector, sensor and solar cells.

Superparamagnetic systems: Nanocomposite materials obtained by dispersing iron oxide or cobalt nanoparticles in polymers have been found to possess magnetic properties, which can be appropriately exploited for fabricating devices with charge storage capabilities, stability during cycling, and dynamics of charge propagation. Novel superparamagnetic nanocomposites have been prepared by the incorporation of magnetic nanoparticles in conducting polymers, such as poly (pyrrole) (PPy) or poly (aniline) (PANI). The composite materials combine magnetic/conducting properties of their inorganic and organic components and exhibit modified mechanical and thermal characteristics. Poly (pyrrole) has been modified by the incorporation of Fe_2O_3 nanoparticles, in order to increase the electrochemical storage of the electroactive material. The influence on the morphology and electrochemical properties of the nanocomposite material has been studied by modifying synthetic parameters such as composition of iron nanofillers and a type of counter-anion incorporated in the nanohybrid, aiming at the optimization of performances for electrochemical storage applications. In the presence of p-toluene sulfonate (PTS) anions an improvement of charge storage capacity has been observed, due to a higher specific surface area attributed to a morphology modification. The incorporation of a high content of PPy-PTS/Fe_2O_3 nanocomposite in an electrode has resulted in improved performances with respect to the bare conductive polymer.

Magnetic polymer nanocomposites constituted by methacrylate capped magnetite NPs chemically bounded to both PMMA and PS polymer chains have also been reported. The increasing of *Tg* of the nanocomposite material has been observed, indicating the positive influence of the NP incorporation and chemical bonding on the mechanical properties of the hybrid nanostructures. Iron NP modified vinyl ester resin nanocomposites have been recently synthesized by direct binding NP surface to monomers, which act as stabilizers for preventing NP aggregation and promote copolymerization. The mechanical and magnetic characterization of the prepared nanocomposites have shown an enhancement of tensile strength and Young's modulus as compared with those of cured pure resin, increased thermal stability and room temperature ferromagnetic behaviour.

Sensors and electrochromic devices: Recently, a large area of research has focused on the incorporation of nanoscale metals in polymers to design and fabricate advanced optoelectronic and sensor devices. The combination of the thermal induced tunable absorption properties of gold nanorod with polymers can be exploited to fabricate nanoscale devices for sensing applications. Uniform dispersion of Au nanoparticles in the polymer matrix has been achieved and the non-covalent Au-S interaction as well as the chemical composition of the polymer has been preserved in the nanocomposite formation. The optical properties based on the plasmon resonance of noble metal nanoparticles are extremely relevant in electro-chromic applications. Since surface plasmon absorption depends also on the dielectric properties of the nanoparticle host matrix, the incorporation of metal nanoprticles in conducting polymers can be used to fabricate tunable electrochromic devices. Ag and Au NPs have been blended with conducting polymer poly(3,4-ethylenedioxythio-phene) : poly styrene sulfonate (PEDOT : PSS) to fabricate electrochromic devices. Both the tunable absorption and time response properties exhibited by the prepared nanocomposites make such materials exploitable in the manufacture of full color electrochromic displays. The optical properties and influence of the addition of Ag NPs on thermal properties of PMMA matrix has also been investigated. Spectroscopic measurements of the optical properties have shown that Ag NPs preserve their plasmon absorption when embedded in the polymer matrix.

Furthermore, metal NPs have been found to enhance thermo-oxidative stability and glass transition temperature of PMMA matrix.

(ii) Polymer-nano clay composites

Clay is referred to a part of soil fraction with the particle size of less than 2 μm and a layer thickness of ~1 nm. Basically clays are alumino-silicates where part of aluminium atoms is substituted with other elements such as magnesium. Important members of clay include Montmorillonite, Saponite, Illite, Vermiculite Kaolinite, etc., which differs in their composition, structure and basal spacing. The interaction between layered silicates and polymer arises from the penetration of polymer chains into the interlayer region of the clay, resulting in an ordered multilayer structure with alternating polymer/inorganic layers at a repeated distance of a few nanometers. Montmorillonite (MMT) clay consisting of an edge-shared octahedral sheet of aluminum hydroxide between two silica tetrahedral layers is characterized by a moderate negative surface charge (cation exchange capacity, CEC). MMT is an effective reinforcement filler due to its high surface area and large aspect ratio (50–1000). Organic modification of clay is achieved through the exchange of sodium ions with organic alkyl ammonium ions. This modification turns the hydrophilic clay to hydrophobic and potentially more compatible with the polymer for preparation of polymer–clay nanocomposites. **Figure 2.24** illustrates the formation of conventional, intercalated and exfoliated composites derived from interaction between layered nanoclays and polymers. The homogeneous dispersion of organically modified clays in polymer matrices leads to enhanced thermal, mechanical, gas barrier, dielectric properties and makes it solvent resistant. Polysaccharide/clay nanocomposites are important for the food packaging industry. These composites make use of naturally occurring polymers, such as starch, mixed with clay to make biopolymer film with enhanced properties, in particular permeability to water vapour.

(iii) Polymer-carbon nanomaterials composites

Among polymer carbon nanocomposites, CNT and graphene are the most promising nanofillers that imparts excellent physical properties such as mechanical, optical, electrical to the nanocomposites. For instance excellent electrical conductivity of CNT's and graphene can be exploited to make traditionally insulating polymer matrices into electrically conductive materials for various applications in conductive adhesives, antistatic coatings, and films. There is a critical loading (percolation threshold) when the composite transitions from an insulator to a conductor due to the formation of a continuous conducting network.

A variety of CNT-polymer composite has been prepared using in-situ polymerization, where the nanotubes are dispersed in the monomer which is then polymerized. Dispersants may be added to assist in the de-agglomeration of the nanotubes. Alternately functionalization of CNTs or polymer adsorption techniques can be used to aid dispersion. This technique can be used to produce both thermoset and thermoplastic materials. Properties of CNT based polymer nanocomposites are particularly interesting due to their flexibility under mechanical and high temperature stresses; and electrical properties. Modified nanotubes *i.e.* doped with B, N, and derivatives with organic groups, give novel electrical and optical properties to the composites. SWCNT is considered to be a better filler as compared to MWCNTs.

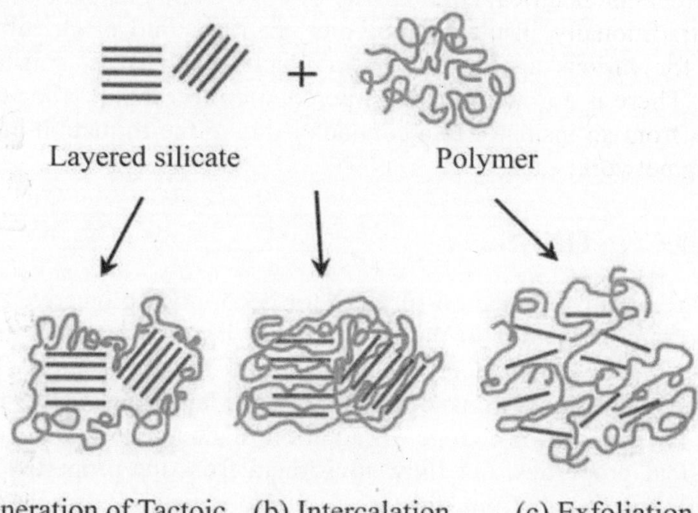

Fig. 2.24. Types of composites derived from interaction between clays and polymers

Graphene sheets provide percolated pathways for electron transfer, making the composites electrically conductive. Graphene-based polymer nanocomposites show improved optical, thermal, electrical, gas barrier and water permeability characteristics as compared to other nano carbon material based reinforcement. Defect-free graphene sheets are impermeable to all gas molecules. Combining graphene with polymeric hosts makes possible large scale barrier membranes with mechanical integrity. Graphene without functionalities interacts with polymer through van der Waals force, $\pi - \pi$ stacking, and hydrophobic interactions. The functional groups on graphene oxide open the possibility of interaction with polymer during processing. Hydroxyl terminated polymers can react with the carboxyl groups on the graphene oxide through esterification. GO can form nanocomposites with hydrophilic polymers such as poly(vinyl alcohol) (PVOH), and poly(ethylene oxide) (PEO). The carboxyl and hydroxyl group of the GO reacts with organic isocyanates. These isocyanate modified graphene oxides have been known to form composites with polystyrene (PS), acrylonitrile-butadiene-styrene (ABS), and styrene-butadiene rubbers (SBR). The functional groups on graphene oxide open the possibility of interaction with polymer during processing. Hydroxyl terminated polymers can react with the carboxyl groups on the graphene oxide through esterification. GO can form nanocomposites with hydrophilic polymers such as poly(vinyl alcohol) (PVOH), and poly(ethylene oxide) (PEO). The carboxyl and hydroxyl group of the GO reacts with organic isocyanates. These isocyanate modified graphene oxides have been known to form composites with polystyrene (PS), acrylonitrile-butadiene-styrene (ABS), and styrene-butadiene rubbers (SBR).

The excellent electrical conductivity of CNT's and graphene can be exploited to make traditionally insulating polymer matrices into electrically conductive materials for various applications in conductive adhesives, antistatic coatings, and films. There is a critical loading (percolation threshold) when the composite transitions from an insulator to a conductor due to the formation of a continuous conducting network.

2.8 NANO COATINGS

Coatings are mainly applied on surfaces for decorative, protective and functional purposes or a combination of these. The major issues concerning the production of coatings are to ensure good mechanical and wear resistance characteristics. Nano coatings are a type of nanocomposites. The layer thickness of a nanocoating is usually in the 1-500 nm range. In addition, nanocomposites coatings can also show original properties, not fully envisioned from the properties of the single components, deriving from the local micro-structural arrangements of the nanosized objects in the matrix. Nanocomposite coatings synthesized by plasma assisted deposition processes have high potential for new applications such as protective and functional coatings in automotive, aerospace, tooling, electronic or manufacturing industry. Functional nanocoatings may induce changes in properties in variety of ways such as to control friction, lubrication and wear or to change the transmission of visible and IR radiation in glass or to introduce new properties such as "self-cleaning" effects. Important functionalities along with associated applications of nanocoatings are briefly discussed in the following sections.

2.8.1 Functional nanocoatings

Chemically active functional nanocoatings perform their activities either at film-substrate interfaces (anticorrosive coatings), in the bulk of the film (fire-retardant or luminescent coatings), or at air-film interfaces (antibacterial, self-cleaning). Some of the important applications of functional nanocoatings are briefly discussed.

Photo luminescent coatings: A novel multilayer composite can be prepared by photo cross-linking layer-by-layer process with functionalized polystyrene matrix reinforced with semiconductor and metal NPs (CdSe@ZnS, Au, and Pt). The resulting nanocomposite multilayers exhibit enhanced and prolonged photo luminescent durability, facile color tuning in accordance to the layer and number thickness and are used as structural materials for manufacturing nanoscale optical devices.

Anti-reflection coatings: Inorganic nanoparticle modified polymers have been extended for the production of high refractive index materials, for highly reflective and antireflection coatings and advanced optoelectronic systems. In particular, optically transparent TiO_2- based nanocomposites with high refractive index are potential materials for UV filter and high refractive coatings. Variation in the titania NP concentration, size, morphology and surface functionalization, as well as a change in the dielectric constant of the matrix result in the possibility of tuning the optical properties of the materials for practical applications. The optimization of the preparative conditions has led to the improvement of both thermal and mechanical properties of the coatings. In another example, ZnS nanoparticles have been used to prepare polymer nanocomposites with tunable refractive index by varying the nanostructures content in the matrix. The nanocomposite coatings exhibit optical transparency in the visible range, thermal stability and good mechanical properties. Fluorescent and highly transparent nanocomposite thin films with high refractive index have been recently prepared by blending o-phenylene diamine (o-PDA) functionalized ZnS NC solutions with poly (vinylpyrrolidone) (PVP). Compared to the bare polymer, the ZnS NC polymeric films have exhibited enhanced refractive index, high optical transparency and preservation of the NC band-edge emission. Nanocomposites with highly dispersed ZnS nanophase contents have shown that ZnS NPs and the polymer matrix are linked by covalent-bonds, thus resulting in the enhancement of thermal and mechanical properties. The small size and the homogeneous dispersion of ZnS NPs contribute to the transparency and the linear behaviour of the refractive index with the volume fraction of the polymer matrix.

Self-Cleaning and Biocidal Coatings: Another example of functional nanocoating is photocatalytic coatings (commercialized as "self–cleaning" glass) which use the catalytic properties of titanium dioxide (TiO_2). Superhydrophobic coatings have surfaces that mimic the surface found in the lotus leaf and are being developed for many applications that require resistance to dirt and ease of cleaning. There is a great interest in the design and development of surfaces that not only are easy to clean and even self-cleaning but also provide biocidal activity. Most of such coatings acquire their biocidal/self-cleaning capacity by incorporating specific nanoparticles: basically silver (Ag) and titanium oxide (TiO_2). Nano TiO_2 is used for developing anti-UV, anti-bacterial and self-cleaning paints. This possesses self-cleaning hydrophobic properties, which causes water droplets to bead-off of a fully cured surface picking up dirt and other surface contaminants along the way.

This self-cleaning action (**Fig. 2.25**) helps clean and maintain important surfaces and to accelerate drying, leaving the surface with minimal spotting. A recent study utilizes corona treatment technique, inert sol-gel coating and anatase TiO_2 layer. With the coronatreatment, an organic surface was activated to allow auniform TiO_2 sol-gel coating. Nanoparticles of surface treated Al_2O_3 molecules help increase hydrophobicity and increase scratch resistance. Microbial evolution on a wide variety of surfaces can cause corrosion, dirt, bad odour and even serious hygiene and health problems. Although no alternatives to the use of biocides are available at present, creation of nanostructured surfaces could offer an innovative and environment-friendly solution to the problem of biofouling. Research has developed new biocidal coating systems that prolong biocidal activity by immobilizing such additives on nanoparticles; the embedded biocides are designed to be released into the environment only when needed, thus extending the lifetime of the biocidal activity

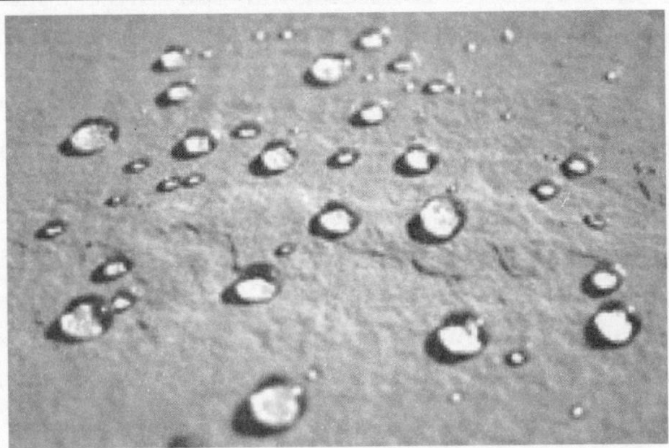

Fig. 2.25. Nano coatings with super hydrophobic properties

Heat dissipation: The coatings of SWNT-ceramic composites with their anisotropic thermal properties have potential for a variety of applications such as, in microprocessors and integrated circuit packages (miniature scale devices) and on high temperature components in turbine engines, reactors and space vehicles (larger scale devices) for effective heat dissipation and spreading. The composite coatings consisting of CNTs and HA/TiO_2 ceramic particles fabricated by EDP on metallic substrates (*e.g.* stainless steel, titanium alloy) are suited for biomedical applications, in particular for the attachment and proliferation of cells).

Tribological nanocoatings: Tribology is the science and technology of interacting surfaces in relative motion. Tribological properties include friction, lubrication and wear. Tribological coatings are those coatings that are applied to the surface of a component in order to control its friction and wear. In this area the term "thin films" is often employed as an alternative to nanocoatings due to the fact that it is an area of innovation that started years ago and has now reached the nanoscale. Tribological coatings play a key role in the performance of internal mechanical components of a vehicle, such as the engine and powertrain.

They are also key elements in cutting tools and machinery in general. By reducing wear and friction these coatings increase the lifetime of the working material while also reducing the dissipation of energy as heat, thus increasing the efficiency of the moving part. When applied to machinery and tools, tribological coatings can reduce (or eliminate) the need for lubricants, increase cutting speed, increase the rate of material removal, reduce maintenance costs or reduce processing cycle times.

Traditional materials used in coatings for tribological applications are carbides, cemented carbides, metal ceramic oxides, nitrides and carbon-based coatings. Since the microstructure controls many of the physical properties of the coating, having a nano-scaled microstructure may lead to significant improvements in the coatings' mechanical properties (*e.g.* hardness), chemical properties (*e.g.* corrosion resistance) and electrical properties. Thus nanocom-posite coatings are now being investigated as alternatives to the traditional approach of using specific alloying elements in single-phase coating materials to improve, for instance, properties such as hardness. One type of nanocomposite coatings is multilayered thin-films made of different layers in the order of nanometers. These films are mostly used for their enhanced hardness and elastic moduli, which is higher in multilayered films than in homogeneous thin films of either component, and for their wear properties. Commercial multilayer coatings made with multilayer periods in the nanoscale range already exist, such as WC/C coatings used in the cutting tool industry. Other examples are films of alternating layers of TiN and NbN, or TiAlN/CrN multilayers, which are more efficient than TiAlN films.

2.8.2 Protective nanocoatings

Ceramic coatings produced by physical vapour deposition (PVD) demonstrate attractive functional properties of high hardness and chemical inertness, they are not easy to deposit at a thickness adequate to provide satisfactory corrosion protection of a metal substrate however, metallic coatings in general do not suffer such thickness restrictions. Thick (>10 μm), lightly-stressed coatings, with insignificant porosity can, in principle, be produced at a rate of 10 μm/hr or higher. Exploring the possibilities of pairing certain low-miscibility transition metals, *e.g.* titanium/chromium/zirconium with copper/nickel - and the ability, with the selection of appropriate deposition parameters, to preferentially supersaturate the ‘early’ transition metal (*i.e.* Ti, Cr, Zr) with nitrogen (or other interstitially-located elements), thick metallic nano-composite coatings can be produced. These coatings exhibit excellent wear resistance owing to their high hardness/elastic modulus ratio (H/E) and, may also possess many other potentially desirable properties (*e.g.* toughness and corrosion protection – sacrificial as well as ‘barrier’) inherent to metallic coatings - especially with the addition of aluminium as a base. These nanocomposite coatings which comprise a hard nanocrystalline phase embedded in an amor-phous matrix exhibit improved properties over a range of varying conditions with enhanced thermal stability and improved physical and mechanical properties. Such coatings can be deposited on austenitic stainless steel (304), low alloy steel (4145) and silicon wafers using closed-field unbalanced magnetron sputter equipment in reactive (N_2) atmosphere. These coatings can act as alternatives to electroplated cadmium, IVD-aluminium, electroplated hard chromium, electr-oless nickel and other 'traditional' coatings used in aerospace, automotive and other engineering applications – where environmental legislation is restricting the use of toxic processes and materials. These metallic coatings with multifunc-tional and adaptive properties, will suit a range of technically challenging environments.

2.8.3 Smart nanocoatings

A functional coating is termed as smart coating when a system is able to respond to a stimulus based on predefined functionalities and providing a smart feedback. In other words, an active coating sensitive to an external stimulus changing its physical or chemical properties, preferably in a reversible way, and able to work as a sensor and/or actuator. Examples are coatings, switching from hydrophilic to hydrophobic according to the environmental humidity or supplying a mechanism of damage prevention which is activated in response to environmental conditions. Responsive nanocoatings react to the environmental conditions, such as light or heat, either in a passive or an active way. These coatings allow changing the properties of some materials, such as glass, by conferring new or improved properties. As an example, the use of glass is very common in modern buildings to allow the construction of transparent and seemingly lightweight structures. However, the relative high transmittance of visible and infrared (IR) light is a major disadvantage, since this leads to a large heat transfer which is particularly undesirable in summer. The problem is reverse in winter, when heat is dispersed through the glass. In order to address these problems various types of nano-coatings that modulate light transmission in glass are under investigation and being commercialized. The aim is to reduce indoor heating in summer, so less air-conditioning is required to keep the atmosphere cool, with consequent energy saving. One type of coating, referred to as "low-e" meaning "low emissivity", is based on a thin silver film, about 10 nm thick, surrounded by dielectric layers. Metallic layers have been widely used to increase the reflectivity of light (and reduce transmittance) for years, but they have the disadvantage of giving a mirror-like appearance. Silver loses its metallic appearance when scaled to a nano-film, thus eliminating this problem. "Low-e" coatings are a type of passive nano-coating, since the properties of the layers are unperturbed during its operation. Another class of coatings used in glasses are those often described as dynamic or "smart coatings". In this case the environmental conditions, such as radiation intensity or temperature, induce a change in the properties of the coating (*e.g.* darkening of windows). When the effect is a change in the colour (meaning also transparency) they are called "chromogenic smart materials". The change can be induced actively, by pressing a button. This is the case of electro chromic coatings where applying a small voltage induces a change in transmittance, and in the case of gas chromic coatings, which change their transmittance in the presence of specific gases. Gas chromic glazing makes use of the properties of tungsten oxide thin films ($W0_3$) which go from colourless to blue in the presence of hydrogen with a suitable catalyst. Gas chromic windows follow a double pane model: on one pane a film of WO_3 is deposited with a thin layer of catalysts on top. Hydrogen gas is fed into the gap producing coloration (windows colour with 1.1-10% hydrogen, which is below the flammability concentration). To switch the colour of the window back, another gas is purged in (oxygen). Smart coatings can also be passive in the sense of changing their optical properties due to a change of external temperature (thermo chromic) or light incidence (photo chromic). Another example of nanotechnology applied to smart coatings is the use of a family of wavelength-selective films for manufacturing "heat mirrors". One of these materials is indium-tin-oxide (ITO), an infrared absorber. A 300 nm ITO coating on glass provides more than 80% transmission for the wavelengths predominant in sunlight. The transmission properties of the window can be varied by changing the thickness and material composition of the coating, so that a combination of materials could be used to produce smart windows that reflect solar energy in summer but transmit solar energy in winter.

Suggested further readings

1. R. Sardar, A. M. Funston, P. Mulvaney and R.W. Murray, "Gold Nanoparticle: Past, Present, and Future", *Langmuir*, 25 (2009) 13840–13851.
2. M. Fernández-Garcia and J.A. Rodriguez, "Metal oxide nanoparticles", Nanomaterials: Inorganic and Bioinorganic Perspectives, Encyclopedia of Inorganic Chemistry, Wiley VCH, 2009.
3. N. R. Jana, Y. Chen and X. Peng, "Size and Shape Controlled Magnetic (Cr, Mn, Fe, Cu, Ni) Oxide Nano Crystals via a Simple and General Approach", *Chem. Mater.*, 60 (2004) 3931–3938.
4. P. A. Ajibade and J. Z. Mbese, "Synthesis and Characterization of Metal Sulfides Nanoparticles/Poly(methyl methacrylate) Nanocomposites", *Int. J. Polym. Sci.*, 8 Article ID752394 (2014).
5. C. Giordano, C. Erpen, W. Yao et al., "Metal Nitride and Metal Carbide Nanoparticles by a Soft Urea Pathway", *Chem. Mater.*, 21 (2009) 5136–5144.
6. T. Pradeep, *Nano: The Essentials*, Tata McGraw Hill, New Delhi, 2007.
7. P. R. Sajanlal, T. S. Sreeprasad, A. K. Samal, T. Pradeep, "Anisotropic nanomaterials: structure, growth, assembly, and functions", *Nano reviews*, **2:** 5583 (2011), 1-62.
8. F. Hussain, M. Hojjati, M. Okamoto and R. E. Gorga, "Polymer-Matrix Nanocomposites, Processing, Manufacturing, and Application: An Overview", *J. Compos. Mater.*, 40 (2006) 1511–1575.

9. Carbon Nanotechnology: Recent Developments in Chemistry, Physics, Material Science and Device applications, L. Dai, Ed., Elsevier, 2006.
10. J. Meyer, A. K. Geim, M. I. Katsnelson, K. S. Novoselov, T. J. Booth and S. Roth, "The Structure of Suspended Graphene Sheets", *Nature* 446 (7131), (2007) 60–63.
11. A. Hirsch, "The Era of Carbon Allotropes", *Nat. Mater.*, 9 (2010) 868.
12. Krueger, "The Structure and Reactivity of Nanodiamonds", *J. Materials Chem.*, 18 (2008) 1485.
13. N. Z. Logar and V. Kaucic, "Nanoporous Materials: From Catalysis Hydrogen Storage to Wastewater Treatment", *Acta Chim. Slov.*, 53 (2006) 117–135.
14. Y. Zheng, X. Li and P. K. Dutta, "Exploitation of Unique Properties of Zeolites in the Development of Gas Sensors", *Sensors* 12 (2012) 5170–5194.
15. M.K. Patra, K. Manzoor, S.R. Vadera, N. Kumar, "Functional Nanomaterials for Sensors and Display Applications", *Encyclopaedia of Nanoscience and Nanotechnology*, Ed. H.S. Nalwa, **14** (2011) 385-435.
16. A.B. Morgan, Polymer-clay nanocomposites: Design and application of multi-functional materials, *Material Matters*, **2** (2007) 20-23.
17. S. Abbott and N. MacDermid, *Nanocoatings: Principles and Practice: From Research to Production*, DEStech Publications, Inc., 2013, Lancaster.
18. J. R. Potts, D. R. Dreyer, C. W. Bielawski and R. S. Ruoff, Graphene-Based Polymer Nanocomposites, *Polymer*, 52 (2011) 5–25.
19. A. Krueger, "The Structure and Reactivity of Nanodiamonds", *J. Materials Chem.*, 18 (2008) 1485.

Review questions

1. Classify nanomaterials based on dimensions and morphology, give example in support of your answer.
2. What are the basic similarities and difference between nanofiber, nanorods, nanowires and nanotubes?
3. What are inorganic nanotubes and discuss their role in nanotechnology?
4. Discuss different types of organic nanomaterials and their significance.
5. What is the significance of cellulose and how can it be converted into nano crystalline cellulose?
6. What are dendrimers? How do they differ from a polymer in general?
7. Define quantum dots. How does band gap of QDs affect luminescent properties?
8. What is the difference between electronic magic numbers and structural magic numbers?
9. Name natural and synthetic allotropes of carbon. Give the structure and dimensions of carbon nano allotropes.

10. How does spherical structure of fullerene facilitate its addition reaction? Give some examples.
11. What are the parameters responsible for giving rise the metallic or semiconductor properties in CNTs?
12. How are the following pairs different in their structure: graphite and graphene; graphene and carbon nanotubes; CNT and carbon nanofiber; carbon onion and fullerenes?
13. Which functional group differentiates graphite oxide from graphene oxide? How can graphene oxide be obtained from graphite oxide?
14. Discuss the unique properties of graphene based on its electronic and band structure.
15. Write a short note on nano diamonds and carbon nano onion.
16. Differentiate self- assembled monolayers from supramolecules.
17. Differentiate between metal organic framework (MOFs) and supra-molecular organic framework (SOFs) structures. Mention significant application areas of SOFs.
18. What are the short range attractive forces? Discuss their implications in self-assembled structures?
19. How does entropy of molecules help in the formation of self-assembled structures?
20. What are liquid crystals and how does self- assembly help in their existence?
21. Define the terms: composite, matrix, reinforcement and nanocomposite. Classify nanocomposites on the basis of matrix material.
22. 'Nature has mastered the synthesis of nanocomposites' justify the statement giving few examples.
23. What are ceramics? Give few examples of nano ceramic reinforced metal matrix composites and their properties.
24. Describe different types of polymer-based nanocomposites.
25. Discuss the significance of nanocomposites reinforced with carbon nanotubes.
26. Discuss significant properties and applications of polymer-clay nanocomposites.
27. What do you understand by 'functional coatings'? Suggest suitable compositions for different functional coatings?

■■■■

Synthesis of Nanomaterials 3

3.1 STRATEGIES FOR SYNTHESIS OF NANOMATERIALS

Synthesis of a wide variety of nanomaterials, over the last three decades, in different sizes and structural forms, by using different methods and techniques has led to unprecedented growth in the area of nanotechnology. Over the time materials scientists and engineers have made significant developments in the improvement of methods in synthesis of solid nanomaterials.

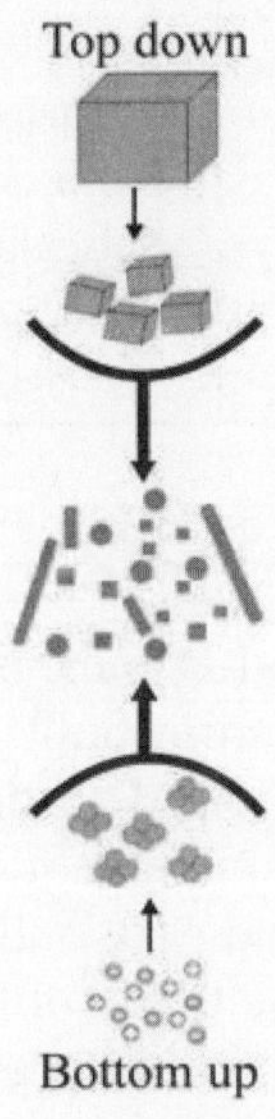

Fig. 3.1. "Top-down" and "bottom-up" approach to nanomaterials synthesis

In general, most of these techniques can be classified into two main categories viz. bottom-up and top-down (**Fig. 3.1**), while many of the strategies have elements of both the approaches. Attrition or milling is a typical top-down method, whereas the colloidal dispersion (sol-gel method) is a good example of bottom-up approach in the synthesis of nano-particles. Lithography may be considered as a hybrid approach, since the growth of thin films is bottom-up whereas etching is top-down. There are advantages and limitations in both approaches. Both the approaches can be performed in either gas, liquid or solid states; accordingly **Table 3.1** enlists some of the important synthetic procedures for nanomaterials. The most important aspect of any method lies in its ability to control the particle size, shape, size distribution, composition as well as degree of agglomeration. Green routes involving plant saps/microbes and bio-mimetic processes of self-organization and self-assembly have also been suggested for for nanomaterials. Innovations and modifications of fabrication/synthesis procedure is a continuous process which results in innumerable recipes and techniques. However, synthetic strategies are maturing every day with the new postulates and procedures.

Table 3.1. Common strategies for synthesis of nanomaterials

Top-down approaches	Bottom-up approaches
Solid phase techniques:	**Vapor phase technique:**
Mechanical attrition	Chemical vapor deposition
Sputtering	Chemical vapor Condensation
Etching	**Solution phase techniques**:
Laser ablation	Sol-gel
Lithography	Reduction
Aerosol- based techniques:	Exchange reaction
Electro spraying	Matrix mediated synthesis
Flame pyrolysis	Solvothermal synthesis
Liquid phase techniques:	Sonochemical synthesis
Electrospinning	Self-assembly techniques

3.2 TOP-DOWN APPROACHES

In the top-down approach a material in bulk is broken down into smaller pieces. Methods to produce nanoparticles from bulk materials include high-energy ball milling, sputtering, and laser ablation. Top down processes are executed in an inert atmosphere or in vacuum. Immediately after processing, the nanoparticles are very reactive to agglomerates and may undergo some additional reactions if a reactive gas is present. To overcome this problem, freshly prepared nanoparticles can be coated with a material that would prevent further interaction with other particles or the environment.

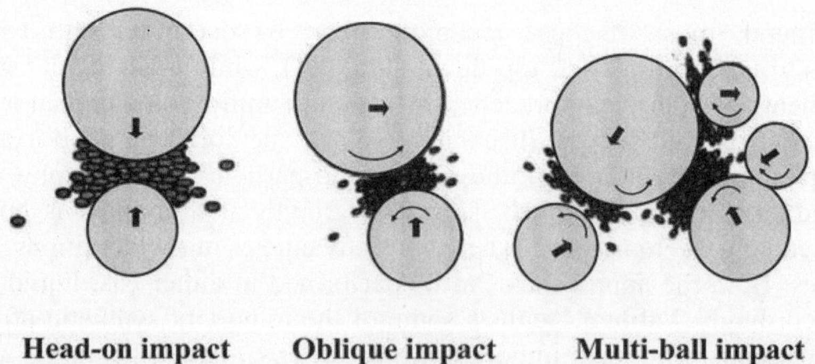

Fig. 3.2. Different forms of impact in mechanical ball milling

3.2.1 Mechanical attrition

Mechanical attrition or milling is a process routinely used in powder metallurgy and mineral processing industries. In this process, mixtures of elemental or prealloyed powders are subjected to grinding under protective atmosphere in equipment capable of high-energy compressive impact forces such as attrition or shaker mills. A variety of ball mills including tumbler mills, shaker mills, vibratory mills, planetary mills, etc. have been developed for different purposes. In a typical mechanical milling powders with typical particle diameters of about 50 μm are placed together with a number of hardened steel or tungsten carbide (WC) coated balls in a sealed container which is shaken or violently agitated. Since the kinetic energy of the balls is a function of their mass and velocity, dense materials are preferable to ceramic balls.

When a single-phase elemental powder or intermetallic compound powder is milled, the grain size of the powder particles continues to decrease until it reaches a minimum level – in the range of 3-25 nm. For some intermetallic compounds, the powder becomes amorphous beyond this point. For intrinsic brittle powders, such as silicon powder or carbide and oxide powders, the reduction of the grain size is a natural outcome of the trans granular fracturing and cold welding. The advantage of the mechanical milling process is that the processing temperature is low, so the newly formed grains grow very slowly. Mechanical attrition methods allow the preparation of a broad range of alloys, intermetallics, ceramics and composites which cannot be synthesized via conventional casting routes. Uniform dispersion of ceramic particles in a metallic matrix and alloys of metals with quite different melting points with the goal of improved strength and corrosion resistance is achieved through this technique. High-energy mechanical milling is a very effective process for synthesizing metal-ceramic nanocomposite powders as it allows incorporation of the metal and the ceramic phases into each powder particle. The optimized conditions for production of iron nanoparticles by high energy ball milling is given in **Scheme 3.1**.

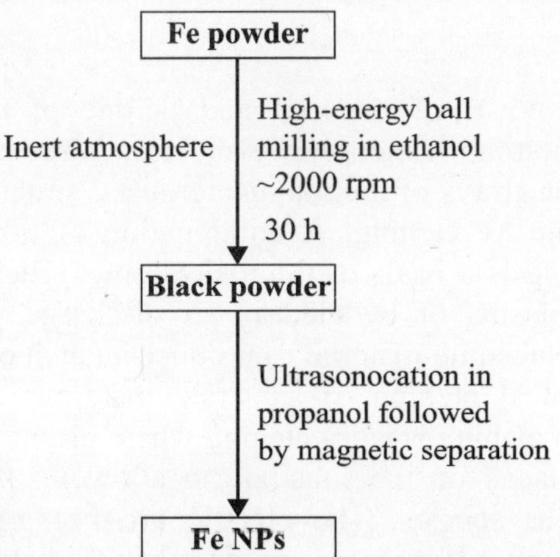

Scheme 3.1. Steps involved in synthesis of iron nanoparticles by high energy milling

Conventional ball mills can be modified as a low temperature chemical reactor allowing the reaction to occur during the actual milling. This process is known as **mechanochemical processing (MCP)** which turns out to be a cost effective method for manufacturing a wide range of nanopowders. Here the milling increases the reaction kinetics of precursors at relatively lower temperatures and results in refinement of the grain structure to the nanometer scale. Oxides, carbonates, sulphates, chlorides, fluorides, hydroxides or other compounds are used as the precursor material. The chosen precursor is then milled with an appropriate reactant. The resulting product phase is formed as individual single nanometer-sized grain in a by-product matrix. One simple example is high-energy milling of a mixture of $FeCl_3$ powder and Na pieces induces a reaction between $FeCl_3$ and Na, forming Fe nanoparticles mixed with NaCl. The NaCl can be easily leached out from the powder by using water, and Fe nanopowder is produced.

3.2.2 Sputtering

Sputtering, also known as Sputter Deposition, is a conventional thin film deposition technique, useful for high melting point material. It involves eroding material from a "target" source onto a "substrate" *e.g.* a silicon wafer. Sputtered atoms ejected into the gas phase are not in their thermodynamic equilibrium state, and tend to deposit on all surfaces in the vacuum chamber. Unlike many other vapor phase techniques there is no melting of the materials. Sputtering is done at low pressure on a cold substrate. The process involves three steps, (i) glow discharge to produce energetic particles (ii) momentum transfer from an indirect energetic projectile to a solid or molten target and (iii) condensation of sputtered species. Sputtering observed to occur below the threshold energy of physical sputtering is often called chemical sputtering.

3.2.3 Etching

Etching refers to removal of material and is one of the common transfer techniques used in microfabrication, however, it has also been used to synthesize nanoparticles. Regular arrays of the nanometer-sized structures can be produced on a planner substrate by etching. A combination of lithographically defined patterning with etching is a basis of microelectronics. Etching can be achieved through chemicals, plasma or by electric arc-discharge. In chemical etching, chemicals react with substrate material to produce an etch profile. Processing gas is selected for chemical etching of substrate materials. Plasma etching also known as reactive ion etching is a dry etching where plasma is used to ionize gas. A negative bias is placed on the substrate to allow for physical etching from positively charged gas species. The electric arc-discharge method has been shown to have potential for mass production of gold nanoparticles. The nanoparticles thus produced could be stabilized by just water molecules. The underlying mechanism of stabilization is that the electrons are injected into the gold nanoparticles by the cathode during arc discharge and negatively charged surface of gold nanoparticles are saturated by atomic oxygen, which may result in hydrogen bonding with water molecules in an aqueous environment. Without any stabilizers and surfactants, negatively charged gold nanoparticles surrounded by water molecules can create a stable suspension. Such a gold colloid can be heated up to 100ºC without destruction of its structure. Electrochemical etching can also be used to produce regular arrays of shapes within nanometer range. For example, layers of porous silicon are formed by electrochemically etching the crystalline silicon wafers, employing a mixture of hydrofluoric acid and ethanol as an electrolyte.

3.2.4 Laser ablation

Laser ablation is the process of removing material from a solid (or occasionally liquid) surface by irradiating it with a laser beam. At low laser flux, the material is heated by the absorbed laser energy and evaporates or sublimates. At high laser flux, the material is typically converted to plasma. Usually, laser ablation refers to removing material with a pulsed laser, but it is possible to ablate material with a continuous wave laser beam if the laser intensity is high enough. The depth over which the laser energy is absorbed and thus the amount of material removed by a single laser pulse depends on the material's optical properties, laser wavelength and pulse width. The total mass ablated from the target per laser pulse is usually referred to as ablation rate. Laser pulses can vary over a very wide range of duration (milliseconds to femtoseconds) and fluxes, and can be precisely controlled. This makes laser ablation a very attractive process in the fabrication of a broad range of nanomaterials. Typically a focused laser beam irradiates on the target (precursor) material and evaporates the target under high vacuum conditions. Atoms and clusters are subsequently being ejected from the target and deposited on to a substrate placed opposite the target. Laser ablation has been extensively used for the preparation of a variety of metallic and semiconductor nanoparticles and particulate films. The distinctive advantage of laser ablation over other deposition techniques is the capability of producing multi-component materials with well-controlled stoichiometry. Laser ablation technique can be complemented by other techniques such as ultrasonication and solution chemistry.

3.2.5 Lithography

Lithography is an important method for fabrication of nanodevices such as integrated circuits. It involves the use of a light or electron beam to selectively remove micro-scale structures from a precursor material called resist. Lithographic patterning methods capable of structuring material on a fine scale are referred as microlithography and nanolithography. Typically, features less than

10 micrometers are considered microlithographic, and less than 100 nanometers are considered nanolithographic. A typical lithographic process consists of three successive steps:

i. Coating a substrate (Si wafer or glass) with a sensitive polymer layer (the resists),

ii. Exposing the resist to light, electrons or ion beams,

iii Developing the resist image with a suitable chemical (developer), which reveals a positive or negative image on the substrate depending on the type of resist used (*i.e.* positive tone or negative tone resist).

Lithographic techniques can be subdivided in two broad categories:

1. **Photolithography:** Photolithography uses UV, deep-UV, extreme-UV or X-ray light source to expose a layer of radiation-sensitive polymer (photoresist) through a mask. The mask is a nearly optically flat glass (or quartz) plate which contains the desired pattern *i.e.* opaque areas (the pattern, made of an absorber metal) on a UV-transparent background. Photolithography is commonly used for fabricating semiconductor microchips and micro-electro-mechanical systems (MEMS) devices.
2. **Soft lithography:** A scanning beam irradiates the surface of the resist sequentially, point by point, through a computer-controlled program where the mask pattern is defined. Methods that uses software mask are collectively called Soft lithography.

Soft lithography is a term used for a number of techniques that fabricate and use a soft mould prepared by casting a liquid polymer precursor against a right master. These methods have been developed specifically for making large-scale micro and nanostructures with equipment that is easier to operate as compared to those used in conventional lithography. The master is normally fabricated via a conventional lithographic method. Various polymers (*e.g.* polyurethanes, epoxides and polyimides) are used for molding; most common is poly (dimethl-siloxane) (PDMS). A PDMS mould is fabricated by pouring its liquid precursor over a lithographically-made master (*e.g.* a photoresist or silicon master), cured to induce cross-linking, and then peeled off from it. The stamp can then be used for printing a desired material from the stamp to a suitable surface. **Figure 3.3** shows the overall principle of soft lithography.

The resolution of soft lithography is mainly determined by van der Waals interactions, by wetting and by kinetic factors such as filling the capillaries on the surface of the master, but not by optical diffraction. This is an important advantage over conventional lithographic techniques. Soft lithography provides access to three-dimensional and curved structures, tolerates a wide variety of materials, generates well-defined and controllable surface chemistries, and is generally compatible with biological applications. It is also low in cost, experimentally convenient and has emerged as a technology useful for a number of applications that include cell biology, microfluidics, lab-on-a-chip, microelectro-mechanical systems and flexible electronics/photonics. The commonly used soft lithographic techniques are: (i) μ-contact printing of alkanethiols and proteins on gold-coated and glass substrates; (ii) replica molding for fabrication of microfluidic devices in poly(dimethyl siloxane), and of nanostructures in polyurethane or epoxy; and (iii) solvent-assisted micro molding of nanostructures in poly(methyl methacrylate).

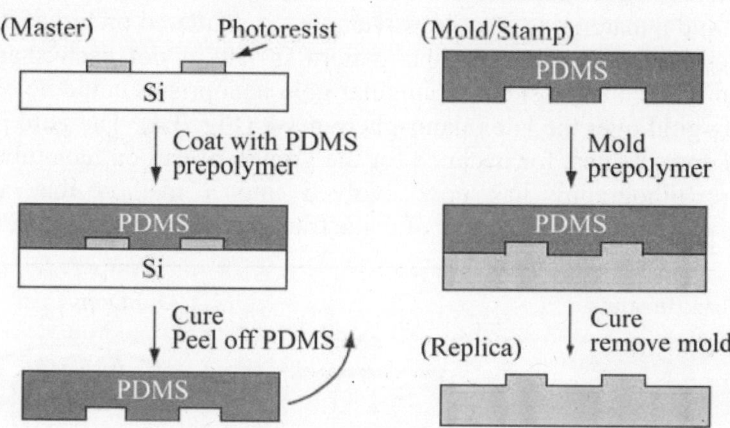

Fig. 3.3. Soft lithography with polydimethyl siloxane (PDMS) molds

3.2.5.1 Electron beam and focused ion beam lithography

Energetic particles such as electrons and ions can be used to pattern appropriate resist films leading to features with nanometer resolution. When using electrons, the technology is called electron beam lithography (e-beam), whereas when using ions the technology is called focused ion beam lithography. In a typical e-beam lithography process, a focused beam of electrons scans across the surface an electron-sensitive resist film, such as poly (methyl methacrylate) (PMMA). The main advantage of e-beam lithography over photolithography is its high resolution patterns with features as small as 50 nm. The resolution of this technology is mainly determined by the scattering of the electrons in the resist film and the substrate. Focused ion beam lithography works on the same principle as of e-beam lithography, but ions such as H^+, He^{++}, Li^+ and Be^{++} are used. Both techniques provide a resolution much higher than photolithography but share a main disadvantage of being very slow; therefore, their use is mostly limited in producing photomasks in optical lithography.

3.2.5.2 Nanosphere lithography and colloidal lithography

In nanosphere lithography the nanospheres are dispersed in a liquid (*i.e.* a colloid) and a droplet placed on a surface is left to dry. Depending on the surface properties (*e.g.* charge) and media used in the colloid (*e.g.* presence of electrolytes), the nanosphere will self-assemble in an ordered pattern. In some conditions, a colloidal crystal is obtained and each nanoparticle is surrounded by six other nanospheres. This regular arrangement (which is a 2D colloidal crystal) can be used to create ordered structures on surfaces. In the regular arrangement of nanospheres there will be an empty space between them, which is regularly repeated in the entire surface. In the simplest case, this space is employed to create relatively flat nanopatterns on the surface. The nanosphere pattern is used as a mask, and a material (*e.g.* gold, silver, etc.) is sputtered on top of it. Once the nanospheres are removed, a regular pattern is left of dot each shaped like a triangle but with concave sides. Triangular gold nanoprisms could be obtained by evaporating gold over the latex nanosphere mask **(Fig. 3.4)**. The gold pattern can also act as growth sites, for instance for the growth of carbon nanotubes or ZnO. Nanosphere lithography has now evolved into a method that allows the fabrication of very complex arrays of nanostructures, including 3D features.

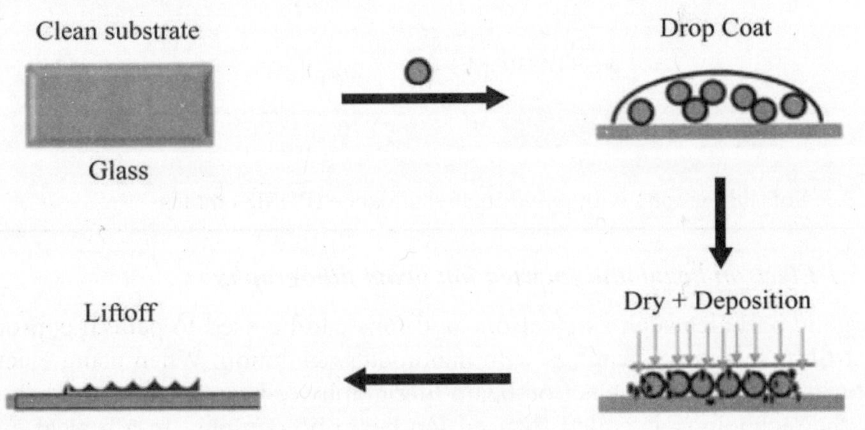

Fig. 3.4. Nanoshpere lithography

Colloid lithography shares the same principle of nanosphere lithography of using a colloid as a mask for the fabrication of nanostructures on surfaces. In this method however, electrostatic forces are employed to obtain short-range ordered arrays of nanospheres on the surface. The electrolyte concentration affect the size of the colloidal particles in the formation of the ordered array. The array can then be used to create a number of different nanostructures, through various processes like etching, liftoff, etc. It is interesting to note the different types of nanostructures that can be formed: holes, cones, rings, sandwiches made of different material, etc. In case of metals, these nanostructures present a localized surface plasmonic resonance effect (LSPR) which can be used for sensing *e.g.* in medical diagnostic devices.

3.2.5.3 Nanoimprint lithography (NIL)

Nanoimprint lithography (NIL) is a revolutionary approach to low-cost and high throughput nanolithography. Nanoimprint Lithography (NIL) in its original version was developed as an alternative technique for the embossing of high resolution patterns in thermoplastic materials. The concept of nano-imprint lithography is to use a hard master with 3D nanostructure to mould another material, which assumes its reverse 3D structure. Since the master has a fine nano-structure, to be successful the process must be done under pressure; a coating must first be placed on the master to avoid catastrophic adhesion to the mould, and the mould must be heated (above its Tg temperature) in order to be soft enough to completely enter the fine master nanostructure and be effectively replicated. The method is the equivalent of embossing at the nanoscale. Nanoimprint lithography is imprinting using a mold to create a thickness contrast in a resist and pattern transfer using anisotropic etching to remove residue resist in the compressed areas. NIL patterns a resist by physically deforming the resist shape with a mold (*i.e.* embossing), rather than modifying the resist chemical structures with radiation as in the case of conventional lithography. This fundamental difference in principles frees NIL from many problems suffered in conventional lithography, such as diffraction limit, scattering, and chemistry. As a result NIL can achieve ~10 nm structures, over large areas with low cost and high throughput, a feat currently unachievable using existing lithography. NIL is believed to impact not only future integrated circuit development, but also many other disciplines, such as biology, chemistry, medicine, and materials. The main advantages of NIL are:

a. *High resolution fast process*: Due to the stamping nature of the process, NIL is inherently parallel process and therefore fast. It may take hours, days or weeks to fabricate the master stamp, but to replicate the patterns only takes seconds or minutes.

b. *Direct 2.5D patterning*: Different height levels or curved features in the stamp can be replicated.

c. *Direct structuring of functional materials*: After the imprinting step, it is not mandatory to follow the etching step to transfer the pattern into the substrate; in many cases, the structured polymer is the final device.

Hot embossing or thermal NIL (using a thermoplastic polymer), UV-NIL (utilizing a UV-curing polymer) and micro-contact printing (μ-CP) or nano-contact printing (n-CP), where an ink is transferred from the stamp to the substrate are three basic variants of the same nanoimprint lithography. Each of these three processes appears in a different form *i.e.* roller NIL, reversal NIL, combined photo- and nanoimprint lithography. The imprint polymer can be droplet dispensed or spin coated. Typically hot embossing processes require higher pressures than UV-NIL processes with μ-CP processes using the lowest pressure. Micro contact printing (μCP) is useful for pattering features with lateral dimension of 500 nm or larger. One of the major challenges for μCP has been to achieve the capability to print with high resolution, *i.e.* with lateral dimension lower than 100 nm. This has been recently achieved by improving the stability of the PDMS, which being soft and highly compressible, has tendency to deform and collapse. One way to improve the stability of the patterns is to affix a backplane to the stamp or to change the chemical formulation of the stamp itself, in order to obtain a harder polymer. With these modifications, it is now possible to print features as small as 50 nm. This printing method, which uses harder stamps, is called nano-contact printing (nCP).

3.2.5.4 Scanning Probe Lithography: Dip-Pen Nanolithography

Scanning probe microscopy (*i.e.* STM, AFM, etc.) uses <50 nm tips to image surfaces with atomic resolution. In this form they are referred to as scanning probe lithography (SPL), which uses the tip of an AFM to selectively remove certain areas on a surface and dip-pen nanolithography (DPN), which similarly uses the AFM tip to deposit material on a surface with nanometer resolution. Both are direct writing techniques and their main advantages are high resolution and the ability to generate patterns with arbitrary geometries. Like e-beam and ion-bean lithography, SPL and DPN are serial techniques whose main limitation is speed.

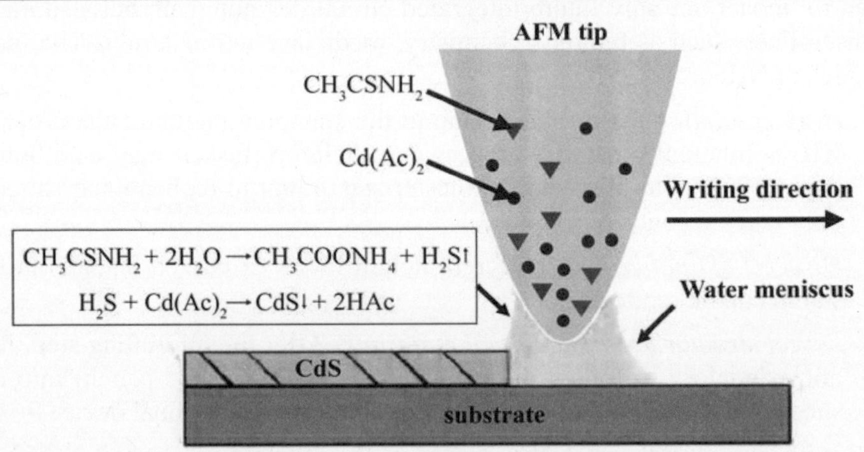

Fig. 3.5. Illustration of dip-pen nanolithography

This ability has opened up opportunities for their use in generating nanostructures and nano devices. One of the more recent techniques for nanocontact printing consists of the direct writing of a molecular ink onto an appropriate substrate *via* the ultra-fine tip of an atomic force microscope. **Figure 3.5** illustrates the process of dip-pen nanolithography for writing nano features of CdS on mica and SiO_2 substrate. The AFM tip is coated with a liquid containing the molecules to be deposited. The use of such a "nano fountain pen" is known as dip-pen nanolithography (DPN) was first demonstrated by Mirkin and coworkers in the late 1990s. Though the earliest examples of DPN featured alkylthiols as the ink onto Au surfaces, there are now an increasingly large number of other ink/ substrate combinations that have been reported for DPN (**Table 3.2**). Top-down lithographic methods will, of course continue to be the best for the important, specialized task of making planar structures with arbitrary patterns (*e.g.* circuits). Top-down assembly methods are currently utilized for interconnecting of electronic circuits. A variety of approaches have been demonstrated for the fabrication of nanostructures on solid supports for such plasmonic applications. These includes various lithographic methods such as electron-beam lithography, nanosphere lithography and dip-pen nanolithography.

Table 3.2. Substrate-ink combinations for DPN

Substrate	Suitable molecular ink
Gold (Au)	Alkylthiols, Proteins, DNA
Silica (SiO_2)	Silazanes, proteins, conjugated polymers, DNA, thioacetamides, cadmium acetate, Colloidal particles
Gallium arsenide (GaAs)	Silazanes
Mica	Thioacetamide, cadmium acetate
Silicon (Si)	Alkynes, alkoxysilanes, metal salts
Germanium (Ge)	Metal salts

3.2.6 Aerosol based processes

Aerosol based process is a common method for the industrial production of nanoparticles. Aerosols can be defined as solid or liquid particles in a gas phase, where the particles range, from molecules up to 100 μm in size. Aerosol based techniques involve atomization of a liquid precursor into finely divided sub micro meter liquid droplets distributed in gas medium. Aerosols were used in industrial manufacturing long before the basic science and engineering of the aerosols were understood. For example, carbon black particles used in pigments and reinforced car tires are produced by hydrocarbon combustion; titania pigment for use in paints and plastic is made by oxidation of titanium tetrachloride; fumed silica and titania formed from respective tetrachloride by flame pyrolysis; optical fibers are manufactured by similar process. Aerosols can be generated by different techniques of which most convenient are flame assisted ultrasonic methods and electrospray which are described here briefly.

3.2.6.1 Ultrasonic spraying

Ultrasonic spray pyrolysis may be employed to generate an aerosol from a dilute aqueous metal salt solution, resulting in the production of particles with a narrow size distribution. **Figure 3.6** shows a setup for ultrasonic spray pyrolysis, a technique used to synthesize fine powders by aerosol decomposition. Silver nanoparticles are produced by flash pyrolysis of a liquid feed solution of silver nitrate nebulized by an ultrasonic atomizer. Aqueous silver nitrate solution is gravity fed from a burette to an atomizer, and the atomized spray is discharged into the reaction chamber, maintained above 650°C in a tube furnace. Silver nanoparticles are deposited on the substrate placed in the flow path, while the water vapor and off-gases are discharged through an exhaust line. A vacuum pump is used to put the reactor under slight vacuum to discharge the water vapor and exhaust gases to vent.

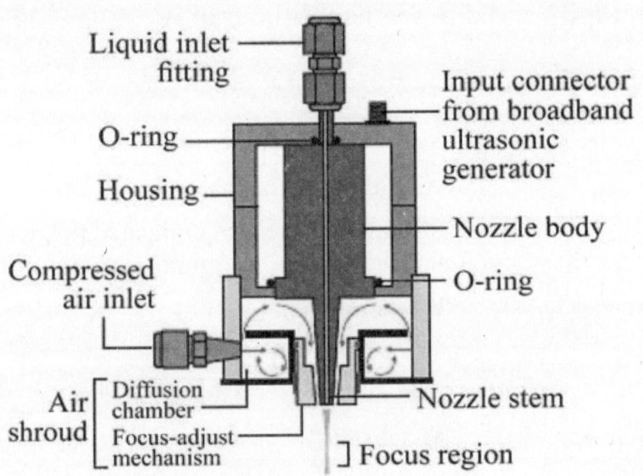

Fig. 3.6. Schematic of ultrasonic spray pyrolysis set-up

3.2.6.2 Electrospray deposition

Electrospraying is a method of liquid atomization of spraying solution or suspensions by electric field, which allows production of fine particles, down to nanometer size. This technique has successfully demonstrated the ability to disperse masses of liquids into very fine mists and has revolutionized the nanotechnology fields in fabricating bio-macromolecules without losing their functionalities, metal oxides, polymeric nanoparticles etc.

When a sample liquid stored in a thin capillary and a high voltage is applied, a strong electric field is generated at the tip of the capillary, due to the effect of electric field concentration. As the liquid begins to exit from the capillary, it forms a conical shape with the electrically charged ions gathered on its surface (This is called Taylor cone). Subsequently, when the electrostatic force becomes stronger than surface tension, the liquid erupts from the tip of the capillary to form a fine jet. Since the jet is highly charged, the liquid immediately turns into fine droplets to generate spray with each droplet spilt from the next by electrostatic force (This is called the Coulomb Explosion). The droplets formed by means of electrospray are tiny. Therefore, the solvent evaporates and dries in a very short period of time, and as a consequence, it forms very fine nanoparticles. The Electro spray deposition method facilitates nano structures in a larger area. Sprayable substances include solution and dispersion liquid, such as organic/inorganic compounds, biomacromolecules and/or synthetic polymers, and deposit various nano structures. The basic set-up for electrospraying is shown schematically in **Fig. 3.7**.

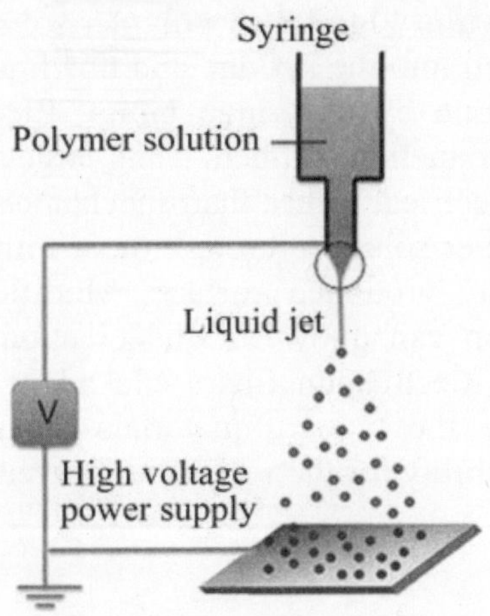

Fig. 3.7. Schematic of a basic set-up for electrospray

A typical one step preparation of polyacrylonitrile (PAN) nanoparticles by electro spraying is given as under:

1-3% (w/w) solution of polyacrylonitrile in DMF is placed into a syringe connected to a high voltage source and syringe pump. The electrical field is formed between the collector and the tip of the syringe tube, the electrospray of solution resulted in formation of nanoparticles of PAN.

3.2.7 Electro-Spinning

It is an emerging technology for the manufacture of nanopolymer fibers and is based on the principle of spinning the dilute polymer solutions in a high voltage electric field. In this process a suspended drop of polymer is charged with thousands of volts. At a characteristic voltage, the droplet forms a Taylor cone and a fine jet of polymer releases from the surface in response to the tensile forces generated by interaction of an applied electric field with the electrical charge carried by the jet. This produces a bundle of polymer fibers. The jet can be directed to a grounded surface and collected as a continuous web of fibers ranging in size from a few μm to less than 100 nm. Electrospinning is a variation of electrospray technique, where application of high voltage to a polymer solution results in the formation of a conejet geometry which turns into very fine fibers instead of breaking into small droplets as in case of Electrospray. Electrospinning shares characteristics of both electrospraying and conventional solution dry spinning of fibers. The standard laboratory setup for electrospinning consists of a spinneret (typically a hypodermic syringe needle) connected to a high-voltage (5 to 50 kV) direct current power supply, a syringe pump, and a grounded collector as shown in **Fig. 3.8**. A polymer solution, sol-gel, particulate suspension or melt is loaded into the syringe and this liquid is extruded from the needle tip at a constant rate by a syringe pump. Electrospinning process is particularly suited to the production of fibers using large and complex molecules. The technique relies on electrical rather than mechanical forces to form fibers. The charge pulls the polymer solution through the air into a tiny fiber, which is collected on an electrically grounded surface. Modification of the spinneret and/or the type of solution can allow for the creation of fibers with unique structures and properties. Electrospun fibers can adopt a porous or core-shell morphology depending on the type of materials being spun as well as the evaporation rates and miscibility for the solvents involved.

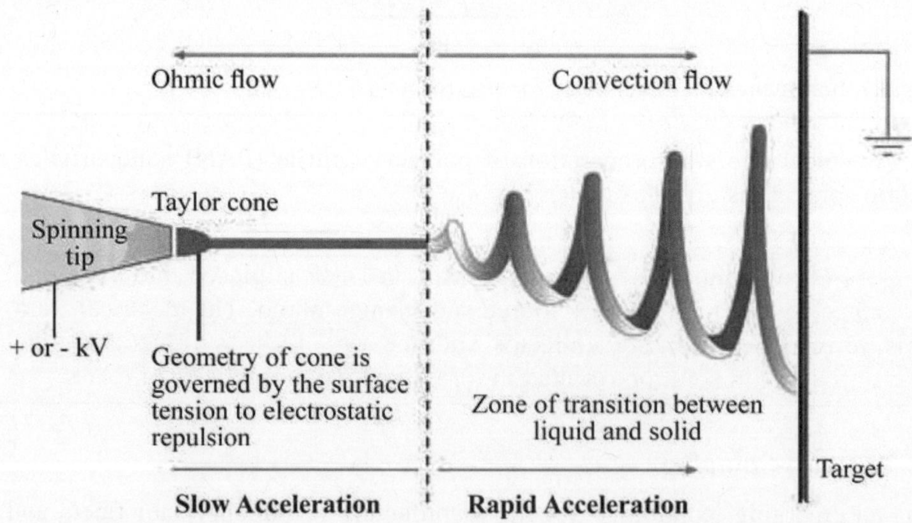

Fig. 3.8. Schematics of electrospinning process for nanofiber synthesis

Synthesis of hierarchical TiO_2 comprising nanorods on-nanofibers is a typical example of combined application of electrospinning techniques together with sol-gel and solvothermal processes.

3.3 BOTTOM-UP APPROACH

The bottom-up approach involves synthesis of the nano materials from atomic or molecular species via chemical reactions or self-assembly, allowing for the precursor particles to grow in size or gradually assembling the atomic or molecular precursors until the desired structure is achieved. Bottom-up approach is more often used for preparing most of the nanomaterials with the ability to generate a uniform size, shape and distribution. It effectively covers chemical synthesis and precisely control the reaction to inhibit further particle growth. It plays a significant role in the fabrication and processing of a variety of nanostructures and nanomaterials. Bottom-up approach is basically the chemical process based transformations in gas, liquid or aerosol form. The main bottom-up techniques includes chemical vapor deposition (CVD), chemical vapor condensation (CVC), sol-gel, precipitation, reduction, solvothermal, matrix mediated (template assisted) processing, self-assembly etc. Sol-gel processing differs from other chemical processes due to its relatively low processing temperature.

3.3.1 Chemical Vapor Deposition (CVD)

Chemical vapour deposition or CVD is a generic name for a group of processes that involve depositing a solid material from a gaseous phase and is similar in some respects to physical vapour deposition (PVD). The later differs in that the precursors are solid, with the material to be deposited being vaporised from a solid target and deposited onto the substrate. CVD is a well known process in which a solid is deposited on a heated surface via a chemical reaction from the vapor or gas phase. CVC reaction requires activation energy to proceed. This energy can be provided by several sources *e.g.* thermal, plasma, laser or photon laser.

In thermal CVD the reaction is activated by a high temperature above 900°C. A typical apparatus comprises of a gas supply system, a deposition chamber and an exhaust system is shown in **Fig. 3.9**. Thermal CVD is the simplest and scalable method to produce nanoparticles by heating the desired material in a heat resistant crucible containing the desired material. Heating normally of the precursor is carried by arc heating, electron beam heating or Joule heating. The atoms are evaporated into an atmosphere, which is either inert (*e.g.* He) or reactive (so as to form a compound). To carry out reactive synthesis, materials with very low vapor pressure have to be fed into the furnace in the form of a suitable precursor such as organometallics, which decompose in the furnace to produce a condensable material. The hot atoms of the evaporated matter lose energy by collision with the atoms of the cold gas and undergo condensation into small clusters via homogeneous nucleation. The clusters would continue to grow if they remain in the supersaturated region. To control their size, they need to be rapidly removed from the supersaturated environment by a carrier gas. The cluster size and its distribution are controlled by three parameters *i.e.* rate of evaporation (energy input), rate of condensation (energy removal), and rate of gas flow (cluster removal).

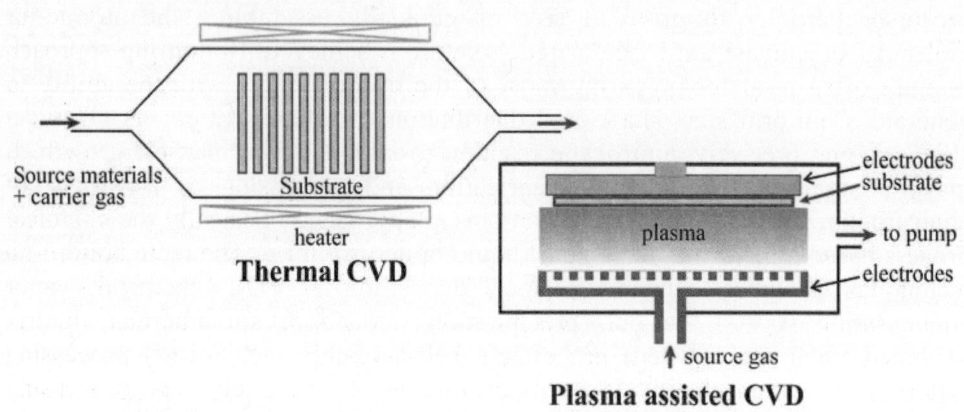

Fig. 3.9. Schematic representation of chemical vapor deposition set up

A plasma is a highly energized gaseous state matter where significant fraction of the matter is ionized and hence is a conductor of electricity. Although there are free charges in plasma, overall negative and positive charges compensate each other. In plasma CVD, the reaction is activated by plasma at temperatures between 300 and 700°C. A typical apparatus is shown in **Fig. 3.9.** The major advantage of the plasma assisted pyrolysis in contrast to the thermal activation is the low temperature reaction which reduces the tendency for agglomeration of the primary particles. Additionally, it has been shown that by introducing another precursor into a second reaction zone of the tubular reactor, *e.g.* by splitting the microwave guide tubes, the primary particles can be coated with a second phase. For example, it has been demonstrated that ZrO_2 nanoparticles can be coated by Al_2O_3.

Microfabrication processes widely use CVD to deposit materials in various forms, including: monocrystalline, polycrystalline, amorphous, and epitaxial. These materials include: silicon, carbon fiber, carbon nanofibers, filaments, carbon nanotubes, Synthetic nanodiamonds, SiO_2, silicon-germanium, tungsten, silicon carbide, silicon nitride, silicon oxynitride, titanium nitride, and various high-k dielectrics. CVD is also used to produce synthetic diamonds and also to deposit conformal films. Nano composite powders have been prepared by CVD *e.g.* SiC/Si N composite powder was prepared using SiH_4, CH_4, WF_6 and H_2 as a source of gas at 1400°C.

Due to the inherent simplicity, it is possible to scale up this process from laboratory (mg/day) to industrial scales (tons/day). Few examples of commercially important nanomaterials prepared by CVD are given here. CVD for tungsten is achieved from tungsten hexafluoride (WF_6), which may be deposited in two ways:

$$WF_6 \rightarrow W + 3\ F_2$$

$$WF_6 + 3\ H_2 \rightarrow W + 6\ HF$$

Other metals, notably aluminum and copper can also be deposited by CVD. Niobium(V) oxide layers can be produced by the thermal decomposition of niobium(V) ethoxide with the loss of diethyl ether according to the equation:

$$2\ Nb\ (OC_2H_5)_5 \rightarrow Nb_2O_5 + 5\ C_2H_5OC_2H_5$$

3.3.2 Chemical Vapor Synthesis/Condensation (CVS/CVC)

In Chemical Vapor Synthesis (CVS) chemical vapors of precursor materials are brought to react in a reaction chamber. The reaction chamber is typically heated using joule heating. The method is similar to chemical vapor deposition, however instead of deposition of nanomaterial as a thin film, CVS process encourages formation of nanoparticles. Thus, the method is well recognized for the flexibility it provides as a nanomaterial synthesis process. The precursors can be in solid, liquid or gas phase at the ambient temperature but are delivered to the hot wall reactor at the vapor phase. The method has been adopted to fabricate a wide range of nanomaterials from a variety of precursor materials. The main process parameters of the CVS process are the residence time of precursors, gas flow rate, pressure different betweenthec inlet and the main chamber and temperature of the hot wall. In the simplest form of CVS, metal organic precursor is introduced to the hot wall reactor at a controlled rate **(Fig. 3.10)**. In this approach, vapor phase precursors are brought into a hot-wall reactor under conditions that favor nucleation of particles in the vapor phase rather than deposition of a film on the wall. It is called chemical vapor synthesis or chemical vapor condensation (CVC). This method has tremendous flexibility in producing a wide range of materials and can take advantage of the huge database of precursor chemistries that have been developed for CVD processes. The precursors can be solid, liquid or gas at ambient conditions, but are delivered to the reactor as a vapor (from a bubbler or sublimation source, as necessary).

When a mixture of gas reactants are delivered into a reaction chamber, the chemical reactions among the gas molecules are induced by an input of energy such as resistant heating, laser, and plasma. Chlorides are popular reactants for the formation of oxides because of their generally low vaporization temperature and low cost. The typical reaction is

$$SnCl_4\ (gas) + 2H_2O\ (gas) \rightarrow SnO_2\ (solid) + 4HCl\ (gas)$$

This method is used mainly for metal nanoparticles. A bulk material is heated in vacuum to produce a stream of vaporized and atomized matter, which is directed to a chamber containing either inert or reactive gas atmosphere. Rapid cooling of the metal atoms due to their collision with the gas molecules results in the condensation and formation of nanoparticles. If a reactive gas like oxygen is used then metal oxide nanoparticles are produced. The theory of gas-phase condensation for the production of metal nano-powders is well known. Gas-phase condensation uses a vacuum chamber that consists of a heating element, the metal to be made into nano-powder, powder collection equipment and vacuum hardware.

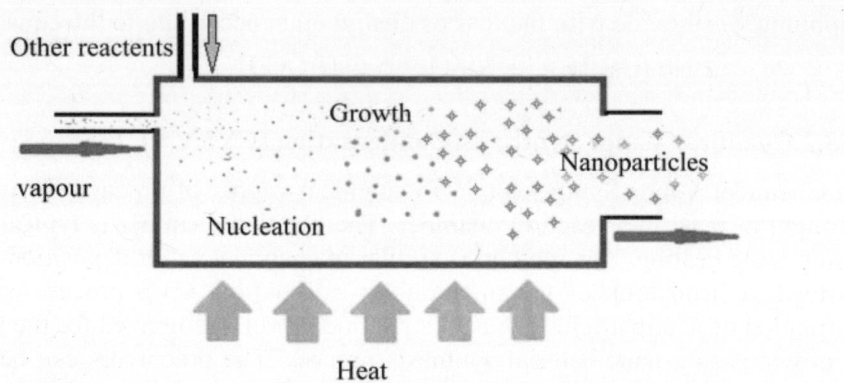

Fig. 3.10. Schematic diagram of CVS reactor for synthesisn of nanoparticles

The reactor chamber can be filled with reactive gases to produce the respective metal oxide, halide, nitride or carbide during the process. Mixtures of nanoparticles of two phases can be fabricated by supplying two or more precursors simultaneously. The same strategy is used to make doped nanomaterials as well. CVS process can also be used to make coated or core shell nanoparticles. This is typically achieved by supplying the second precursor at a later stage of the reactor. CVS is also regarded as a high through-put process as the production is continuous. Even a small-scale reactor can produce considerably high amounts of nanomaterials compared to other manufacturing techniques.

3.3.3 Plasma arcing

This is the most common method for fabricating nanotubes. The method uses plasma, which is an ionized gas. A potential difference is created between two electrodes and the gas ionizes in between the electrodes. A typical arcing device is made of two electrodes an arc that passes from one electrode to the other. The first electrode (anode) vaporizes as electrons are taken from it by potential difference. For instance when a carbon electrode is used, it will be consumed during the reaction producing carbon cations. These positively changed ions pass to the other electrode, pick up electrons and are deposited to form nanotubes. Plasma arcing can also be used to deposit nanolayers on surfaces rather than making new structures. The deposit can be as little as a few atoms in depth. In this sense plasma arcing is complementary to chemical vapor deposition.

3.3.4 Wet chemical methods

Wet chemical methods are the most exploited bottom-up approach for synthesis of a variety of nanomaterials. A number of methods included under this category are discussed in the following sections.

3.3.4.1 Chemical reduction method

Chemical reduction of metal salts is the most common method for synthesis of metal nanoparticles and has the potential to be scaled up even at room temperature. The basic scheme for synthesis of metal nanoparticles by chemical reduction method is shown in **Scheme 2**.

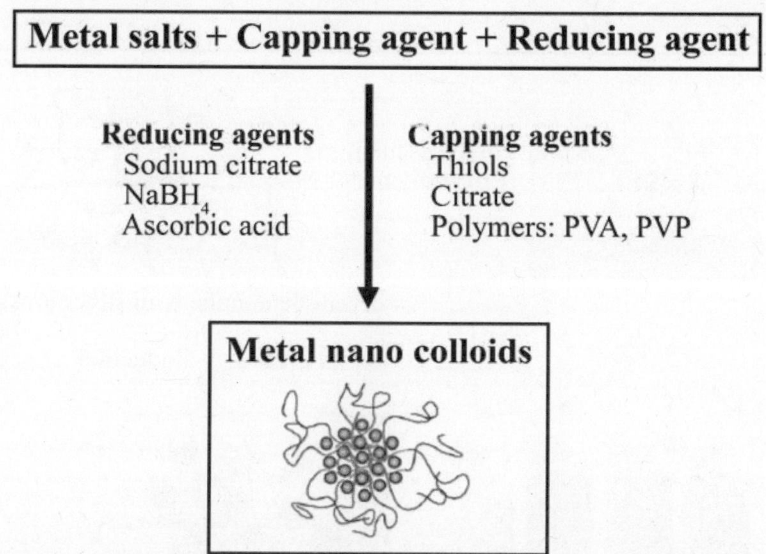

Scheme 2. Schematic representation of chemical reduction method for nanosynthesis

Several reducing agents such as ascorbic acid, citric acids and their salts, alkali hydrides namely $NaBH_4$, $NaB(C_2H_5)_3H$, $LiB(C_2H_5)_3H$, etc. along with various capping agents such as polymers *e.g.* polyvinyl pyrrolidone; organic ligand, *e.g.* quaternary amines, thiols; and surfactants have been successfully used to prevent aggregation of metal nano particles. Typical preparations of silver and gold nanoparticles are given in following sections.

Gold nanoparticles can be prepared in a simple and two steps process (nucleation and growth). In a typical preparation of gold nanoparticles, an aqueous solution of Au (HAuCl4, 0.001M) is mixed with mercapto succinic acid (MSA, 0.005M in methanol) to give a transparent solution. Freshly prepared aqueous solution of sodium borohydride (NaBH4, 0.010M) is then added under vigorous stirring conditions. The solution immediately turns dark brown but remain transparent (which indicates that the nanoparticles have not yet formed) until enough reductant was added till a flocculent dark brown precipitate is generated. The pH of the solution increase gradually with the addition of reactant and finally reaches to 8.6. The solvent can be removed by decantation and dried by rotary evaporation.

Optically clear and stable colloidal solutions of silver nanoparticles coordinated with dendrimer polymer, can be synthesized at room temperature by chemical reduction method (Scheme 3). In a typical experiment, 0.85 gm silver nitrate mixed with 100 mL 0.85% polyethleneimine (PEI) is added to 1 ml, formaldehyde. Apperance of light yellow colour, indicates nucleation and growth of Ag NPs.

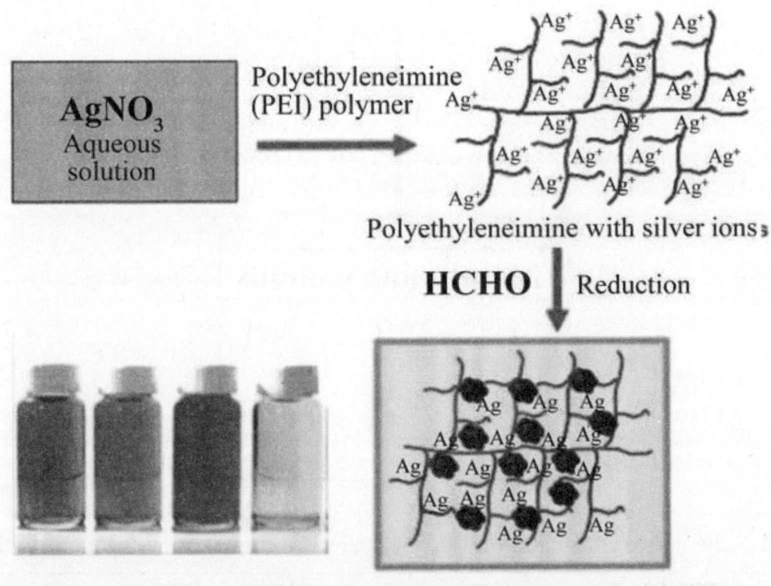

Scheme 3. Preparation of Silver nanoparticles by reduction method

3.3.4.2 Exchange reactions

The exchange reactions leading to precipitation of solids from a solution with metal ions is one of the most common processes for the synthesis of nanomaterials. In a supersaturated solutions, particles are formed via homogeneous or heterogeneous nucleation. The growth of particles usually occurs via diffusion or Ostwald ripening. Particle size and particle size distribution, crystallinity as well as morphology are controlled by applying reaction kinetics. Important controlling parameters are concentration, temperature, pH of the solution, order of adding the reactants and mixing processes. For the synthesis of quantum dots, the formation of nuclei has to be fast and the particle growth slow. For the synthesis of monodisperse particles, all nuclei have to form at about the same time. Metal oxides, chalcogenides, and metallic nanoparticles can be produced by using precipitation methods. For example the synthesis of metallic nanoparticles, reducing agents are added to the acidic solution of metal salts; metal oxides by adding alkalis to metal salts; metal chalcogenides nanoparticles can be formed by mixing together appropriate reagents, *e.g.*, metal sulfides can be produced by purging or *in-situ* generation H_2S to soluble metal salts (halides, sulphates, nitrates, etc.). To avoid the formation of agglomerates during the precipitation process and to produce stable dispersions of nanoparticles, surfactants (especially for high particle concentrations) or electrolytes (especially for diluted suspensions) can be added. Co-precipitation enables the production of nanoparticles containing various elements. In order to avoid inhomogeneity and phase separation, all species have to be precipitated simultaneously.

3.3.4.3 Hydrothermal/Solvothermal

This is another solution- based method to prepare a wide range of nanomaterials having diverse shapes and sizes. Hydrothermal and solvothermal methods are very similar to each other in that in both the cases reaction is done in a stainless pressure vessel, known as autoclave. The only difference being that the precursor solution is usually not aqueous in case of solvothermal method. Using the solvothermal route one gains the benefits of both the sol-gel and hydrothermal routes. Thus solvothermal synthesis allows for the precise control over the size, shape distribution, and crystallinity of metal oxide nanoparticles or nanostructures. These characteristics can be altered by changing certain experimental parameters, including reaction temperature, reaction time, solvent type, surfactant type, and precursor type. In a typical synthesis, titanium (IV) alkoxide, 10 g, was dissolved in a 70 cm^3 portion of alcohol in a test tube which was then set in a 200 cm^3 autoclave The autoclave was thoroughly purged with nitrogen, heated to the desired temperature (523-573 K) at a rate of 2.5 $Kmin^{-1}$, and kept at that temperature for 2 h. After the autoclave treatment, the resulting powders were washed repeatedly with acetone and dried in air.

3.3.4.4 Reverse micelle method

Reverse micelles are water-in-oil droplets stabilized by a surfactant. The surfactant most often used is sodium salt of bis (2-ethylhexyl) sulfosuccinate, usually called Na(AOT). These droplets are dispersed randomly and subjected to Brownian motion. They exchange their water content and reform two distinct micelles.

Furthermore, the size of the water-in-oil droplets increases linearly, *i.e.* the micellar concentration decreases, with increasing water content, defined as $w=[H_2O]/[AOT]$. This involves the preparation of two separate micro-emulsions, incorporating the different reactants. Upon mixing, nucleation occurs on the micelle edges as the water inside them becomes supersaturated with reactants. Growth then take place around this nucleation point, with the arrival of more reactant fed via inter-micellar exchange. The reverse micelle permits synthesis of a very wide range of nanomaterials such as II-VI group semiconductors, metals, etc. with good control of size and shape. A simple approach is shown in **Scheme 4.**

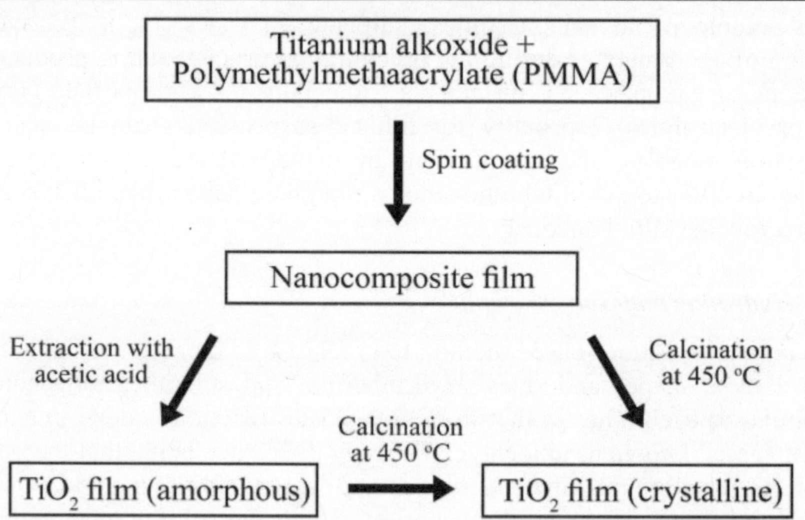

Scheme 4. Illustration of the reverse micelle preparation route for hierarchical structuring of TiO_2

3.3.4.5 Sol-gel method

A "sol" is a type of colloid in which a dispersed solid phase is mixed in a homogeneous liquid medium. An example of a naturally occurring sol is blood. Gel is the formation of a network in continuous liquid phase by a sol. Sol- gel technique is a long-established industrial process for the generation of colloidal nanoparticles from liquid phase; it has been further developed in last years for the production of advanced nanomaterials and coatings. Sol-gel-processes are well adapted for the synthesis of metal oxide nanoparticles and composite nanopowders. The main advantages of sol-gel techniques are low temperature of processing, versatility, and flexible rheology allowing easy shaping and embedding. The reactions involved in the sol-gel chemistry are based on the hydrolysis and condensation of metal alkoxides M(OR) x which can be described as follows:

$$MOR + H_2O \rightarrow MOH + ROH \text{ (hydrolysis)}$$

$$MOH + ROM \rightarrow M\text{-}O\text{-}M + ROH \text{ (condensation)}$$

The typical sol-gel process involves four steps:

1. Formation of stable solutions (**sol**) of alkoxide or solvated metal precursor by hydrolysis reaction, in which the –OR group is replaced with an –OH group. The hydrolysis reaction can occur without a catalyst but is more rapid and complete when a catalyst (*e.g.* a base NaOH, NH_3 or an acid HF, CH_3COOH) is used.
2. After hydrolysis, the sol starts to condense and polymerise (**gel**). This leads to a growth of particles to dimensions of a few nanometers. Gelation resulting from the formation of an oxide- or alcohol-bridged networks by a polycondensation reaction results in a dramatic increase in the viscosity of the solution.
3. Syneresis, the aging of the gel during which the polycondensation reactions continue until the gel transforms into a solid mass, accompanied by contraction of the gel network and expulsion of solvent from gel pores. Ostwald ripening (also referred to as coarsening, is the phenomenon by which smaller particles are consumed by larger particles during the growth process) and phase transformations may occur concurrently with syneresis. The aging process of gels can exceed 7 days and is critical for the prevention of cracks in gels.
4. Drying of the gel, when water and other volatile liquids are removed from the gel network. If the solvent (such as water) is extracted under supercritical or near supercritical conditions, the product is an *aerogel*. Gels are stabilized by dehydration, during which surface-bound M-OH groups are removed. This is normally achieved by calcining the monolith at temperatures up to 800°C.

All the four steps, illustrated above are affected by the initial conditions for the hydrolysis reaction and the condensation/polymerization. These conditions include pH, temperature and time of reaction, nature of the catalyst, etc. **Figure 3.11** summarizes the different steps involved in sol-gel processes.

The sol-gel process is very commonly used to make silica gels. Other types of gels can also be formed; aluminosilicate gels are special because they form tubular structures. One such product is imogolite which has an external diameter of about 2.5 nm and internal tube diameter of 1.5 nm. These types of nanostructures are known to be good adsorbents for anions such as chloride, chlorate, sulphate and phosphates. The imogolite structure can be dissolved away with HF. Therefore these nanostructures can be used for template synthesis: the tube can be filled with atoms and then dissolved away, leaving rows of atoms. Number of strategies based on sol-gel can be adopted to prepare nanocoatings. The gel are placed on a surface, a greater bulk-area ratio is obtained. Another strategy is to form an aerogel. These are three-dimensional continuous networks of particles with are trapped at their interstices. Aerogels are characterized by being porous and very light yet able to withstand 100 times their weight. A versatile way to create ordered surface nanostructures is to perform the sol-gel synthesis in a liquid which is itself ordered. For example, liquid crystals; they have a crystalline structure but exist in a liquid phase. Nanostructured silica with controlled pore size, shape and ordering can be made in this way. The liquid crystalline casting method described above can also be used to produce nanostructured metals. This development is very useful for making nanostructured catalytic surfaces, like platinum or palladium surfaces. Since these metals are very rare and expensive, it is highly advantageous to have surfaces where nearly all metal atoms can take part in the catalytic reaction and not just surface atoms as in conventional solids. The interest in this synthesis method arises due to the possibility of synthesizing nonmetallic inorganic materials like glasses, glass ceramics or ceramic materials at very low temperatures compared to the high temperature process required by melting glass or firing ceramics. The major difficulties to overcome in developing a successful bottom-up approach are controlling the growth of the particles and secondly stopping the newly formed particles from agglomerating. Other technical issues are ensuring the reactions are complete so that no unwanted reactant is left on the product and completely removing any growth aids that may have been used in the process. Also, production rates of nanopowders are very low by this process. The main advantage is that one can get mono dispersed sized nanoparticles by any bottom-up approach.

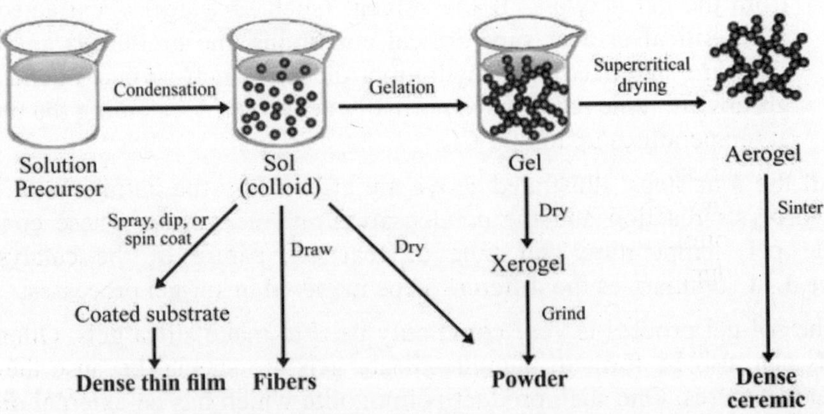

Fig. 3.11. Schematic representation of sol-gel process for synthesis of a variety of nanostructures

3.3.4.6 Sonochemical method

High intensity ultrasound can be used for the production of novel materials and provides an unusual route to known materials without bulk high temperatures, high pressures, or long reaction times. Several phenomena are responsible for sonochemistry and specifically the production or modification of nanomaterials during ultrasonic irradiation. The most notable effects are consequences of acoustic cavitation and can be categorized as primary sonochemistry (gas-phase chemistry occurring inside collapsing bubbles), secondary sonochemistry (solution-phase chemistry occurring outside the bubbles), and physical modifications (caused by high-speed jets or shock waves derived from bubble collapse). Acoustic cavitation (the formation, growth, and implosive collapse of bubbles in a liquid as shown in **Fig. 3.12**) creates extreme conditions inside the collapsing bubble and serves as the origin of most sonochemical phenomena in liquids or liquid-solid slurries. Nebulization (the creation of mist from ultrasound passing through a liquid and impinging on a liquid-gas interface) is the basis for ultrasonic spray pyrolysis (USP) with subsequent reactions occurring in the heated droplets of the mist. In both cases, we have examples of phase-separated microreactors for sonochemistry; it is a hot gas inside bubbles isolated from one another in a liquid, while for USP it is hot droplets isolated from one another in a gas. Cavitation-induced sonochemistry provides a unique interaction between energy and matter, with hot spots inside the bubbles of ~5000K, pressure of~1000 bar, heating and cooling rates of $>10^{10}$ Ks^{-1}; these extraordinary conditions permit access to a range of chemical reaction space normally not accessible, which allows for the synthesis of a wide variety of unusual nanostructured materials. Complementary to cavitational chemistry, the micro droplet reactors created by USP facilitate the formation of a wide range of nanocomposites.

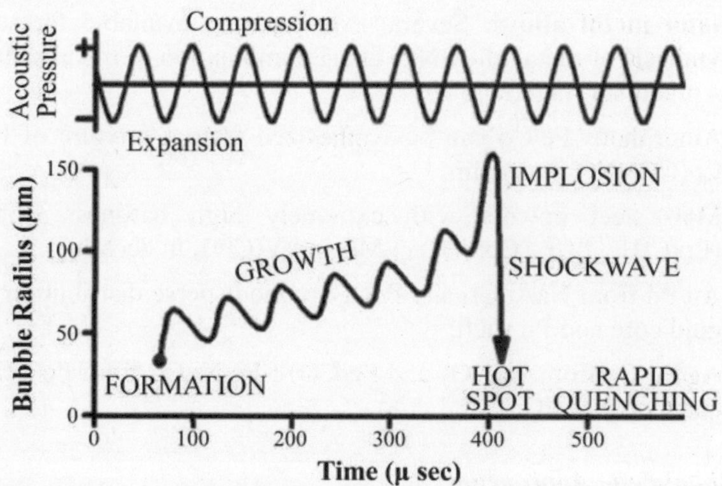

Fig. 3.12. Transient cavitation during sonolysis

Synthesis of inorganic nanoparticles by sonolysis: Molecules can undergo chemical reactions due to the application of powerful ultrasound radiation (20 kHz to 10 MHz). In almost all sonochemical reactions nanomaterials are obtained; they may vary in size, shape, structure, and in their solid phase (amorphous or crystalline) depending on the reaction conditions. Examples of some typical formation of nano metallic powders, metallic colloids and amorphous alloys are given below.

i. **Nanometallic powder** can be obtained from the sonication of metal carbonyls (*e.g.* Fe from $Fe(CO)_5$; Ni from $Ni(CO)_4$; Co from $Co(CO)_3NO$; Pd from $Pd(O_2CCH_3)_2$).

   In a typical preparation, 2.0 mL $Fe(CO)_5$ (0.15 M) and 1.5 mL $Co(NO)(CO)_3$ (0.15 M) are dissolved in 100 ml diphenylmethane. The sonolysis of the solution is performed for 3 h at 20–30°C under argon atmosphere by means of a ultrasonic device with a direct immersion titanium horn (working frequency 20 kHz, electric power of generator 600 W, irradiation surface area of the horn 1 cm^2). The black solid product can be separated by centrifugation, washed with pentane inside the N_2-filled glove box, and dried under vacuum at room temperature. The annealing of the solids is performed under an argon flow (99.996%) for 5 h at 600°C, yielding ~60% $Fe_{40}Co_{60}$ alloy by weight.

ii. **Metal colloids**: Sonication of Au, Pd, Ag in the presence of a surfactant, which largely stabilizes the colloid have resulted in Fe NPs with 8 nm average size from $Fe(CO)_5$ in the presence of oleic acid; Co NPs stabilized by oleic acid (magnetic fluids); elongated Cu NPs using a zwitterionic surfactant; Pt NPs (3nm) from K_2PtCl_6 solution.

iii. **Nano metal alloys:** Several examples are available for sonochemical synthesis of metal alloy NPs using combination of metal salts/complexes as precursor material.

- Amorphous Fe/Co can be synthesized from a mixture of $Fe(CO)_5$ and $Co(CO)_3NO$ in decalin;
- M50 steel powder with extremely high hardness from $Fe(CO)_5$, $(EtxC_6H_{6-x})_2Cr$, $(EtxC_6H_{6-x})_2Mo$ and $V(CO)_6$ in decalin;
- Au/Pd from $NaAuCl_4$ and $PdCl_2$: monodisperse distribution (8 nm) with gold core and Pd shell;
- Ag/Fe_2O_3 from $AgNO_3$ and $Fe(CO)_5$; Fe/Ni/Co from $Fe(CO)_5$, $Ni(CO)_4$ and $Co(CO)_3NO$.

3.3.4.7 Biomimetic Approaches

Nature through millions of years of evolution has selected organisms that can create diversified inorganic material structures in nano- or microscale, known as biomineralization. Examples include calcification and silification found in diatoms, and the formation of magnetite crystals in magnetotactic bacteria. These organisms are capable of organizing nanoscale building blocks into large-scale hierarchical structures to form hard tissues serving different functions, such as mechanical support, filtration, light harvesting, gravity sensing, and locomotion. Scientists have long been exploring the approaches and mechanisms adopted by nature to create functional structures for technical applications, which is usually referred to as biomimetics. Biomimetics refers to applying biological principles for materials formation. Initially bacteria were used to synthesize nanoparticles and this was later succeeded with the use of fungi, actinomycetes and more recently plants and specific peptides. Number of plant products extracts, predominately medicinal plants, has been used for extracellular synthesis of nanoparticles. **Figure 3.13** shows the AFM image of silver nanoparticles synthesized using root extract of *Curculigo orchioides*.

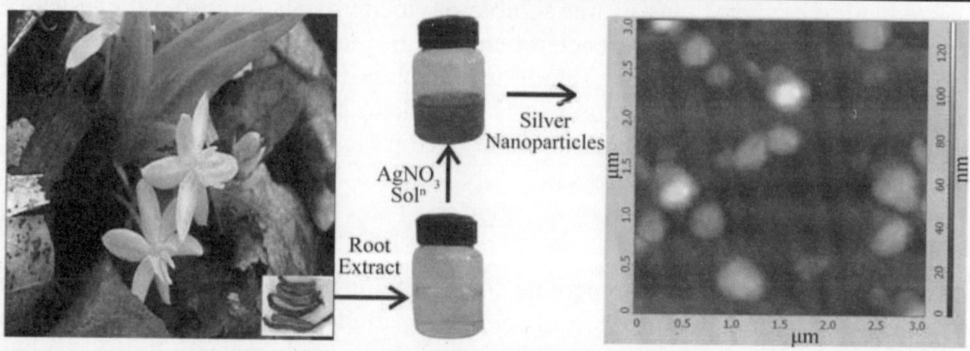

Fig. 3.13. Plant *Curculigo orchioides* (inset shows dried root extract used for nano-synthesis) and AFM image of resultant silver nanoparticles

Bacteria: Microorganisms are considered as a potential biofactory for the synthesis of nanoparticles like gold, silver and cadmium sulphide. Some well-known examples of bacteria synthesizing inorganic materials include magneto-tactic bacteria (synthesizing magnetic nanoparticles) and *S. layer* bacteria which produce gypsum and calcium carbonate layers. The mechanisms involve: efflux systems, alteration of solubility and toxicity via reduction or oxidation, bioso-rption, bioaccumulation, extracellular complexation or precipitation of metals and lack of specific metal transport systems *e.g. Pseudomonas stutzeri* AG 259 isolated from silver mines has been shown to produce silver nanoparticles.

Actinomycetes: *Actinomycetes* are microorganisms that share important characteristics of fungi and *Prokaryotes* such as bacteria. Even though they are classified as prokaryotes, they were originally designated as ray fungi. Focus on actinomycetes has primarily centered on their exceptional ability to produce secondary metabolites such as antibiotics which now finds application in biosynthesis of metal nanoparticles.

Fungi: Fungi have been widely used for the biosynthesis of nanoparticles. In addition to monodispersity, nanoparticles with well defined dimensions can be obtained using fungi. Compared to bacteria, fungi can be used as a source for the production of large amount of nanoparticles. This is due to the fact that fungi secrete more amounts of proteins which directly translate to higher productivity of nanoparticle formation. Instead of fungi culture, isolated proteins from them have also been used successfully in nanoparticles production.

Plants: The advantage of using plants for the synthesis of nanoparticles is that they are easily available, safe to handle and possess a broad variability of metabolites that may aid in reduction. A number of plants have been investigated for their role in the synthesis of nanoparticles.

While fungi and bacteria require a comparatively longer incubation time for the reduction of metal ions, water soluble phytochemicals do it in a much lesser time. Therefore, compared to bacteria and fungi, plants are better candidates for the synthesis of nanoparticles. Taking use of plant tissue culture techniques and downstream processing procedures, it is possible to synthesize metallic as well as oxide nanoparticles on an industrial scale once issues like the metabolic status of the plant etc. are properly addressed.

Specific peptides

A genetically engineered polypeptide for inorganics (GEPI) defines a sequence of amino acids that specifically and selectively binds to an inorganic surface. In molecular biomi-metics, inorganic-binding proteins could potentially be used as (i) linkers for nan-oparticle immobilization; (ii) functional molecules assembled on specific subst-rates; and (iii) hetero bifunctional linkers involving two (or more) binding prot-eins linking several nano inorganic units.

Protein, bovine serum albumin (BSA) has been used as a scaffold to make Ag and Au nano crystals (<1 nm) with yellow and red emission respectively via a simple, one-pot, solution-phase, green synthetic route. In a typical experiment, aqueous $HAuCl_4$ solution (5 mL,50 mM) is added to BSA solution (10 mL, 3 mg/mL) with vigorous magnetic stirring at room temperature. The mixed solution was vacuumized and kept static under nitrogen protection for 2 h. Then, 0.2 mL of $N_2H_4{\cdot}H_2O$ was injected into the vacuumed solution under magnetic stirring. After reaction, the resulting mixed solution was aged under ambient conditions for 24 hrs. Controlled binding and assembly of proteins onto inorganics is at the core of biological materials science and engineering with wide-ranging applications such as implants, hard-tissue engineering, genomics, pharmogenetics and proteomics.

3.3.5 Molecular Self-Assembly

Nanomaterials can, in principle, be made using both top-down and bottom-up techniques. Self-assembly bridges these two techniques and allows materials to be designed with hierarchical order and complexity that mimics those seen in biological systems. Self-assembly of nanostructured materials holds promise as a low-cost, high-yield technique with a wide range of scientific and technological applications. Nanoparticles of a wide range of materials including a variety of organic and biological compounds as well as inorganic oxides, metals, and semiconductors can be synthesized using chemical self-assembly techniques which cannot be synthesized by the above mentioned top-down methods, the most important and obvious of which are molecules. Materials developed by self-assembly are also useful to:

Develop newer methodologies for the synthesis of metal nanoparticles in aqueous medium under ambient conditions without using reducing agents or elevated temperatures.

Formation of stable supramolecular assembly, by bringing metal nanoparticles close to each other, results in plasmon coupling. This strategy can be effectively utilized for the development of metal nanoparticle based sensors.

Supramolecular organization of molecules on surfaces.

Major challenges lies in the application of organic molecules in electronic devices are (i) precise positioning of organic molecules at predefined locations and the creation of well defined structures and (ii) interconnecting various functional devices and further integrating into the macroscopic world. Self assembly approach for nanofabrication addresses many of such requirements and challenges. Detail discussion on these approaches is given in chapter 6.

3.3.6 Langmuir-Blodgett (LB) Film Formation

The Langmuir-Blodgett Films are thin organic films of a thickness of a few nanometers. An organic thin film can be deposited on a solid substrate by various techniques such as thermal evaporation, sputtering, electrodeposition, molecular beam epitaxy, adsorption from solution, Langmuir-Blodgett (LB) technique, self-assembly, etc. The LB-technique is one of the most promising techniques for preparing such thin films as it enables (i) the precise control of the monolayer thickness, (ii) homogeneous deposition of the monolayer over large areas and (iii) the possibility to make multilayer structures with varying layer composition. An additional advantage of the LB technique is that monolayers can be deposited on almost any kind of solid substrate. Langmuir balance is used for depositing LB films. It consists of a Teflon trough to hold the sub phase. The trough is thermostated by circulating water in channels placed underneath. The surface area of the trough can be varied by sweeping movable barriers over the surface of the trough. The barrier is made of Delrin, a hydrophilic material, and is heavy enough to prevent any leakage of the monolayer beneath the barrier. The surface pressure and the mean molecular area are continuously monitored during the compression.

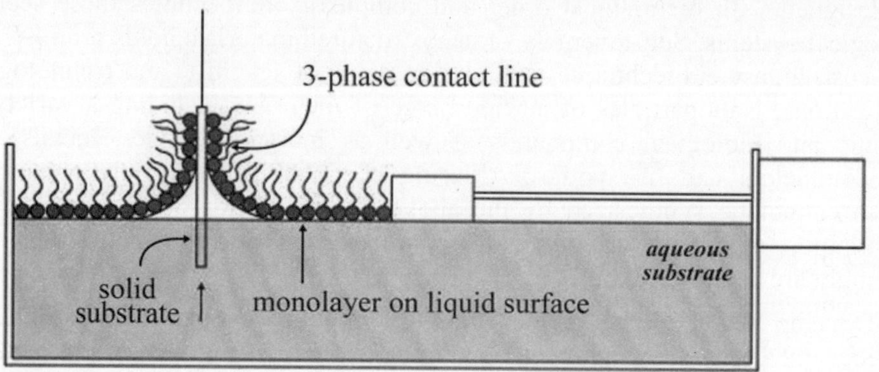

Fig. 3.14. Schematic illustration of a floating monolayer on a solid substrate

The Langmuir film balance can also be used for building up highly organized multilayers of the amphiphiles. This is accomplished by successively dipping a solid substrate up and down through the monolayer while simultaneously keeping the surface pressure constant by a computer controlled feedback system between the electro balance measuring the surface pressure and the barrier moving mechanism. Consequently the floating monolayer is adsorbed to the solid substrate. In this way multilayer structures of hundreds of layers can be produced. These multilayered structures are commonly called Langmuir-Blodgett or simply LB films. The deposition process is schematically shown in **Figure 3.14.** The LB deposition is traditionally carried out in the "solid" phase. The surface pressure is then high enough to ensure sufficient cohesion in the monolayer, *e.g.* the attraction between the molecules in the monolayer is high enough so that the monolayer does not fall apart during transfer to the solid substrate. This also ensures the buildup of homogeneous multilayers. The surface pressure value that gives the best results depends on the nature of the monolayer and is usually established empirically. However, amphiphiles can seldom be successfully deposited at surface pressures lower than 10 mN/m, and at surface pressures above 40 mN/m collapse and film rigidity often pose problems. When the solid substrate is hydrophilic (glass, SiO_2 etc.) the first layer is deposited by raising the solid substrate from the sub phase through the monolayer, whereas if the solid substrate is hydrophobic (HOPG, silanized SiO_2, etc.) the first layer is deposited by lowering the substrate into the sub phase through the monolayer. There are several parameters that effect on what type of LB film is produced. These are the nature of the spread film, the sub phase composition and temperature, the surface pressure during the deposition and the deposition speed, the type and nature of the solid substrate and the time the solid substrate is stored in air or in the sub phase between the deposition cycles.

3.4 STABILIZATION AND FUNCTIONALIZATION OF NANOPARTICLES

Due to their high reactivity nanoparticles have a high tendency to build aggregates resp. agglomerates, which could lead to a loss of the desired properties. Therefore, it is often necessary to stabilize the nanoparticles with additional treatments. The commercial success or failure of nanoparticles in a particular application usually depends upon the ability to prepare stable dispersions in water or organic fluids with controlled rheology. In turn, the ability to prepare stable nanoparticle dispersions with controlled rheology is enabled by tailoring nanoparticle coatings. On the other hand, coating nanoparticles with another material of nanoscale thickness is a simple way to alter the surface properties of nanoparticles. Core-shell structured nanoparticles have been shown to display advanced optical, mechanical and magnetic properties. One common method to stabilize or modify the reactivity of the nanoparticle is the encapsulation with a molecular or polymeric layer. A thin polymeric shell enables compatibility of the particles with a wide variety of fluids, resins and polymers. In this way, the nanoparticles retain their original chemical and physical properties, but the coating can be tailored for a wide variety of applications. One example is the Discrete Particle Encapsulation (DPE) method patented by Nanophase Technologies Corporation. Collecting nanoparticles in liquid suspension to prevent agglomeration is another way to ensure the stability of the collected nanoparticle powders. For semiconducting particles, stabilization of the liquid suspension has been demonstrated by the addition of a polar solvent; surfactant molecules have been used to stabilize the liquid suspension of metallic nanoparticles.

There are two general ways of stabilizing nanoparticles in aqueous solutions. Firstly by adjusting the pH of the system the nanoparticle surface charge can be manipulated in such way that an electrical double layer is generated around the particle. Overlap of two double layers on different nanoparticles causes repulsion and hence stabilization. The magnitude of this repulsive force can be measured via the zeta potential. The second method involves the adsorption of polymers onto the nanoparticles in such a way that the particles are physically prevented from coming close enough for the van der Waals attractive force to dominate. This is termed steric stabilization. A combination of these two mechanisms is called electrostatic stabilization and occurs when polyelectrolytes are adsorbed on the nanoparticle surface.

Suggested further readings

1. L.L. Hench, J.K.West, The sol-gel process, *Chem. Rev.*, **90** (1990) 33.
2. P. Milani, S. Iannotta, *Cluster Beam Synthesis of Nanostructured Materials*, Springer, 1999.

3. Z. Wang, Y. Liu, X. Zen, One step synthesis of Fe_2O_3 nanoparticles by laser ablation. Powder Tech.**6** (2006) 65-68.

4. J. M. Deitzel, J Kleinmeyer, D Harris, NC Beck Tan, The effect of processing variables on the morphology of Electrospun nanofibers and textiles, *Polymer*, **42**, (2001) 261-272.

5. V. Sáez and T. J. Mason, "Sonoelectrochemical Synthesis of Nanoparticles", *Molecules,14* (2009) 4284-4299.

6. H. Xu, B. W. Zeiger and K. S. Suslick, Sonochemical synthesis of nanomaterials, *Chem. Soc. Rev.*, **42 (**2013**)** 2555-2567.

7. S.W. Kang, Applications of soft Lithography for nano functional devices, *Lithography* Ed. M. Wang, InTech. 2010.

8. A. Shukla, M.K. Patra, M. Manoth, S. Songara, V.K. Singh, S.R. Vadera and N Kumar, Preparation and chracterization of biocompatible and water-dispersible superparamagnetic iron oxide nanoparticles (SPIONS), *Adv Sci. Lett.*, **3** (2010)1-7.

9. M. Manoth, K. Manzoor, M. K. Patra, P. Pandey, S.R. Vadera and N. Kumar, Dendrigraft polymer-based synthesis of silver nanoparticles showing bright blue fluorescence*", Mater. Res. Bull.*, **44** (2009) 314-317.

10. Y. Liu, J. Goebl and Y. Yin, Templated synthesis of nanostructured materials, **42** (2013) 2610-2653.

11. M. Sarikaya, C. Tamerler, A.K.Y. Jen, K. Schulten, and F. Baneyx, Molecular biomimetics: Nanotechnology through biology", *Nature Materials*, **2**(2003) 577-585.

12. S. Dave and S. Kumbhat, "Green synthesis of silver nanoparticles", Indian Journal of Chemical Technology, March 2018.

REVIEW QUESTIONS

Q1. What do you understand by "top-down" and "bottom-up" approach for nanosynthesis?

Q2. How is mechanochemical processing (MCP) method for nano-synthesis different from mechanical milling?

Q3. What are the basic advantages of electro-expulsion method to produce nano metallic powders?

Q4. Explain the basis of thin film deposition by sputtering technique.

Q5. What do you understand by reactive ion etching?

Q6. How can you obtain sputtering of insulating materials?

Q7. Why do nano metallic powders produced by electro-expulsion method have greater chemical and metallurgical reactivity?

Q8. What is the main advantage of scanning lithography over photolithography?

Q9. Describe the basic principles of electrospinning and its applications in production of nanofibers.

Q10. How will you differential chemical vapor condensation and chemical vapor deposition technique for synthesis of nanoparticles.

Q11. What are the essential parameters required to obtain NPs of nano size distribution by chemical vapor deposition (CVD) method?

Q12. What is plasma enhanced CVD and how is it different from thermal CVD?

Q12. Illustrate versatility of CVD technique giving one example each for preparation of metals, metal oxide, and silicon carbide nanoparticles.

Q13. How does liquid phase synthesis differ from gas phase synthesis for production of nanoparticles?

Q14. What are the precursors and appropriate solvent systems to obtain (a) metals, (b) alloys, and (c) metal oxides using sonochemical method?

Q15. Nature is known to be the best laboratory to synthesize a variety of nanomaterials. What are the routes it may be following, illustrate with some examples?

Q16. What is self-assembly? How are the self-assembly approaches useful in preparing nano-structured materials?

Q17. Distinguish between precipitation and sol-gel phenomena. Explain how these techniques are useful for nanosynthesis.

Q18. Enumerate with suitable examples the various steps involved in chemical reduction method for synthesis of nanomaterials.

Q19. How can you obtain stable gold nanoparticles without adding any stabilizer or surfactant?

Q20. Suggest a suitable method for preparing polymer composites.

■■■■

Characterization of Nanomaterials Part I: Spectroscopic and Scattering Techniques

4

4.1 INTRODUCTION

Characterization of nanomaterials for their molecular structure and chemical composition is carried out in the same manner as for bulk materials. However, characterization of nanomaterials emphasise on parameters such as size, size distribution, porosity, surface area, shape, wettability, zeta potential, adsorption potential, aggregation, distribution of conjugated moieties and impurities. Spectroscopy, microscopic and scattering techniques are employed for the determination of molecular structure, size, size distribution and chemical composition. The present chapter deals with the spectroscopic and scattering techniques.

The term "spectroscopy" defines a large number of techniques that use absorption and emission of electromagnetic radiation, ranging from UV to visible, to infrared, to obtain information on the structure and properties of matter. **Figure 4.1** shows the classification of electromagnetic radiation along with range of wavelength and frequency and associated phenomenon as an outcome of the interaction of radiation with matter.

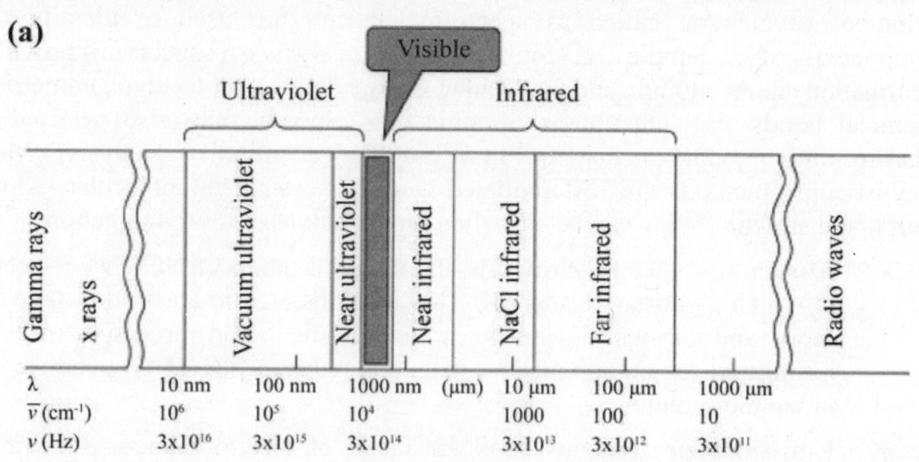

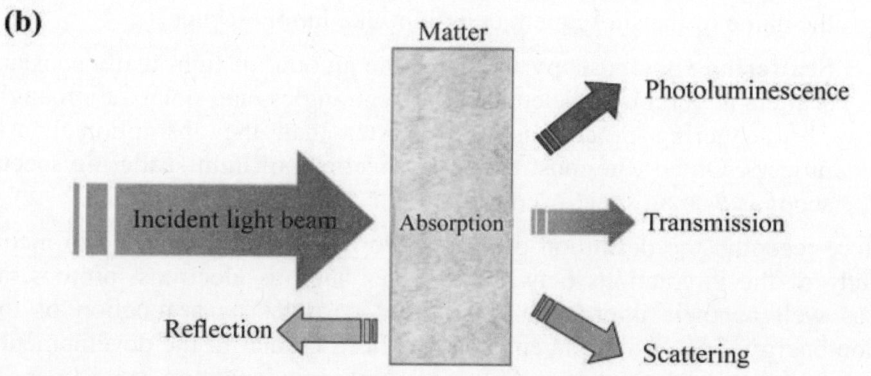

Fig. 4.1. (a) Electromagnetic radiation (b) associated spectral phenomena

Electromagnetic radiation is composed of oscillating electric and magnetic fields that have the ability to transfer energy through space. The energy propagates as a wave, such that the crests and troughs of the wave move in vacuum at the speed of 3 x 10^8 meters per second. Components of electro-magnetic radiation appear different to an observer; visible light is perceived to the human eye, while X-rays, UV, infrared, microwave and radio waves are not. The splitting of electromagnetic radiation into its component wavelengths is fundamental to spectroscopy. Evolving from the first crude prism spectrographs that separated sunlight into its constituent colors, modern spectrometers have provided ever-increasing wavelength resolution. Large-grating spectrometers are capable of resolving wavelengths as close as 10^{-3} meter, while modern laser techniques can resolve optical wavelengths separated by less than 10^{-10} meter. The basic principle shared by all spectroscopic techniques is to expose the sample to electromagnetic radiation and observe the response as a function of radiation wavelength, called as spectrum. Spectra are used to identify the components of a sample *i.e.* for qualitative analysis. A spectrum provides information about atomic and molecular energy levels, molecular geometries, chemical bonds and interactions of molecules. Spectra may also be used to measure the amount of material in a sample (quantitative analysis). Most spectroscopic methods are differentiated as either atomic or molecular. Along with that distinction, they can be classified on the nature of their interaction:

- **Absorption spectroscopy** uses the range of the electromagnetic spectra in which a substance absorbs. This includes atomic absorption spectroscopy and various molecular techniques, such as infrared spectroscopy in infrared region and nuclear magnetic resonance (NMR) spectroscopy in the radio region.
- **Emission spectroscopy** uses the range of electromagnetic spectra in which a substance radiates (emits). The substance first must absorb energy. This energy can be from a variety of sources, which determines the name of the subsequent emission, like luminescence.
- **Scattering spectroscopy** measures the amount of light that a substance scatters at certain wavelengths, incident angles, and polarization angles. The scattering process is much faster than the absorption/emission process. One of the most useful applications of light scattering spectroscopy is Raman spectroscopy.

More recently, the definition of spectroscopy has been expanded to include the study of the interactions between particles such as electrons, protons, and ions, as well as their interaction with other particles as a function of their collision energy. Spec-troscopic analysis has been crucial in the development of the most fundamental theories in physics, including quantum mechanics, the special and general theories of relativity, and quantum electrodynamics. Spectroscopy, as applied to high-energy collisions, has been a key tool in developing scientific understanding not only of the electromagnetic force but also of the strong and weak nuclear forces.

4.2 UV-VISIBLE AND PLASMON ABSORPTION SPECTROSCOPY

The ultraviolet region falls in the range between 190 and 380 nm and the visible region falls between 380 and 750 nm. Ultraviolet and visible light interact with matter and causes transition of electron from one filled orbital to another of higher energy unfilled orbital. When a molecule absorbs light of an appropriate wavelength and an electron is promoted to a higher energy molecular orbital, the molecule is then in an excited state. The possible electronic transitions are: $\pi - \pi^*$; $n - \pi^*$; $\sigma - \sigma^*$; $n - \sigma^*$; as shown in **Fig. 4.2a**. The σ to σ* transition requires an absorption of a photon with a wavelength which does not fall in the UV-Vis range, therefore, allowed transitions are π to π*, n to π*and d to d' (transition metal ions). UV-Vis spectroscopy is one of the most common and powerful technique used in the analysis of organic compounds and metal complexes.

UV-vis spectroscopy becomes useful to characterize and identity most of the organic and biological molecules with conjugated π systems. In these molecules, the energy gap for π -π* transitions is smaller than for isolated double bonds, and thus the wavelength absorbed is longer. Molecules or parts of molecules that absorb light strongly in the UV-vis region are called chromophores. The color of a molecule may be intensified by groups called auxochromes which generally do not absorb significantly in the 200-800 nm region, but will affect the spectrum of the chromophore to which it is attached. The most important auxochromic groups are OH, NH_2, CH_3 and NO_2 and their properties are acidic (phenolic) or basic. **Figure 4.2b** shows a typical UV-Vis spectra of tetraphenyclopentadienone, showing n to π* and π to π* transitions.

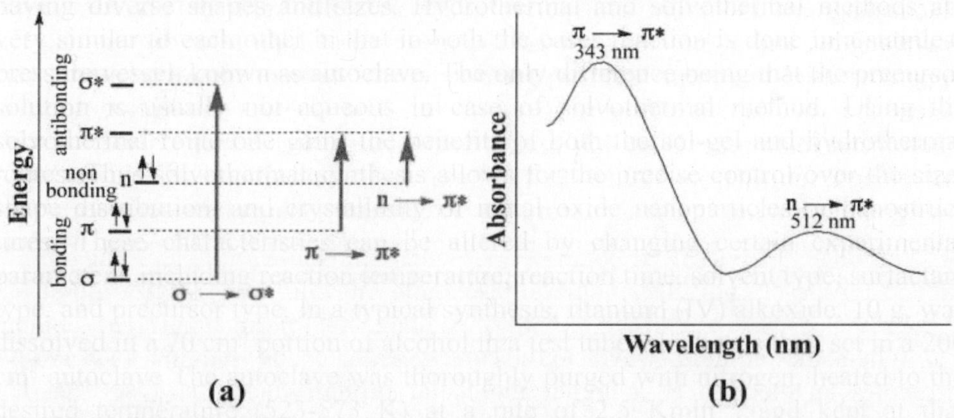

Fig. 4.2. Possible electronic transition with energy scale (*a*); UV-Vis spectrum of tetraphenyclopentadienone, showing n to π* and π to π* transitions

According to Beer-Lambert law, the absorbance (*A*) at a given wavelength is the logarithmic ratio of intensity of incident radiations entering the sample to the intensity of emitted radiation emerging from the sample (Eqn. 4.1).

$$A = \log I0/I \qquad 4.1$$

At a given wavelength, absorption is proportional to the concentration of absorbing molecules and the path length of the light through the sample (Eqn. 4.2).

$$A = \varepsilon cl \qquad 4.2$$

where, I_0 = intensity of incident radiation

I = intensity of emitted radiation

l = length of the light path through the sample, in centimetres

c = concentration of the sample, in moles/liter

ε = molar absorptivity (M^{1}.cm^{-1}; where M = mol.L^{-1})

The molar absorptivity (also called the extinction coefficient) of a compound is a constant and is characteristic of the compound at a particular wavelength. It is possible to calculate the concentration of the sample if the molar absorptivity for the sample is known. The molar absorptivity can be calculated by recording the absorbance of a standard solution of 1 mol/dm^3 concentration in a standard cuvette where the light travels a constant distance of 1 cm. Once the molar absorptivity and the absorbance of the sample are known, with the length the light travels being fixed, it is possible to determine the concentration of the sample solution

To obtain and analysis of a spectrum requires a source of light or other electromagnetic radiation; a disperser to separate the light into its component wavelengths, and a detector to sense the presence of light after dispersion. The apparatus used to accept light, separate it into its component wavelengths, and detect the spectrum is called a spectrometer. The UV-Vis absorption spectrophotometer instrument **(Fig. 4.3)** uses a Xenon pulse lamp which provides a wide wavelength range (from 190 nm to 1100 nm) and a dual silicon diode detector. A holographic grating is used to scan the wavelength. The reference beam in the spectrophotometer travels from the light source to the detector without interacting with the sample. The sample beam interacts with the sample exposing it to the UV light of continuously changing wave length.

When the emitted wavelength corresponds to the energy level which promotes an electron to a higher molecular orbital, energy is absorbed. The detector records the ratio between reference and the sample beam intensities (I_o/I); a computer aided plot of intensity of absorption vs wavelength, *i.e.* the absorption spectrum is recorded to determines the wavelength at which sample absorbs maximum (λ_{max}) which is the characteristic property of an analyte. Peaks in UV-Vis spectra tend to be quite broad, often spanning well over 20 nm at half-maximal height. It may be attributed to the fact that in a molecule many-fold vibrational and rotational energy levels are superimposed on an electronic energy level.

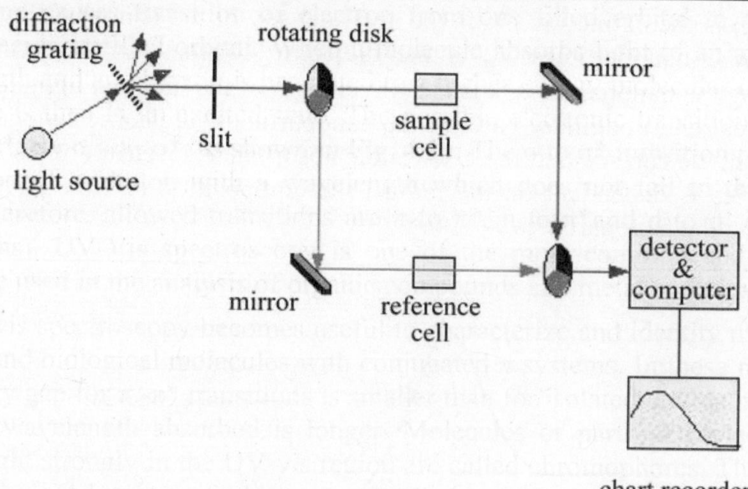

Fig. 4.3. Schematic of a UV-Vis spectrophotometer

UV-Vis Spectroscopy of Nanomaterials

Nanoparticles have optical properties that are sensitive to size, shape, concentration, agglomeration state, and refractive index near the nanoparticle surface. All these effects makes UV/Vis/IR spectroscopy a valuable tool for identifying, characterizing, and studying these materials. UV-Vis spectroscopy offer the possibility to characterise nanoparticles, especially quantum dots metallic and organic based nanoparticles like fullerenes and carbon nanotubes. UV-Vis spectroscopy have been used to compare colloidal aqueous suspensions of C_{60} as well as to study the relationship between absorbance spectra and particle size distributions for quantum-sized nanocrystals. UV-visible absorbance spectroscopy is a powerful tool for detecting noble metal nanoparticles, because the localized surface plasmon resonance (LSPR) of metal nanoparticles allows for highly selective absorption of photons. The LSPR of metal nanoparticles is governed by the Mie theory.

Mie theory describes the interaction of light with a homogenous sphere and can be used to predict the UV-visible absorbance spectrum of spherical metallic nanoparticles. The calculated absorption spectrum for 50 nm gold nanoparticles is shown in **Fig. 4.4**, along with the expected contributions from scattering and absorption. The relationship between the extinction coefficient, the sum of absorption and scattering of light, of spherical nanoparticles described by Mie theory is given by Eqn. 4.3.

$$E(\lambda) = \frac{24\pi N_A a^3 \varepsilon_m^{3/2}}{\lambda \ln(10)} \left[\frac{\varepsilon_i}{(\varepsilon_r + 2\varepsilon_m)^2 + \varepsilon_i^2} \right] \quad 4.3$$

where, $E(\lambda)$ is the extinction, N_A is the areal density of the nanoparticles, a is the radius of the nanoparticles, ε_m is the dielectric constant of the environment surrounding the nanoparticles, λ is the wavelength of the incident light, and ε_r and ε_i are the real and imaginary parts of the nanoparticles' dielectric function. From this relation, we can see that the UV-visible absorbance spectrum of a solution of nanoparticles is dependent on the radius, the composition and the environment surrounding the nanoparticles

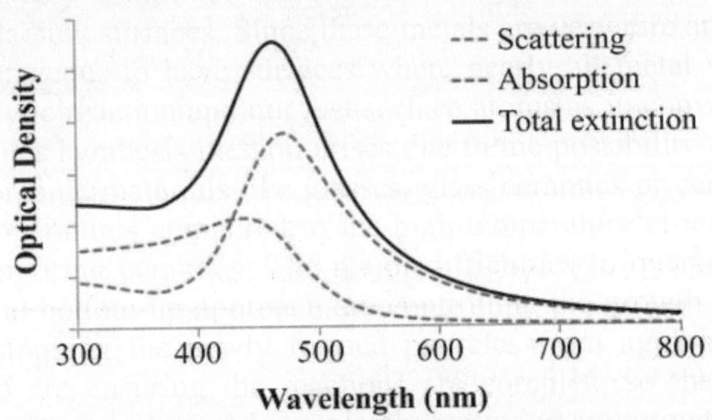

Fig. 4.4. The calculated spectrum for 50 nm gold nanoparticles in water (a); contributions from scattering (b); contributions from absorption (c)

Mie theory is limited to spherical nanoparticles, but there are other theoretical techniques that can be used to predict the UV-visible spectrum of more complex noble metal nanostructures. These techniques include surface-based methods such as the generalized multipole technique and T-matrix method, as well as volume-based techniques such as the discrete dipole approximation and the finite different time domain method.

Metal nanoparticles, in particular gold and silver, are characterized by a plasmon resonance absorption that gives rise to intensely colored solutions. The absorption band is due to electrons confined at the particle surface that collectively oscillate at a specific frequency, commonly referred to as the surface plasmon resonance frequency. Absorption can be in the visible and UV region of the spectrum. Particles can be visualized by absorbance in solution at nanomolar and picomolar concentrations. Plasmon resonance in metals like gold and silver is responsible for their unique optical properties. Nanoparticles of plasmonic metals scatter light elastically with remarkable efficiency. The position and shape of the plasmon absorption of plasmonic nanoclusters are strongly dependent on the particle size, dielectric medium, and surface adsorbed species. The absorption spectrum of metal nanoparticles shifts to longer wavelengths with increasing particle size. For examples, the plasmon absorption band of 20 nm silver particle is centred at 395 nm, resulting in a yellow solution, while 20 nm gold particles absorbs at 520 nm resulting in a red solution. The plasmon absorption effect occurs for particles up to approx. 50 nm in diameter and scales with particle volume.

In larger metal nanoparticles (>30 nm) another effect, light scattering, is observed. When illuminated with white light, metal nanoparticles in the 50-120 nm diameter size range scatter light of a specific color at the surface plasmon resonance frequency. This effect is called plasmon resonance light scattering. As in the case of plasmon absorbance, light scattering scales with particle volume, but the scattered light can be detected at much lower concentrations than absorbed light. For example, light scattered by a solution of 80 nm diameter gold particles is detectable down to 5 *fM* concentration ($fM = 10^{-15}$M). For this reason metal nanoparticles are interesting materials for use in techniques which needs labeling.

UV-visible absorbance spectrum is highly dependent on geometry of nano-particles and can be used to determine the geometry of nanoparticles. **Figure 4.5** shows UV-visible spectrum of gold NPs in different geometrical shapes (spheres, decahedra and rods).

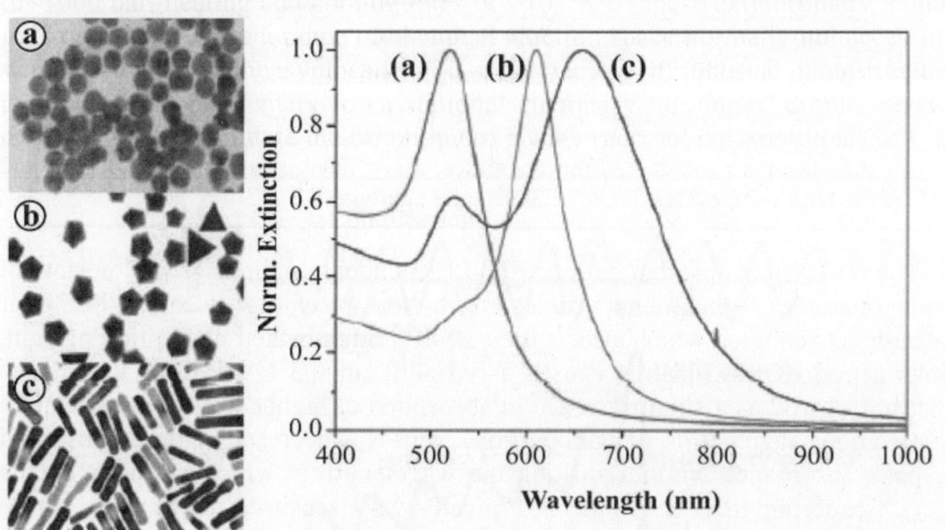

Fig. 4.5. UV-VIS spectrum of gold nanoparticles with different geometries: spheres (a); decahedra (b); and rods (c)

The λ_{max} of the UV-visible absorbance spectrum of noble metal nanoparticles is highly dependent on the environment surrounding the nanoparticles. Because of this, shifts in λ_{max} can be used to detect changes in the surface composition of the nanoparticles. One potential application of this phenomenon is using UV-visible absorbance spectroscopy to detect the binding of biomolecules to the surface of noble metal nanoparticles. For example, the 'red shift', a shift to longer wavelengths in the λ_{max} of the UV-visible absorbance spectrum will be observed on binding the proteins at the surface of the nanoparticles.

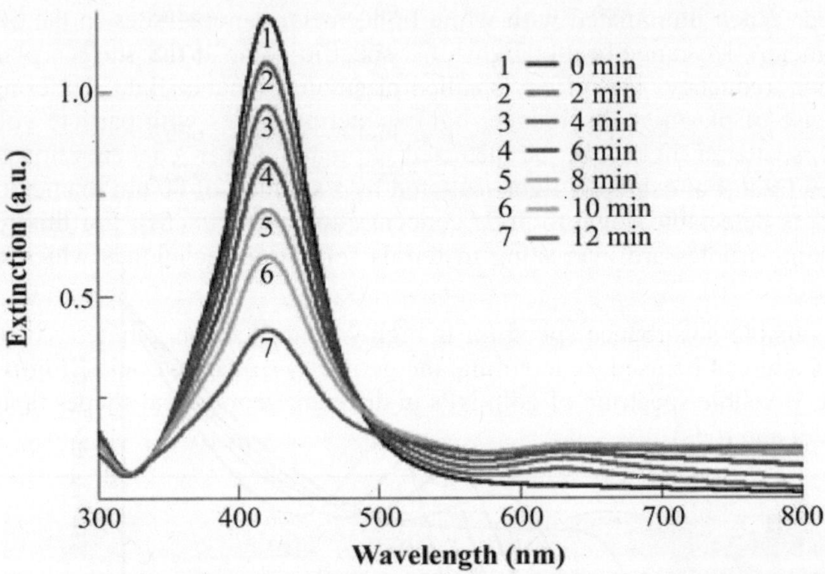

Fig. 4.6. The absorption spectrum of silver nanoparticles. On addition of 20 mM Na_2CO_3 the plasmon peak at 420 nm decreases, a secondary peak at 620 emerges, and the baseline elevates due to scattering by aggregates

The UV-visible absorbance spectrum is dependent on the aggregation state of the nanoparticles. When nanoparticles are in close proximity to each other, their plasmon gets coupled which affects their LSPR and thus their absorption of light. Dimerization of nanospheres causes a red shift, in the UV-visible absorbance spectrum as well as a slight increase in absorption at higher wavelengths. Unlike dimerization, aggregation of nanoparticles causes a decrease in the intensity of the peak absorbance without shifting the wavelength at which the peak occurs (λ_{max}). Scattering from a sample is typically very sensitive to the aggregation state of the sample, with the scattering contribution increasing as the particles aggregate to a greater extent. UV-visible spectroscopy can be used as a simple and reliable method for monitoring the stability of nanoparticle solutions. As the particles destabilize, the original extinction peak will decrease in intensity (due to the depletion of stable nanoparticles), and often the peak will broaden or a secondary peak will form at longer wavelengths (due to the formation of aggregates). **Figure 4.6** shows change in the absorption spectrum of 50 nm nano silver on addition of a salt solution. The absorption spectrum of silver nanoparticles changes as the transition of particles from a well-dispersed state to an aggregated state following the addition of a concentrated salt solution. As aggregation proceeds, the plasmon peak at 420 nm decreases, a secondary peak at 620 emerges, and the baseline elevates due to scattering by aggregates. The final aggregated state is readily seen by eye, as the original dispersed particles form a yellow solution and the aggregated particles become grey.

The band gap of the semiconductor quantum dots can be tuned with the size and size distribution of the particles. The minimum energy for an electron to get excited from the ground state is the energy to cross the band gap. In an absorption spectra, this is given by the first exciton peak at the maximum wavelength (λ_{max}). The size of quantum dots can be approximated corresponding to the first exciton peak wavelength. Emperical relationships have been determined relating the diameter of the quantum dot to the wavelength of the first exciton peak. The Group 12-16 semiconductor quantum dots such as cadmium selenide (CdSe), cadmium telluride (CdTe) and cadmium sulfide (CdS) have been studied in detail using UV-Vis spectroscopy. In case of a very narrow size distribution, the first exciton peak will be very sharp **(Fig. 4.7)**. This is because the differences in band gap between different sized particles will be very small and hence most of the electrons will get excited over a smaller range of wavelengths.

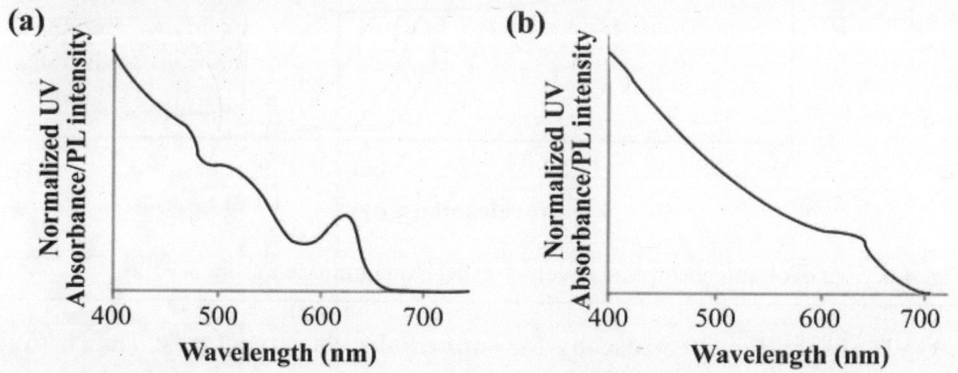

Fig. 4.7. Narrow emission spectra (a); broad emission spectra (b) of CdSe QDs

Bulk cadmium telluride (CdTe) has a band gap of 1.44 eV (860 nm) and as such it absorbs in the infrared region. Sizes of CdTe, CdSe etc. can be engineered to have different band edges and thus, different absorption spectra as a function of wavelength. A typical CdTe spectra is shown in **Figure 4.8.** Due to the small bandgap energy of CdTe, it can be used in tandem with CdSe to absorb in a greater part of the solar spectrum. **Table 4.1** shows the bulk band gap of other group 12-16 semiconductor systems as determined from UV-Vis spectra of respective samples. The band gap of ZnS falls in the UV region, while those of ZnSe, CdS, and ZnTe fall in the visible region.

Table 4.1. Bulk band gaps of different Group 12-16 semiconductors

Material	Band gap (eV)	Wavelength (nm)
ZnS	3.61	343
ZnSe	2.69	460
ZnTe	2.39	518
CdS	2.49	497
CdSe	1.74	712
CdTe	1.44	860

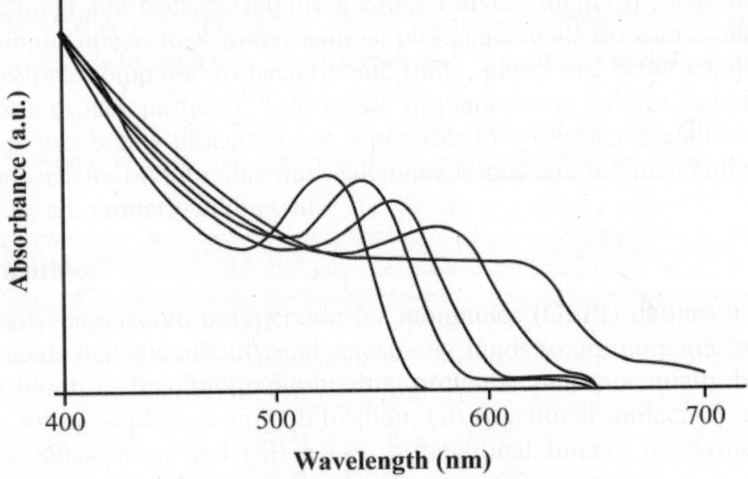

Fig. 4.8. Size evolving absorption spectra of CdTe quantum dots from 3-7 nm

UV-Vis absorption spectroscopy for supramolecular structures: The UV-Vis absorption spectroscopy method is used extensively in biology to detect the presence and relative amount of DNA or proteins. For example, this technique is sensitive to the π-bonding in the amine bases of DNA. The π-bonding absorption line occurs at a wavelength around 260 nm for the various nucleotides. The UV-Vis absorption technique is also sensitive to the presence of two amino acids forming the proteins: Tryptophan and Tyrosine. The absorption signal from proteins (around the 280 nm absorption line) is 40 times smaller than that from DNA (around the 260 nm absorption line) for comparable concentrations.

4.3 INFRARED SPECTROSCOPY

The term "infra red" covers the range of the electromagnetic spectrum between 0.78 and 1000 μm; the energy of the infrared radiation usually lie in the range of

0.8 - 1.6 ε_v corresponds to molecular vibrations. In infra red spectroscopy, wavelength is measured in "wave numbers", which have the units cm^{-1}. The most useful IR region lies between 4000 and 670 cm^{-1}. For a molecule to show infrared absorptions it must possess an electric dipole moment which change during the vibration. This is the *selection rule* for infrared spectroscopy. **Figure 4.9** illustrates an example of an 'infrared-active' molecule, a *heteronuclear* diatomic molecule.

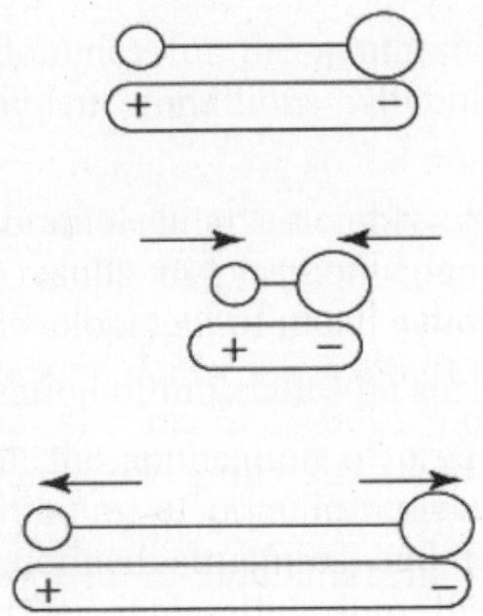

Fig. 4.9. Change in dipole moment of a heteronuclear diatomic molecule

4.3.1. Basic concepts

For a molecule to absorb IR, the frequency of the radiation should match the vibrational frequency of the molecule. In order to begin with a basic model, a molecule can be looked upon as a system of masses joined by bonds with spring-like properties. Taking first the simple case of diatomic molecules, such molecules have three degrees of translational freedom and two degrees of rotational freedom. The positions of atoms in a molecules are not fixed; they are subject to a number of different vibrations. The atoms in the molecules can also move relative to one other, that is, bond lengths can vary or one atom can move out of its present plane. This is a description of stretching and bending movements that are collectively referred to as vibrations.

It can be shown that the number of normal modes of vibration for a given molecule can be determined as follows:

$$\text{Number of normal modes} = 3N - 6 \text{ (nonlinear)}$$
$$= 3N - 5 \text{ (linear)}$$

where N is the number of component atoms in the molecule.

In the near-infrared region (NIR, λ range 0.78-2.5μm), which meets the visible region at about 12500 cm^{-1} and extends to about 4000 cm^{-1}, generally there are many absorption bands that result from harmonic overtones of fundamental and combination bands often associated with hydrogen atoms which as such are not of much significance. The wavenumber, *v* of middle-infrared region (λ range 2.5-15.4 μm) extends in range 4000-650 cm^{-1} and that of far-infrared region (FIR λ range 15-1000 μm) extends in the range 667-10 cm^{-1}.

The infrared absorption spectrum of a substance is sometimes called its 'molecular fingerprint'. Many useful correlations have been found in the mid-infrared region. This region is divided into the 'group frequency' region, 4000-1300 cm^{-1}, and the fingerprint region, 1300–650 cm^{-1}. In the group frequency region the principal absorption bands are assigned to vibration units consisting of only two atoms of a molecule: in the interval from 4000-2500 cm^{-1}, the absorption is characteristic of hydrogen stretching vibrations with elements of mass 19 or less. The intermediate frequency range, 2500-1540 cm^{-1} (unsaturated region) contains triple bond frequencies which appear from 2500-2000 cm^{-1} and double bond frequencies from 2000-1540 cm^{-1}. In the region between 1300 and 650 cm^{-1} there are single bond stretching frequencies and bending vibrations (skeletal frequencies) of polyatomic systems that involve motions of bonds linking a substituent group to the remainder of the molecule.

Change in inter-atomic distance along bond axis is referred as stretching which may be symmetric or nonsymmetric. In symmetric stretching *e.g.*, CO_2 molecule, the net change in dipole moment is very small thus leading to no absorption in infrared. FTIR bands for the stretching modes of some common bonds are:

3700 - 2500 cm^{-1} for H−(C, N, O or S) stretching.

2300 - 2000 cm-^{1} for C≡ (C or N) stretching.

1900 - 1500 cm_{-1} for C = (C, N or O) stretching.

800 - 1300 cm^{-1} for C−(C, N or O) stretching.

A very broad peak in the region between 3100 and 3600 cm^{-1} indicates the presence of exchangeable protons, typically from water alcohol, amine, amide or carboxylic acid groups (see further discussion of this below). First, alkane C-H stretching absorptions just below 3000 cm^{-1} demonstrate the presence of saturated carbons, and signals just above 3000 cm^{-1} demonstrate unsaturation. The frequencies from 2300 - 2000 cm^{-1} are normally void of other absorptions, so the presence of alkyne or nitrile groups can be easily seen here. The region below 2000 cm^{-1}, normally contains many peaks of varying intensities due the presence of C=C, C=N or C=O stretching vibrations in the 1900-1500 cm^{-1} while for C-C, C-N or C-O stretching vibrations in the 1300-800 cm^{-1}. This complex lower region is also known as the "fingerprint region" because almost every organic compound produces a unique pattern in this area. Therefore, identity can often be confirmed by comparison of this region to a known spectrum.

Change in angle between two bonds is termed as 'bending'. Different modes of bending are rocking, scissoring, wagging and twisting as shown in **Figure 4.10**.

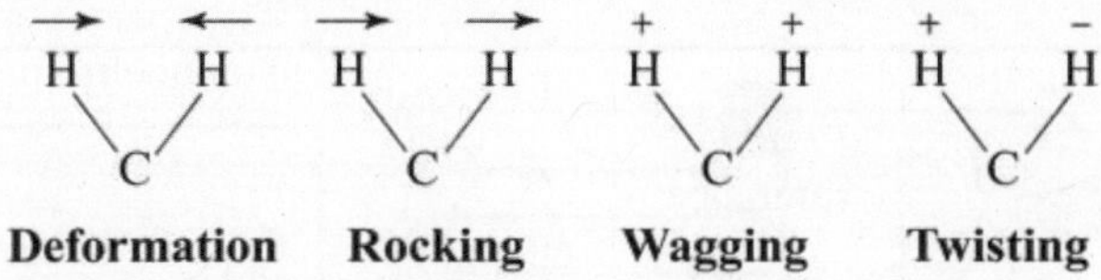

Deformation **Rocking** **Wagging** **Twisting**

Fig. 4.10. Different modes of bending modes

The region 667–10 cm^{-1} contains the bending vibrations of carbon, nitrogen, oxygen and fluorine with atoms (*M*) heavier than mass 19, and additional bending motions in cyclic or unsaturated systems. The low-frequency molecular vibrations in the far-infrared are particularly sensitive to changes in the overall structure of the molecule; thus the far-infrared bands often differ in a predictable manner for different isomeric forms of the same basic compound. The far-infrared frequencies of organometallic compounds are often sensitive to the metal ion or atom, and this can be used advantageously in the study of coordination bonds. Moreover, this region is particularly well suited to the study of organometallic or inorganic compounds whose atoms are heavy and whose bonds are inclined to be weak. In addition absorption, though weak, in IR spectra also occur as a consequence of out-of-plane and in-plane bending vibrations as illustrated in Figure 4.11.

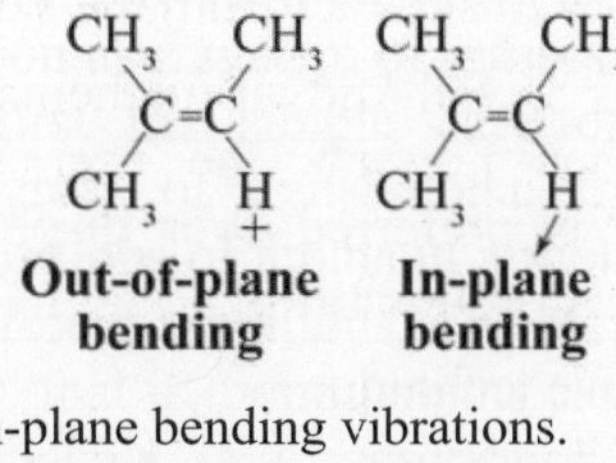

Out-of-plane bending **In-plane bending**

Fig. 4.11. Out-of plane and in-plane bending vibrations.

Infrared spectroscopy has been a workhorse technique for materials analysis in the laboratory for over six decades. An infrared spectrum represents a fingerprint of a sample with absorption peaks which correspond to the frequencies of vibrations between the bonds of the atoms making up the material. Because each different material is a unique combination of atoms, no two compounds produce the exact same infrared spectrum. Therefore, infrared spectroscopy can result in a positive identification (qualitative analysis) of every different kind of material. The "fingerprint zone" (1300~400 cm^{-1}), reveal extremely sensitive changes in molecular structure which provides a reliable determination to analyze unknown composition and structure. In addition, the size of the peaks in the spectrum is a direct indication of the amount of material present. With the aid of modern software algorithms, infrared spectroscopy can be used for quantitative analysis as well.

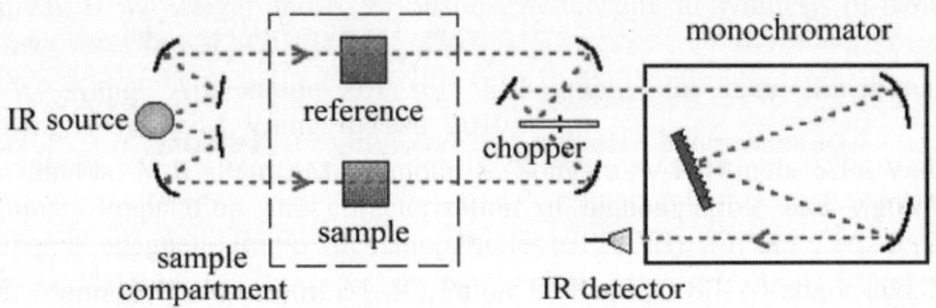

Fig. 4.12. Schematic diagram of a Fourier-transform infrared spectrometer (FTIR)

Traditionally, IR spectrometers were of dispersive type, where a diffraction grating in a monochromator is used to disperse the different wavelengths of light. Modern IR absorption instruments use Fourier-transform techniques with a Michelson interferometer **(Fig. 4.12).** To obtain an IR absorption spectrum, one mirror of the interferometer moves to generate interference in the radiation reaching the detector. Since all wavelengths are passing through the interferometer, the interferogram is a complex pattern. The absorption spectrum as a function of wave number (cm^{-1}) is obtained from the Fourier transform of the interferogram, which is a function of mirror movement (cm). This design does not have the reference cell of a dispersive instrument, so a reference spectrum is recorded and stored in memory to subtract from the sample spectrum.

Fourier transformation is a well-known mathematical technique, performed by the computer which then presents the user with the desired spectral information for analysis. It is preferred over dispersive or filter methods of infrared spectral analysis for several reasons:

- It is a non-destructive technique
- It provides a precise measurement method which requires no external calibration
- It can increase speed, collecting a scan every second
- It can increase sensitivity – one second scans can be co-added together to ratio out random noise
- It has greater optical throughput
- It is mechanically simple with only one moving part.

4.3.2. IR spectroscopy of Nanoparticles

IR spectroscopy provides specific and valuable information due to the excitation of inherently specific fundamental vibrational transitions characteristic of molecular species. Nanoparticles of diverse nature have been characterized and determined using spectroscopic techniques in the infrared range. Nanoparticles with the presence of inherently infrared absorbing functional groups, present at their surface, can be directly characterized via infrared spectroscopy. Furthermore, different ligands attached to nanoparticles may readily be identified according to their vibrational signatures in a rapid, precise, and non-destructive way. Next to the direct characterization of such materials, some nanostructures modify the local optical field, thereby yielding enhanced IR signals facilitating advanced infrared imaging procedures for nanoparticles.

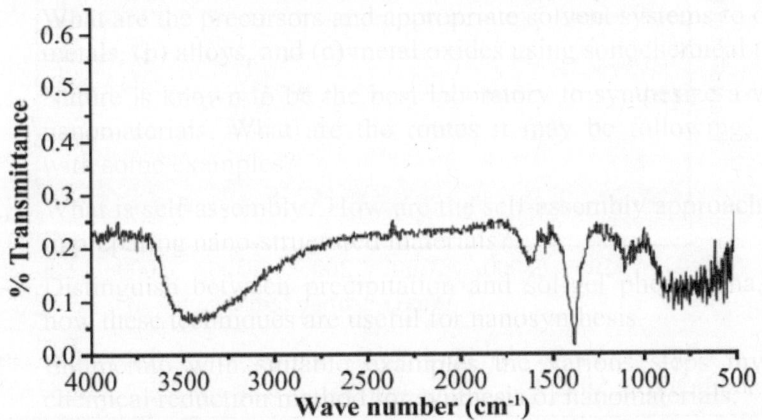

Fig. 4.13. FTIR of silver nanoparticles synthesized by treating silver salt with a plant extract in aqueous media

Infrared spectroscopy is commonly used as a technique for the characterization of nanoparticles of diverse nature including metallic NPs, carbon nanomaterials, as well as core-shell and hybrid nanoparticles. Most commonly, infrared spectroscopy has been applied for characterizing metallic nanoparticles synthesized via a large variety of strategies. IR spectra reveal vibrational signatures of the constituents present in the surrounding media of the nanoparticles, *e.g.*, during synthesis or when applying metallic NPs in complex media such as biological extracts revealing proteins and metabolites via functional groups of amines, alcohols, ketones, aldehydes, and carboxylic acids. For example, the FTIR spectrum of silver nano particles, obtained by green approach, is shown in **Fig. 4.13.** The broad band at 3450 cm^{-1} is assigned for O−H stretching vibration indicating the presence of hydroxyl groups in the reducing medium, the strong intense peaks at 1379 and 1625 cm^{-1} corresponds to C−N stretch vibrations as well as to the amide I bands of proteins in the plant extract.

The result of this FTIR spectroscopic study confirmed that plant extract has the ability to perform dual functions of reduction and stabilization of silver nanoparticles.

Further, the potential of IR spectroscopy for revealing the oxidation state and surface structure of metal-oxide nanoparticles has been demonstrated using CeO_2 NPs by investigating the presence or absence of strongly IR absorbing O-H stretching modes associated with O-H-OH complexes present at fully oxidized NPs.

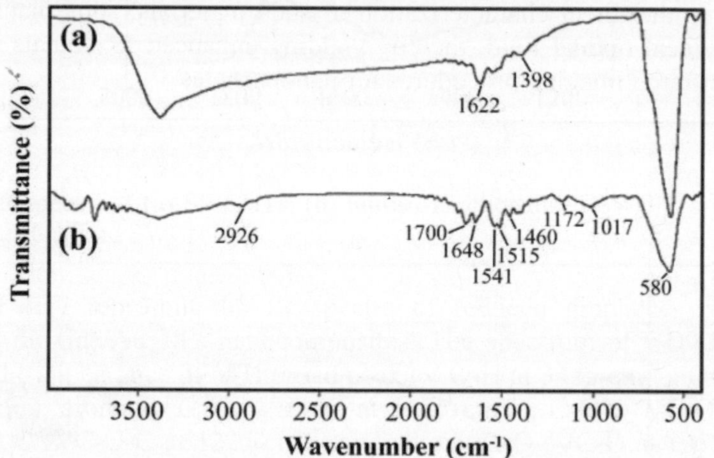

Fig. 4.14. FT-IR spectra of (a) pure Fe_3O_4 nanoparticles; (b) Fe_3O_4 nanoparticles modified with Silan A

In addition, IR spectroscopy is also used to characterize the surface modified metal oxide NPs. A typical example of Fe_3O_4 functionalise with silane-A (3-(trimethoxysilyl) propyl methacrylate) is illustrated in **Fig. 4.14**. The absorption peak at 580 cm^{-1} belongs to the stretching vibration mode of Fe–O bonds in Fe_3O_4. Comparing with the IR spectrum of pure Fe_3O_4 **(Fig. 4.14a)**, the IR spectrum of Silan A-modified Fe_3O_4. **Figure 4.14b** has absorption peaks at 1700 cm^{-1} and 1648 cm^{-1} being relevant to the stretching vibration of conjugated esteric C=O and stretching vibration of C=C bonds, respectively. Absorption bonds of stretching vibrations of esteric C–O bond are appeared at 1172 $cm^{-1}$and 1017 cm^{-1}. Peaks at 2926 cm^{-1} and 1460 cm^{-1} belong to asymmetric stretching vibration and scissoring bending vibration of CH_2 group, respectively. All these peaks reveal the existence of silan-A and indicated that the reactive groups had been introduced onto the surface of nanomagnetite.

IR spectroscopy may help in identification of phase transfer which sometimes XRD cannot differentiate. For example, upon calcination pure Fe_3O_4 nanoparticles transform into γ-Fe_2O_3 and α-Fe_2O_3. To reveal the difference, the

IR spectra of magnetite samples calcined at different temperatures is shown in **Fig. 4.15.** One can find the absorption peak of Fe-O in Fe_3O_4 at 571.1 cm^{-1}, which splits into two peaks at 632.9 and 563.4 cm^{-1}, respectively after calcination at 150°C for 2 h, due to Fe-O bond, are the characteristic peaks of γ-Fe_2O_3. Due to extreme similarity in crystal structure of the two phases, XRD is not sensitive enough to differentiate between Fe_3O_4 and γ-Fe_2O_3. On further heating the peak at 632.9 cm^{-1} becomes increasingly weaker, and disappears, while, another new peak at 463.1 cm^{-1} becomes stronger at 550°C. The two absorption peaks at 532.5 cm^{-1} and 463.1 cm^{-1} can be assigned to the Fe-O absorption in hematite (α-Fe_2O_3). Therefore, one can get that 550°C is the minimal temperature for the complete transformation from magnetite (Fe_3O_4) to hematite (α-Fe_2O_3), consistent with the results of XRD and magnetic properties analysis.

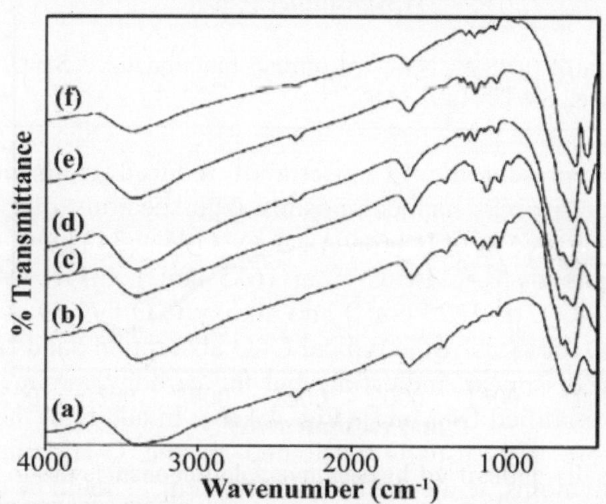

Fig. 4.15. IR spectra of pure Fe_3O_4 nanoparticles at (a) room temperature and after calcination for two hrs at different temperatures (b) 150°C (c) 250°C (d) 350°C (e) 450°C (f) 550°C

IR spectroscopy is also useful for characterizing carbon nanomaterials, *i.e.*, carbon nanotubes, nanodiamonds, graphene quantum dots (GQDs), and carbon dots (CDs) along with their functionalization via ligands. **Figure 4.16** illustrates FTIR spectra of pristine SWCNT and functionalized SWCNTs with nitrosyl group, carboxyl and maleic anhydride group. SWCNTs based on the appearance of a 1577 cm^{-1} mode associated with NO groups **(Fig. 4.16b)** and the 1749 and 1720 cm^{-1} modes associated with MA (Fig. 4.16d), neither of which are present in the pristine (Fig. 4.16a) or COOH functionalized SWCNTs **(Fig. 4.16c)**.

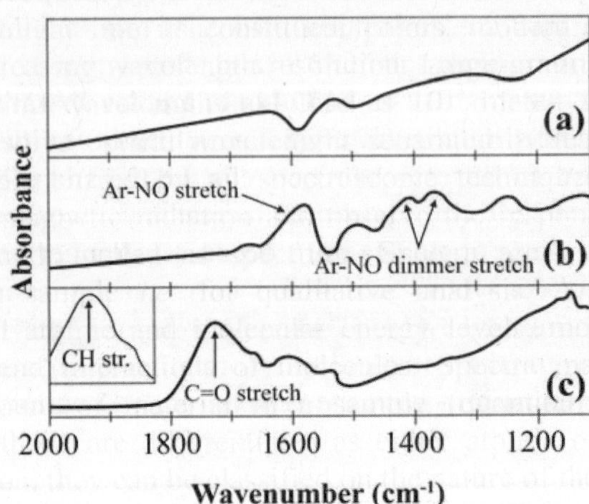

Fig. 4.16. IR spectra of (a) SWCNTs; (b) nitroso-functionalized SWCNTs; (c) COOH-functionalized SWCNTs

Figure 4.17 represent the IR spectra of reduced graphene oxide (GO)/ manganese dioxide/Au NPs nanocomposites. The spectrum of GO **(Fig. 4.17a)** illustrates the appearance of O-H broad peak (V O-H at 3425 cm^{-1}), C=O (VC=O at 1729 cm^{-1}), aromatic C=C (V C=C at 1625 cm^{-1}), C-OH (VC-OH at 1390 cm^{-1}), C-O-C (VC-O-C at 1225 cm^{-1}) and alkoxy C-O (VC-O at 1055 cm^{-1}). In Contrast to the GO, obviously, the typical C=O absorption bands at 1729 cm^{-1} of GO–MnO_2 almost disappear, indicating that the carbonyl groups on the surface of GO sheets are modified by MnO_2 **(Fig. 4.17b)**. In addition, the peaks at 1590 and 1060 cm^{-1} are also usually attributed to the O-H bending vibrations combined with Mn atoms, while the bands at 580 cm^{-1} should be attributed to the Mn -O stretching vibrations. In the case of RGO/MnO_2/AuNPs **(Fig.4.17c)**, it is clearly seen that these characteristic peaks of functional groups have been significantly disappeared after reduction process. Furthermore, the peak at 580 cm^{-1} is reduced in intensity, which may be attributed to the existence of AuNPs that covered absorption peak of MnO_2. The result further confirms the successful synthesis of nanocomposite RGO/MnO_2/AuNPs.

In addition to the chemical structure, IR spectra may also provide information on the size of NPs. For example, diffuse reflectance near-infrared (DR-FTNIR) spectroscopy combined with back-propagation artificial neural network (ANN) multivariate algorithms have been used for estimating the size of TiO_2 NPs. Furthermore, attenuated total reflection infrared (IR-ATR) spectroscopy has been used to determine the surface charge density (SCD) of colloidal SiO_2 NPs in aqueous solution. The frequency position of the Si-(OH) stretching vibration directly correlates with the NPs SCD. In addition, diffuse reflectance infrared Fourier transform (DRIFT) spectroscopy has been used to investigate the surface of SiO_2 NPs relevant for usage in lithium cells. Complementarily, some nanoparticles show emissions in the NIR spectral range, *e.g.*, single-walled carbon nanotubes (SWNTs). NIR fluorescence is a technique widely used for the characterization of SWNTs owing to their chirality, and for their quantitative determination in aquatic environments. Carbon nanotubes (CNTs) are characterized by tunable near-infrared emissions that respond to changes in the local dielectric environment. However, NIR luminescent studies are limited to individually disperse semiconducting nanotubes, since aggregation of nanotubes into bundles quenches the fluorescence via interactions with metallic CNTs, which substantially broadens the absorption spectra. A method for determining the short-wave IR (SWIR) emission wavelengths of individual SWNTs using microscope objectives instead of a conventional spectrometer has also been described. The axial chromatic aberration in the SWIR of the microscope lens has been calibrated such that the depth at which an emissive NP appears brightest and best focused can be used to deduce its peak emission wavelength, thus allowing identification of the CNT structure.

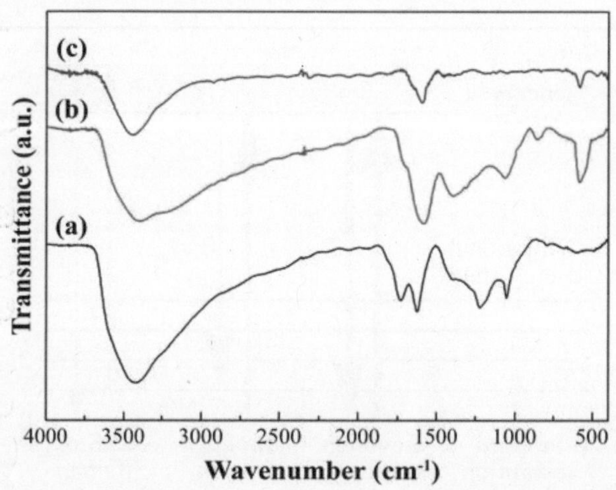

Fig. 4.17 FT-IR spectra of a. GO; b. GO–MnO_2 and c. RGO–MnO_2–AuNPs

4.4 RAMAN SPECTROSCOPY (RS)

Raman scattering of light by molecules is a powerful mean to seek information on a sample's chemical composition and molecular structure. Raman spectroscopy like infrared spectroscopy is a part of vibrational spectroscopy. The vibrational states probed by Raman spectroscopy are similar to those involved in infrared spectroscopy. However, the two vibrational spectroscopy techniques are complementary, in that vibrations that are strong in an infrared spectrum (those involving strong dipole moments) are typically weak in a Raman spectrum.

Likewise, non-polar functional group vibrations that give very strong Raman bands usually result in weak infrared signals. As an example, hydroxyl- or amine-stretching vibrations and the vibrations of carbonyl groups are usually very strong in an FT-IR spectrum and are weak in a Raman spectrum. The stretching vibrations of carbon double and triple bonds and the symmetric vibrations of aromatic groups give a very strong Raman signal. While IR is sensitive to functional groups and to highly polar bonds, Raman is more sensitive to backbone structures and symmetric bonds. Thus, IR and Raman Spectroscopy are complimentary techniques.

Basic Principles: Usually, when light is scattered from an atom or molecule, it has the same energy (frequency) and wavelength as the incident light (Rayleigh scattering). This is an elastic scattering. However, a small fraction of the scattered light (approximately 1 in 10 million photons) is scattered by an excitation in that the scattered photons having energy (frequency) different from the incident photons. This is known as Raman scattering; the phenomenon first observed by Indian scientist Sir C.V. Raman in 1928, who was awarded Nobel Prize in 1930 for this discovery.

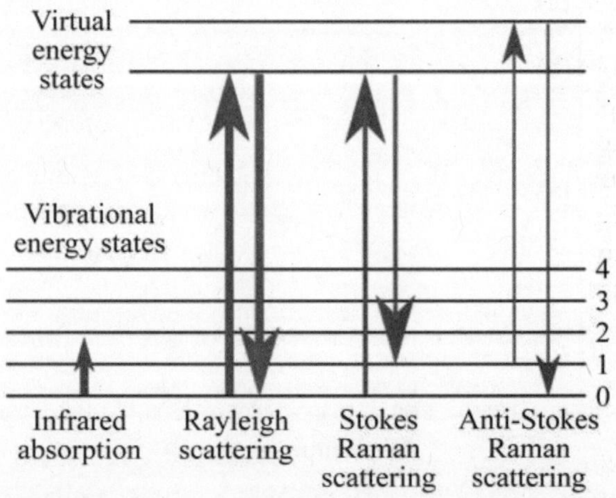

Fig. 4.18. Energy-level diagram showing the states involved in Raman signal. The line thickness is indicative of comparative signal strength from the different transitions

Raman spectroscopy, is used to study vibrational, rotational and other low-frequency modes in a system of condensed matter physics and chemistry. It is based on inelastic scattering (Raman scattering) of monochromatic light usually from laser in the visible, near infrared, or near ultraviolet range. The laser light interacts with phonons or other excitations in the system, resulting in the energy of the laser photons being shifted up or down. The shift in the energy gives information about the phonon model in the system. Raman shifted photons can be of either higher or lower energy, depending upon the vibrational state of the molecule under study. If the final vibrational state of the molecule is more energetic than the initial state, the emitted photon will be shifted to a lower frequency for the total energy of the system to remain balanced (**Fig. 4.18**). This shift in frequency is designated as a Stokes shift. If the final vibrational state is less energetic than the initial state, then the emitted photon will be shifted to a higher frequency, and this is designated as an anti-Stokes shift. The increase or decrease in energy is related to the vibrational energy levels in the ground electronic state of the molecule, and as such, the observed Raman shift of the Stokes and anti-Stokes features are a direct measure of the vibrational energies of the molecule. The difference in energy between the original state and this new state leads to a shift in the emitted photon's frequency away from the excitation wavelength. Accordingly, the signal strength due to Stokes shift is higher then that of anti-Stokes shifts.

The energy increase or decrease from the excitation is related to the vibrational energy spacing in the ground electronic state of the molecule and therefore the wavenumber of the Stokes and anti-Stokes lines are a direct measure of the vibrational energies of the molecule. The pattern of shifted frequencies is determined by the rotational and vibrational states of the sample. This dependence on the polarizability differs from Infrared spectroscopy where the interaction between the molecule and radiation is determined by the dipole moment; this contrasting feature allows one to analyze transitions by Raman spectroscopy that might not be IR active. Raman shifts are typically reported in wave numbers (v) as this value is directly related to energy. Raman shift is expressed as (Eqn. 4.4),

$$\Delta \upsilon = \left(\frac{1}{\lambda_0} - \frac{1}{\lambda_1} \right) \tag{4.4}$$

where, Δv is the Raman shift expressed in wave number (cm^{-1}), λ_0 is the excitation wavelength, and λ_1 is the Raman spectrum wavelength.

The first ever Raman "instrument" was constructed in 1928 by Sir C.V. Raman where monochromatized sunlight was used as a light source and a human eye as a detector was modified using arc lamps and photographic plates and soon became very popular up until the 1950's. Modern instrumentation typically consists of a laser, Rayleigh filter, a few lenses, a spectrograph and a detector (typically a CCD or ICCD). Raman spectrometers as shown schematically in **Fig. 4.19**, a laser light excites the sample, light is scattered in all directions, and some of this light is directed to the detector, which records the Raman spectrum. Raman spectrum shows light at its original laser frequency (Rayleigh) and the Raman spectral features unique to the sample.

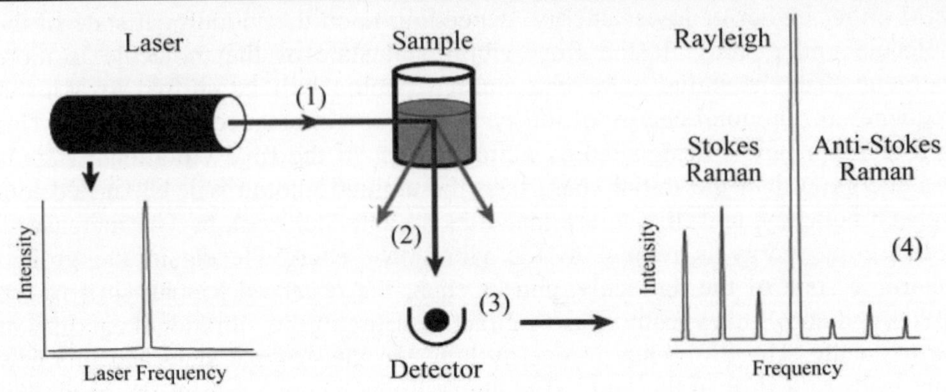

Fig. 4.19. Schematic representation of Raman spectrometer and Raman spectra

Applications: Raman spectroscopy is commonly used in chemistry, since vibrational information is specific to the chemical bonds and symmetry of molecules. Therefore, it provides a fingerprint in the range 500–2000 cm^{-1} by which the molecule can be identified. Raman spectroscopy provides key information about the structure of molecules. The position and intensity of peaks in the spectrum reflect the molecular structure and can be used to determine the chemical identity of the sample. Spectra may also show subtle changes depending on the crystalline form. With the extensive spectral libraries that are now available, it is very straight forward to identify compounds by spectral library searching. In addition, Raman spectroscopy can be used to observe other low frequency excitations of a solid, such as plasmons and superconducting gap excitation.

To be used as a standard, the Raman spectrum of diamond crystals with a regular array of identical carbon atoms, all in the same configuration show just one dominant Raman band (because there is just one molecular environment of the crystal) whereas a polystyrene molecule consisting of different types of atom exhibit complex Raman spectra due to its being less symmetric and presence of different bond types connecting the atoms (**Fig. 4.20**).

The frequencies of vibration depend on the masses of the atoms involved and the strength of the bonds between them. Heavy atoms and weak bonds have low Raman shifts. Light atoms and strong bonds have high Raman shifts. As a consequence of this we see the high frequency carbon-hydrogen (C-H) vibrations in the polystyrene spectrum at about 3000 cm^{-1}. The low frequency carbon-carbon (C-C) vibrations are at around 800 cm^{-1}. It is also to be noted that the vibrations of two carbon atoms linked by strong double bonds (C = C) at around 1600 cm^{-1} is at a higher frequency than two carbon atoms lined by a weaker single bond (C-C, 800 cm^{-1}). These simple rules can be used in the analysis of a variety of organic, inorganic and polymeric or biomolecules through their Raman spectroscopy as a complementary technique to infrared spectroscopy with the distinct advantages in that:

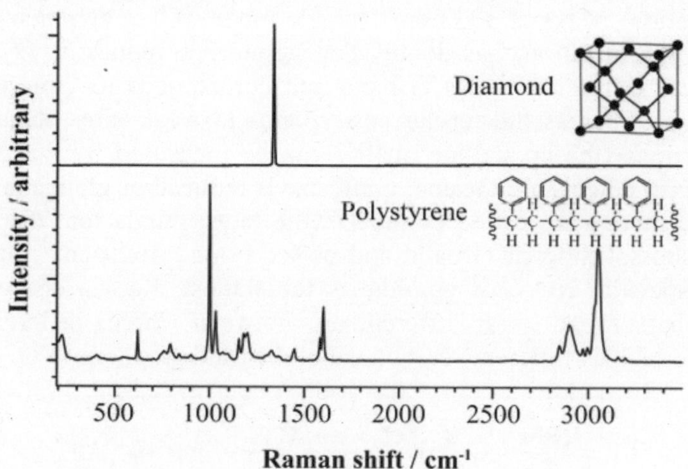

Fig. 4.20. Raman spectra of diamond and polystyrene.

Recently, Raman spectroscopy has proved as an excellent tool to address many issues related to structure and properties of a variety of nanomaterials including carbon nanomaterials, functionalized metals and metal oxides etc. having advantages such as:

- Identification of materials constituting nanostructures
- Number of layers for graphene layers
- Thickness of heterostructure layers
- Diameter and chirality of carbon nanotubes
- Stress/strain characterization
- Electronic properties of doped materials (metallic/semiconductor)
- Localization of separated carbon nanotubes or quantum dots

Raman spectra of carbon nanomaterials: Carbon nanomaterials have made a big impact in the field of the field of material science since individual carbon nanomaterials viz., fullerene, carbon nanotubes, graphene or nanodiamonds offer a myriad of unique electronic, optical, magnetic, mechanical properties in their own right thus making them very interesting materials to a broad range of industries. The great interest in the processing, modification, and customization of these materials has created a strong demand for suitable characterization technique(s) for them. Raman spectroscopy being sensitive to highly symmetric C-C bonds with little or no dipole moment is the technique of choice very well suited to many of the characterization needs with these materials. It is indeed one of the few techniques sensitive to the full range of structural states present in this class of materials, from perfectly crystalline to amorphous.

The conventional allotropes of carbon yield very simple spectra: diamond (sp3 hybridization) peaks at 1332 cm^{-1} (single mode of T_2g symmetry) whereas graphite (sp^2 hybridization) has doubly degenerate E_2g modes at 42 and 1582 cm^{-1}. The latter is referred to as G band and corresponds to vibrations in the graphene planes whereas the former corresponds to weak inter planar Van der Waals interactions (the crystalline quality can be measured by the width of G band). Whenever defects are created, grain size is reduced or graphene planes are curved. These modes are called D and D' (the letter stands for "disorder"). D' results from the splitting of G band and peaks around 1620 cm^{-1}, at the value where the dispersion curve of graphite is the flattest. Raman spectroscopy is shown to characterize and differentiate between three different carbon nanostructures such as fullerene, carbon nanotubes and graphene (**Fig. 4.21**).

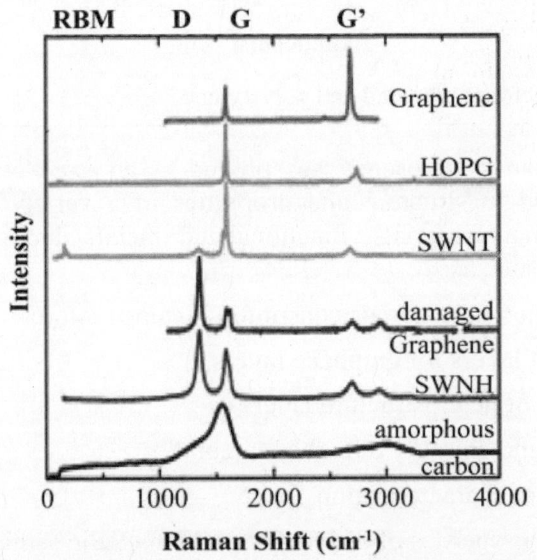

Fig. 4.21. Raman spectra of amorphous carbon and carbon nanomaterials

The presence of disorder in sp^2-hybridized carbon systems results in resonance Raman spectra and thus makes Raman spectroscopy as one of the most sensitive techniques to characterize disorder in sp^2 carbon materials. G-band arises from the stretching of the C-C bond in graphitic materials, and is common to all sp^2 carbon systems. If there are some randomly distributed impurities or surface charges in the graphene, the G-peak can split into two, G-peak (1583 cm^{-1}) and D'-peak (1620 cm^{-1}). The main reason is that the localized vibrational modes of the impurities can interact with the extended phonon modes of graphene resulting in the observed splitting. All kinds of sp^2 carbon materials exhibit a strong peak in the range 2500 - 2800 cm^{-1}. Combined with the G-band, this spectrum is a Raman signature of graphitic sp2 materials and is called 2D(G*)-band. 2D-band is a second-order two-phonon process and exhibits a strong frequency dependence on the excitation laser energy.

Raman spectra of metal oxide: Raman spectroscopy is used to evaluate the crystallinity and composition of material. **Figure 4.22** shows Raman spectrum of WO_3 nanowire networks. All the major peaks can be attributed to tungsten trioxide. More specifically, the peaks at 707 cm^{-1} and 797cm^{-1} are related to the stretching vibration of bridging oxygen in (W–O–W) bonds, and they are typical peaks of monoclinic tungsten trioxide. The other two major peaks at 262 cm^{-1} and 322 cm^{-1} are due to the bending vibration of (O–W–O) bonds.

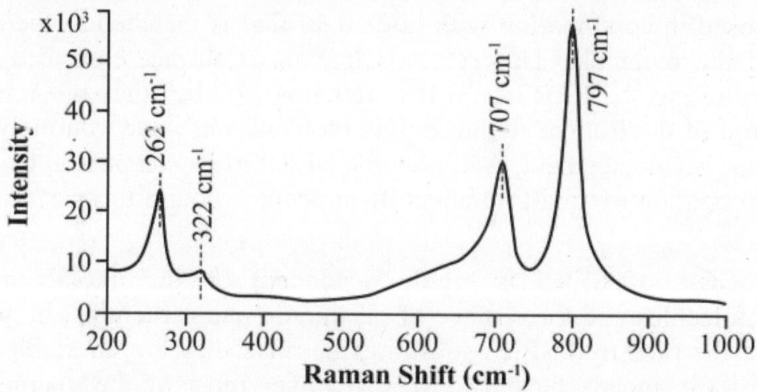

Fig. 4.22. Raman spectra of WO_3 nanowire deposited on alumina substrate

4.5 ENHANCED RAMAN SPECTROSCOPY

Conventional Raman spectroscopy offers a diffraction limited spatial resolution in the order of a few hundreds of nanometers. Despite the difference between this resolution and the actual dimensions of the nanomaterials, Raman spectroscopy still remains an excellent tool to address many issues related to their structure and properties. However, Raman signals are inherently weak, especially when using visible light excitation and so a small number of scattered photons are available for detection. Enhanced Raman spectroscopy methods *viz.,* surface-enhanced Raman scattering (SERS) and Tip Enhanced Raman Spectroscopy (TERS) have overcome limitation in conventional Raman spectroscopy that led to a spectacular increase in sensitivity and resolution very much needed in several fields such as nanotechnology, semiconductor industry, and biomedical applications. Surface-enhanced Raman spectroscopy (SERS) made it possible to achieve high sensiti-vity allowing *e.g.* the detection of a single molecule, while tip-enhanced Raman spectroscopy (TERS) combines the enhancement capability of SERS with the spatial resolution close to 20 nm which with further add on scanning probe methods make out-standing sub-molecular visualization possible; thus, introducing Raman spectroscopy into nanoscale resolution imaging.

4.5.1 Surface Enhanced Raman Spectroscopy (SERS)

SERS uses nanoscale roughened metal surfaces typically made of gold (Au) or silver (Ag). Laser excitation of these roughened metal nanostructures resonantly drives the surface charges creating a highly localized (plasmonic) light field. When a molecule is absorbed or lies close to the enhanced field at the surface, a large enhancement in the Raman signal can be observed. Raman signals several orders of magnitude greater than normal Raman scattering are observed, thereby making it possible to detect low concentrations without the need for fluorescent labeling. The Raman signal can be amplified further when the roughened metal surface is used in combination with laser light that is matched to the absorption maxima of the molecule. This effect is known as surface-enhanced resonance Raman scattering, (SERRS). SERS remains to be the most significant amplification of the Raman signal. Single molecules are now routinely observed due to the large enhancement. Additionally, SERS is an exceptional technique for the characterization of small numbers of molecules bound to or near plasmonic surfaces.

The success of SERS is highly dependent on the interaction between adsorbed molecules and the surface of plasmonic nanostructures. In general, Au and Ag are the preferred SERS substrates because they are air stable materials. Moreover, these metals have LSPRs that cover most of the visible and near infrared wavelength range, where most Raman measurementsare made. By delivering huge enhancement, SERS provides an opportunity to obtain the vibrational and chemical information from nanosized samples. With such a massive enhancement, SERS can even make it possible to measure Raman scattering from a single molecule, provided the sample molecules are properly spread with low density.

The potential of SERS lies in its ability to identify chemical species and obtain structural information in a wide variety of fields including polymer and materials science, biochemistry and bio sensing, catalysis, and electrochemistry.

While SERS came as a revolutionary addition to Raman spectroscopy due to its ability to immensely enhance the light field, it did not solve the restriction on the spatial resolution in Raman measurement imposed by the diffraction limit of light.

4.5.2 Tip Enhanced Raman Spectroscopy (TERS)

Tip Enhanced Raman Spectroscopy (TERS) a combination of Raman spectroscopy with Scanning Probe Microscopy techniques (refer chapter V). Such a combination allows advantages of the two analytical methods within a single platform. The co-localized Raman-SPM measurements give separate information from Raman spectroscopy (chemical characterization and imaging) and SPM (physical properties), acquired from the same region of the sample. As in regular Raman spectroscopy the sample is illuminated with a monochromatic (laser) source and the backscattered radiation is dispersed to obtain the spectral information. It has emerged as a novel method for direct, label-free nondestructive analysis of chemical matter at the nanoscale.

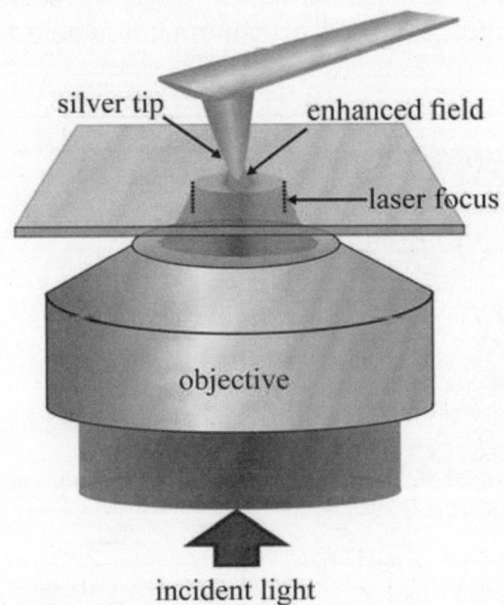

Figure 4.23 Illustration of an AFM-based TERS measurement system

Figure 4.23 describe the setup of a Tip Enhanced System combined with AFM. A thin granular silver layer is deposited via thermal evaporation on the body of the AFM tip. The incident light is focused on a sample placed on a transparent glass substrate, through a high-NA objective lens. When a silver-coated nano tip is brought in the diffraction-limited focus spot, a strongly enhanced and confined nanolight-source is created in the close vicinity of the tip apex. This nanolight-source is utilized to excite Raman scattering from a tiny volume of the sample right under the apex of the tip of approximate size of 20 nm, where the deposited silver particles are disconnected, as illustrated in the inset of **Fig. 4.24a**. The key in TERS is a properly enhancing probe. The functionalized probe acts as a 'plasmonic antenna' which enhances the local electromagnetic field, enhancing the Raman scattering at the very end of the probe. With enough enhancements, the local near-field Raman signal largely overwhelms the far-field signal and Raman signal with nanoscale resolution can be observed. The other near field optical techniques like NSOM always lack in the sensitivity to perform proper Raman detection while the local enhancement obtained from TERS probes truly makes it a reality. The SPM tip thus plays a central role. To serve as effective near field antenna for TERS, it must display a pronounced surface plasmon resonance at the wavelength of interest.

Among others, carbon nano materials such as fullerene (C_{60}), single-walled carbon nanotubes (SWNTs), and graphene provide interesting test samples for TERS measurements. **Figure 4.24 b–d** show examples of TERS measurement from a nanocluster of C_{60}, thin bundles of SWNTs, and a graphene sample, respectively. The lower spectra for all samples correspond to the far-field measurement where the tip was retracted from the sample, while the upper spectra correspond to the near-field measurements where the tip was in contact with the sample.

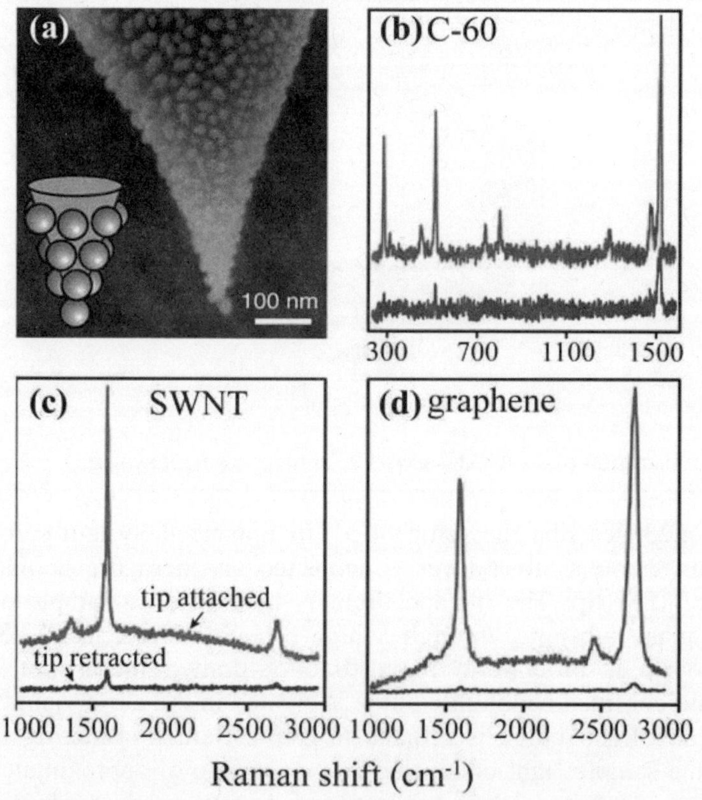

Fig. 4.24. (a) SEM image of the silver-evaporated nano tip; TERS spectra of (b) carbon-60; (c) single-walled carbon nanotubes; and d. graphene. Upper red curves in (b)–(d) represent the near field spectra when tip was in contact with the sample, while the blue curves represent the far-field spectra when the tip was retracted

When the tip is retracted from the sample, there is no confined light near the sample and thus no enhancement is observed in the lower spectra. These spectra represent usual far-field Raman scattering, which is quite weak due to the small volume of the samples. When the tip comes in contact with the sample, the tip apex enhances both incident and Raman scattered light from the sample right under the tip (upper spectra). The far-field Raman scattering in the lower spectra are generated from the sample volume within the diffraction-limited focal spot. On the other hand, the enhanced scattering in upper spectra represents the combination of this far-field Raman scattering together with the near-field Raman scattering generated from a nanometric volume of the sample that is immersed into the confined field at the tip apex. Subtraction of the lower spectra from the upper spectra would provide the intensity of the pure near-field Raman scattering. Enhancement factor in TERS is defined as the ratio between pure near-field and far-field Raman intensities per unit volume. Taking the near-field and the far-field excitation volumes into consideration, one can calculate the enhancement factor from the intensity ratio.

Since one of the most significant features of TERS is the high spatial resolution. SWNTs with one-dimensional structure are the best samples to further explore the extent of this feature **(Fig. 4.25)**

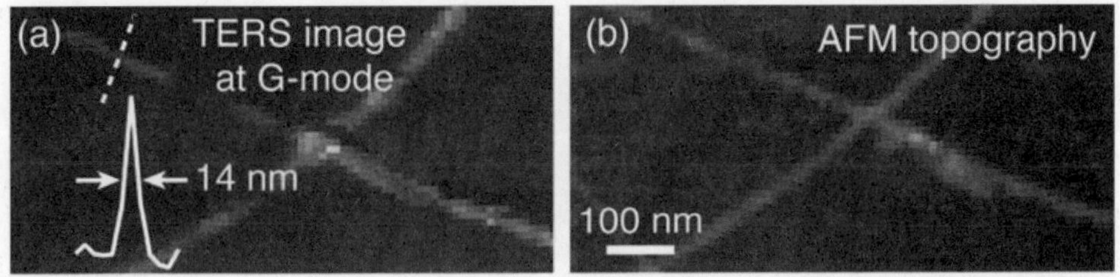

Fig. 4.25 a. TERS image; b. corresponding AFM image of two thin bundles of SWNTs

4.6 ELECTRON SPECTROSCOPY

Electron spectroscopy is mainly used to identify the chemical compositions of the sample. Depending on the source of excitation and the product being detected, there are a variety of spectroscopic techniques including X-ray photoelectron spectroscopy (XPS), auger electron spectroscopy (AES) and secondary–ion mass spectrometry (SIMS).

4.6.1 X-ray Photoelectron Spectroscopy (XPS)

X-ray photoelectron spectroscopy (XPS), also called Electron Spectroscopy for Chemical Analysis (ESCA), was pioneered by the Swedish physicist K. Siegbahn, who subsequently received the 1981 Noble Prize in Physics for this work. X-ray photoelectron spectroscopy enables the determination of the surface chemistry, and the way in which that chemistry changes in the surface and near-surface region. XPS provides quantitative elemental and chemical information with extremely high surface specificity (from outer 10 nm of the sample). Depth profiling extends the technique measuring this film layers. XPS is ideal for comprehensively and quantitatively characterizing the composition, chemical bonding and electronic states of elements at surfaces and interfaces.

The photoelectric effect is the ejection of electrons from the surface of a material upon exposure to electromagnetic radiation of sufficient energy (**Fig. 4.26**). Electrons emitted have characteristic kinetic energies proportional to the energy of the radiation (Eq. 4.5),

$$KE = hv - E_b - \phi \tag{4.5}$$

where, *KE* is the kinetic energy of the electron, *h* is Planck's constant, ν is the frequency of the incident radiation, E_b is the ionization, or binding, energy, and φ is the work function. The work function is a constant which is dependent upon the spectrometer.

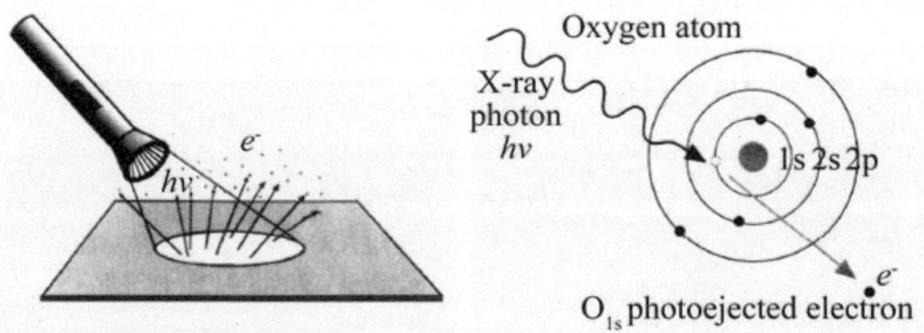

Fig. 4.26. Schematic of photoelectric effect

In X-ray photoelectron spectroscopy, high energy radiation is used to expel core electrons from a sample. The kinetic energies of the resulting core electrons are measured. Using the equation with the kinetic energy and known frequency of radiation, the binding energy of the ejected electron may be determined. By Koopman's theorem, which states that ionization energy is equivalent to the negative of the orbital energy, the energy of the orbital from which the electron originated is determined. These orbital energies are characteristic of the element and its state. **Figure 4.27 (a-d)** is a schematic representation of the physical process involved in various electron spectroscopic techniques.

The three lower lines labeled E_b, E'_b and E''_b, shown in **Fig. 4.27** represent energies of the inner-shell *K* and *L* electrons of an atom. The upper three lines represent some of the energy levels of the outer shell, or valence, electrons. As shown in the illustration, one of the photons of a monochromatic X–ray beam of known energy *hv* displaces an electron e^- from a K orbital of energy E_b. The process can be represented by equation 4.6

$$A + hv \rightarrow A^{+*} + e^- \tag{4.6}$$

Where *A* can be an atom, a molecule, or an ion and A^{+*} is an electronically excited ion with a positive charge one greater than that of A. Various methods are available to determine the value of *v*. The binding energy of an electron is characteristic of the atom and orbital that emits the electron. It is evident from **Fig. 4.27** that XPS provides a means of qualitative identification of the elements present on the surface of solids. Electron spectroscopy records the kinetic energy of emitted electrons, the spectrum thus consists of a plot of the number of emitted electrons, or the power of the electron beam, as a function of the energy (or the frequency or wavelength) of the emitted electrons.

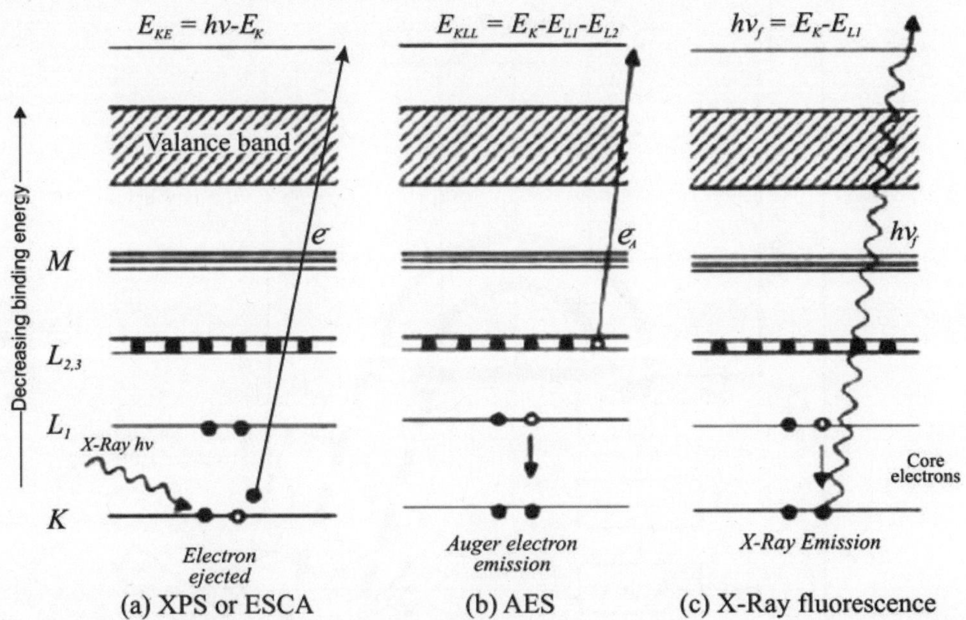

Fig. 4.27. Schematic representation of the physical process involved in Electron Spectroscopy: XPS and Auger spectroscopy

Instrumentation: Electron spectrometers are made up of components whose functions are analogous to those encountered in optical spectroscopic instruments. These components include (1) a source; (2) a sample holder; (3) an analyzer, which has the same function as a monochromator; (4) a detector; and (5) a signal processor and read-out. **Figure 4.28** shows a typical arrangement of these components. Electron spectrometers generally required elaborate vacuum systems to reduce the pressure in all of the components to as low as 10^{-8} to 10^{-10} torr.

Source: The X-ray sources for XPS spectrometers are X-ray tubes equipped with magnesium or aluminum targets and suitable filters. Monochromatic aluminum

(hv = 1486.6 eV) or magnesium (hv = 1253.6 eV) K_{α} X-rays are used to eject core electrons from the sample. The K∝ lines for these two elements have considerably narrower bandwidths (0.8-09 eV) than those encountered with higher atomic number targets; narrow bands are desirable because they lead to enhanced resolution. Non-monochromatic sources typically illuminate a spot a few centimeters in diameter. Relatively sophisticated XPS instruments employ a crystal monochromator to provide an X-ray beam having a bandwidth of about 0.3 eV.

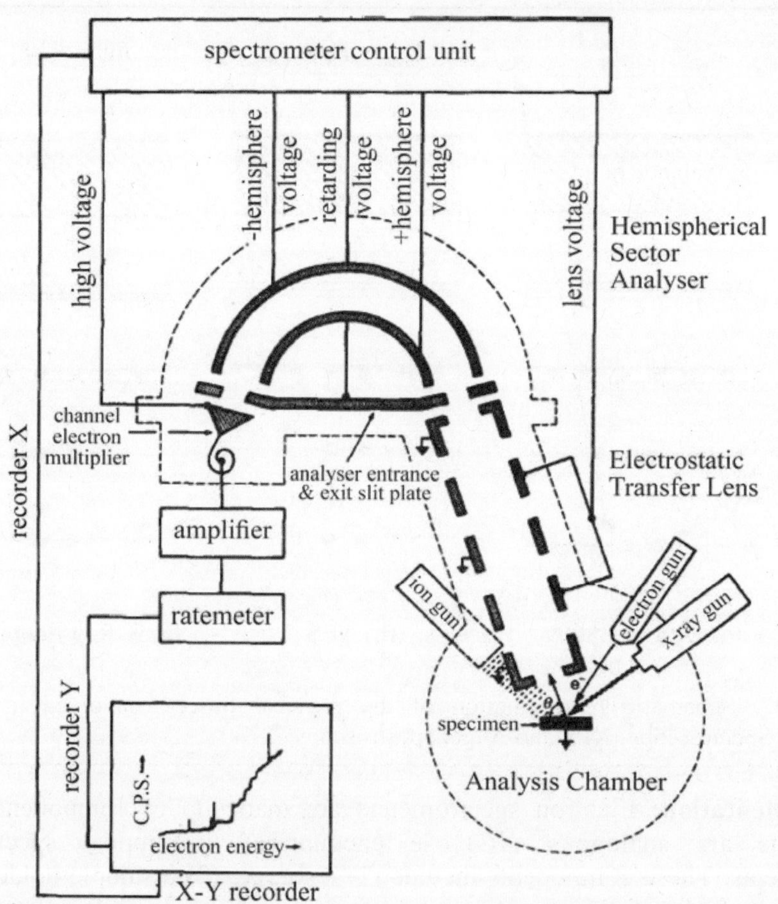

Fig. 4.28. Basic components of XPS / ESCA instrument: a. Electron energy analyzer, b. X-ray source, c. Ar ion gun, d. Neutralizer, e. Vacuum system, f. Electronic controls, g. Computer system

Solid samples are mounted in a fixed position as close as possible to the photon or electron source and the entrance slit of the spectrometer. Ultra-high vacuum conditions are used in order to minimize gas collisions interfering with the electrons before they reach the detector.

Detector: The photoelectrons ejected from the material are detected and their energies measured. The detector consists of the collection lens or lenses and the electron energy analyzer, which disperses the emitted electrons according to their kinetic energy. The lens system usually allows a wide collection angle (–30°) for high efficiency. Typically, photoelectron experiments are carried out in constant analyzer energy mode, in which electrons are accelerated or retarded by the lens system to some user-defined energy as they pass through the analyzer (the pass energy, *E*). Often, pass energies of 5-25 eV will give high-resolution spectra, and 100-200 eV pass energies are used for survey scans. The signal intensity decreases as the pass energy decreases.

The energy analyzers are of the type as illustrated in **Fig. 4.28**, where the electron beam is deflected by the electrostatic field of a hemispherical capacitor. The electrons thus travel in a curved path from the lens to the multi-channel transducer. The radius of curvature depends on the kinetic energy of the electrons and the magnitude of the electrostatic field. An entire spectrum is obtained by varying the filed so as to focus electrons of various kinetic energies on the transducer.

Most modern electron spectrometers are based on solid-state, channel electron multipliers, which consist of tubes of glass that have been doped with lead or vanadium. When a potential difference of several kilovolts is applied across these materials, a cascade or pulse of 10^6 to 10^8 electrons is produced for each incident electron. One of the most important applications of XPS has been identification of oxidation states of elements in inorganic compounds.

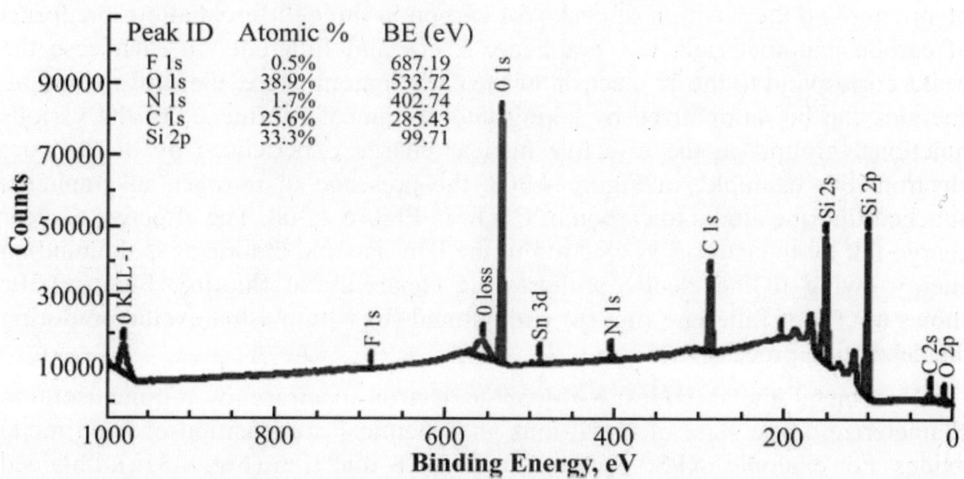

Fig. 4.29. XPS scan survey spectrum for all elements

Data collection: X-ray photoelectron (XP) spectra provide the relative frequencies of binding energies of electrons detected, measured in electron-volts (eV). Detectors have accuracies on the order of ±0.1 eV. The binding energies are used to identify the elements to which the peaks correspond. XPS data is given in a plot of intensity versus binding energy. Intensity may be measured in counts per unit time (such as counts per second, denoted c/s). Often, intensity is reported as arbitrary units (arb. units), since only relative intensities provide relevant information. Comparing the areas under the peaks gives relative percentages of the elements detected in the sample. Initially, a survey XP spectrum is obtained, which shows all of the detectable elements present in the sample. Elements with low detection or with abundances near the detection limit of the spectrometer may be missed with the survey scan. Examples of some of its uses include identification of active sites and poisons on catalytic surfaces, determination of surface contaminants on semiconductors, analysis of the composition of human skin, and study of oxide surface layers on metals and alloys. It is also evident that the method has a substantial potential in the elucidation of chemical structure **(Fig. 4.29).** Information from XPS spectra is comparable to that from nuclear magnetic resonance (NMR) or IR spectroscopy.

Application of XPS to characterize nanomaterials

XPS is a versatile technique for determining the elemental composition of a material's surface. XPS is a powerful surface technique to accurately detect the presence and determine the relative ratios of detectable elements in sample. The technique has been extended for the analysis of structure and composition of surfaces of nanomaterials of different types as discussed below.

Identification of carbon nanomaterials: **Figure 4.30 (a-c)** illustrates the effect of structure on the position of peaks for carbon in three different allotropic forms of carbon nanomaterials viz. graphene, CNTs and fullerene. In each case the peaks correspond to the 1*s* electron of the carbon atom. Here, the shift in binding energies can be rationalized by taking into account the influence of the various functional groups on the effective nuclear charge experienced by the 1*s* core electron. For example, in Figure 4.30a, the presence of nitrogen in graphene; attached fluorine atoms to carbon of CNTs in Figure 4.30b. The effective nuclear charge felt by the carbon 1*s* electron in the later is a maximum, as is the binding energy owing to the electro withdrawing capability of fluorine. Figure 3.30c shows a XPS of fullerene film (a) without and (b) with plasma excitation during the deposition process.

Nanometal /metal oxides: X-ray photoelectron spectroscopy is quite useful to characterize, ionic state of metal ions and chemical composition of nano metal oxides. For example, XPS spectra of TiO_2-WO_3 thin film **(Fig. 4.31a),** obtained from survey scans on the surface of the co-evaporated film between binding energies 0 to 1200 eV. As expected elements of C, O, Ti and W were detected along with impurities of K and P. From high resolution spectra (Figure 4.31b), the core level of Ti $2p_{1/2}$ and Ti $2p_{3/2}$ are found at binding energy of 465.6 eV and 459 eV, respectively with a peak separation of 6.6 eV. **Figure 4.31 b and c** shows the core level of W 4f 5/2 (38.2 eV) and W $4f_{7/2}$ (36 eV) obtained from high resolution measurements and the two peaks are separated by 2.2 eV. The Ti $2p_{3/2}$ peak obtained from the co-evaporated film (459 eV) is in close agreement with the literature value (459.2 eV) which corresponds to the fully oxidized Ti^{+4} in titanium dioxide. The binding energy of W 4f $_{7/2}$ (36 eV) was found to be in reasonable agreement with the corresponding characteristic binding energy of W 4f (35.8 eV) in tungsten trioxide film.

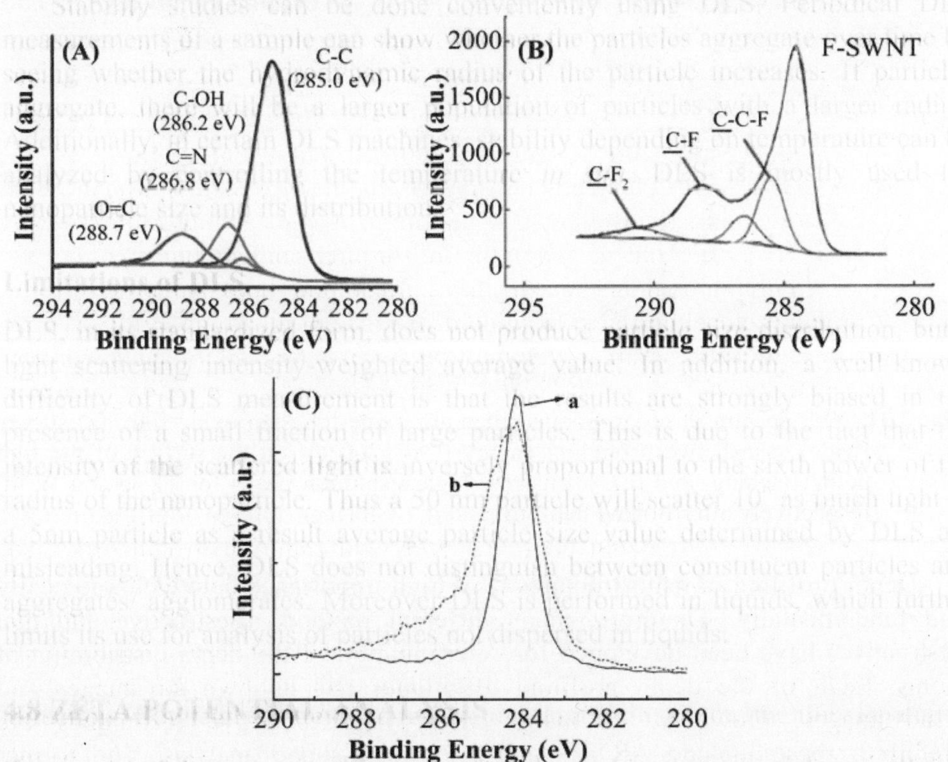

Fig. 4.30. C_{1s} XPS of carbon nanomaterials (A) graphene oxide; (B) fluorinated SWNTs; (C) fullerene film (a) without and (b) with plasma excitation during the deposition process

The peak position of *O* 1s was at 531 eV but this value is shifted, for example by 1.1 eV when compared with the core level of *O*1s (529.9 eV) of the anatase TiO_2 film. This can be due to the nanocrystalline nature of the mixed oxide thin film which may cause a shift of the core levels.

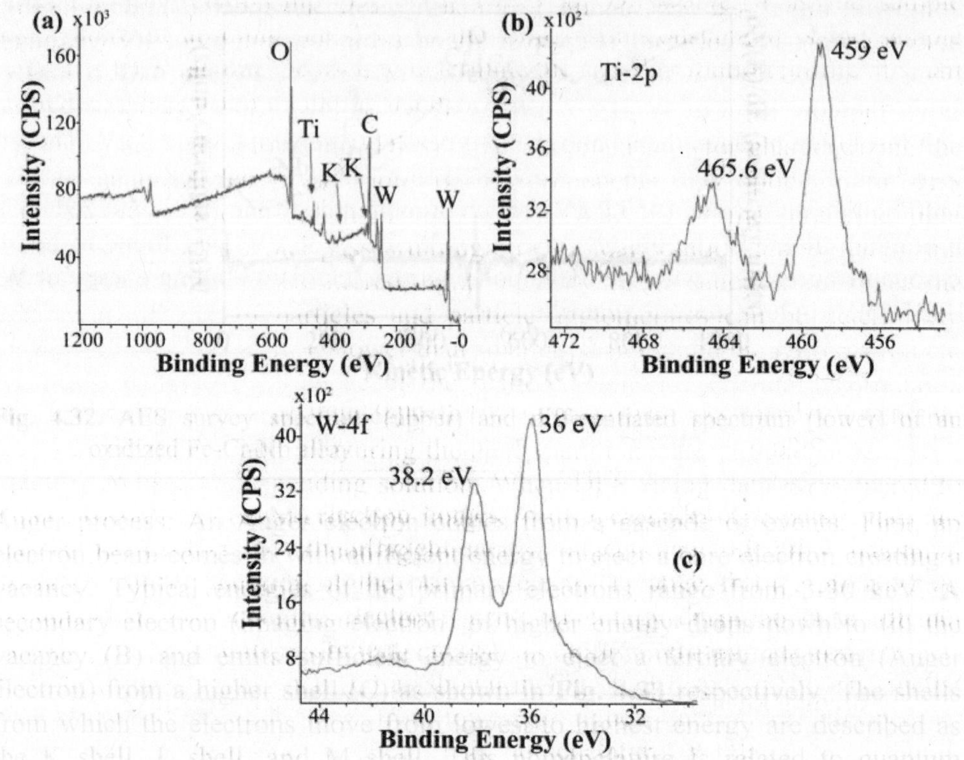

Fig. 4.31. XPS spectra of co-evaporated titanium oxide-tungsten oxide film: (a) survey (wide) scan between binding energy 0 to 1200 eV on the surface of the film before etching, and (b), (c) narrow high resolution peaks for Ti 2p and W 4f, respectively after etching the film

Depth profiling of nanostructures is of high importance both technologically and fundamentally. Therefore, many different methods (destructive and non destructive) have been developed for determination of the depth distribution of atoms. Each of the depth profiling techniques has its own advantages and disadvantages. However, in many cases, non-destructive techniques are preferred; these include ARXPS and XPS peak-shape analysis. The former together with parallel factor analysis is suitable for giving an overall understanding of chemistry and morphology with depth. It works very well for flat surfaces but it fails for rough or nanostructured surfaces because of the shadowing effect. In the latter method shadowing effects can be avoided because only a single spectrum is used in the analysis and this may be taken at near normal emission angle. It is a rather robust means of determining atom depth distributions on the nanoscale both for large-area XPS analysis and for imaging. Traditional XPS analysis relies on XPS peak intensity assuming that the concentration in the surface region is proportional to the XPS peak intensity. The XPS-peak-shape method is non-destructive and therefore also allows one to study the change in surface composition during surface treatment as, *e.g.*, in chemical reactions and gradual annealing. It has been applied in the study of a wide range of systems and physical phenomenon including growth mechanisms and nanostructures of metal/metal, metal/metal oxide, films etc.

Limitations: Lighter elements such as hydrogen and helium cannot be detected using XPS. For this reason, XPS can provide only relative, rather than absolute, ratios of elements in a sample. Also, elements with relatively low atomic percentages close to that of the detection limit or low detection by XPS may not be seen in the spectrum. Furthermore, each peak represents a distribution of observed binding energies of ejected electrons based on the depth of the atom from which they originate, as well as the state of the atom. Electrons from atoms deeper in the sample must travel through the above layers before being liberated and detected, which reduces their kinetic energies and thus increases their apparent binding energies. The width of the peaks in the spectrum consequently depends on the thickness of the sample and the depth to which the XPS can detect; therefore, the values obtained vary slightly depending on the depth of the atom. Additionally, the depth to which XPS can analyze depends on the element being detected.

4.6.2 Auger Electron Spectroscopy (AES)

Auger electron spectroscopy (AES) is one of the most commonly employed surface analysis techniques. It uses the energy of emitted electrons to identify the element present in a sample, similar to X-ray photoelectron spectroscopy (XPS). The main difference is that XPS uses an X-ray beam to eject an electron while AES uses an electron beam to eject an electron. In AES, the sample depth is dependent on the escape energy of the electrons. It is not a function of the excitation source as in XPS. In AES, the collection depth is limited to 1-5 nm due to the small escape depth of electrons, which permits analysis of the first 2-10 atomic layers. In addition, a typical analysis spot size is roughly 10 nm. A representative AES spectrum illustrating the number of emitted electrons, N, as a function of kinetic energy, E, in direct form (red) and in differentiated form (black) is shown in **Fig. 4.32**.

Like XPS, AES measures the kinetic energy (E_k) of an electron to determine its binding energy (E_b). The binding energy is inversely proportional to the kinetic energy and can be found from Eqn. 4.7, where $h\nu$ is the energy of the incident photon and $\Delta\Phi$ is the difference in work function between the sample and the detector material.

$$E_b = h\nu - E_k + \Delta\Phi \tag{4.7}$$

Since the E_b is dependent on the element and the electronic environment of the nucleus, AES can be used to distinguish elements and their oxidation states. For instance, the energy required to remove an electron from Fe^{3+} is more than in Fe^0. Therefore, the Fe^{3+} peak will have a lower E_k than the Fe^0 peak, effectively distinguishing the oxidation states.

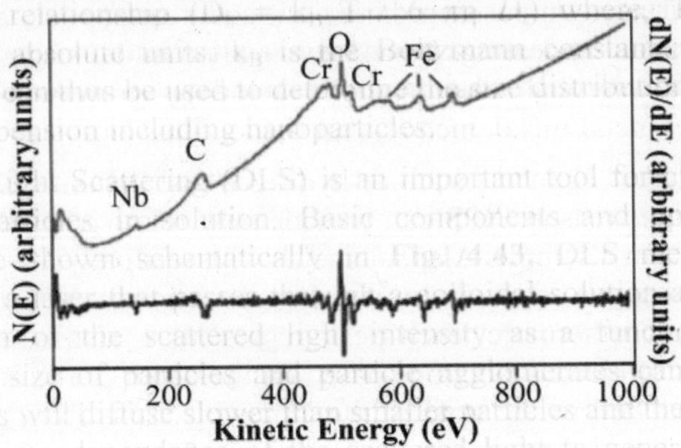

Fig. 4.32. AES survey spectrum (upper) and differentiated spectrum (lower) of an oxidized Fe-Cr-Nb alloy

Auger process: An Auger electron comes from a cascade of events. First, an electron beam comes in with sufficient energy to eject a core electron creating a vacancy. Typical energies of the primary electrons range from 3-30 keV. A secondary electron (imaging electron) of higher energy drops down to fill the vacancy (B) and emits sufficient energy to eject a tertiary electron (Auger electron) from a higher shell (C) as shown in **Fig. 4.33** respectively. The shells from which the electrons move from lowest to highest energy are described as the K shell, L shell, and M shell. This nomenclature is related to quantum numbers. Explicitly, the K shell represents the 1s orbital, the L shell represents the 2s and 2p orbitals, and the M shell represents the 3s, 3p, and 3d orbitals. The cascade of events typically begins with the ionization of a K shell electron, followed by the movement of an L shell electron into the K shell vacancy. Then, either an L shell electron or M shell electron is ejected. It depends on the element, which peak is prevalent but often both peaks will be present. The peak seen in the spectrum is labeled according to the shells involved in the movement of the electrons. For example, an electron ejected from a gold atom could be labeled as Au KLL or Au KLM. The intensity of the peak depends on the amount of material present, while the peak position is element dependent. Auger transitions characteristic of each elements can be found in the literature.

Instrumentation

Important elements of an Auger spectrometer include a vacuum system, an electron source, and a detector. AES must be performed at pressures less than 10^{-3} pascal (Pa) to keep residual gases from adsorbing to the sample surface. This can be achieved using an ultra-high-vacuum system with pressures from 10^{-8} to 10^{-9} Pa. Typical electron sources include tungsten filaments with an electron beam diameter of 3–5 μm, LaB_6 electron sources with a beam diameter of less than 40 nm, and Schottky barrier filaments with a 20 nm beam diameter and high beam current density. Two common detectors are the cylindrical mirror analyzer and the concentric hemispherical analyzer discussed below. Notably, concentric hemispherical analyzers typically have better energy resolution.

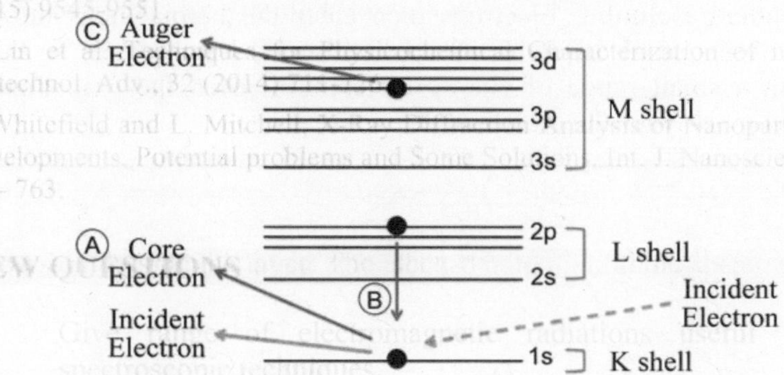

Fig. 4.33. Schematic diagram of the Auger process

Cylindrical mirror analyzer (CMA): A CMA is composed of an electron gun, two cylinders, and an electron detector as shown in **Fig. 4.34.** The operation of a CMA involves an electron gun being directed at the sample. An ejected electron then enters the space between the inner and outer cylinders (IC and OC). The inner cylinder is at ground potential, while the outer cylinder's potential is proportional to the kinetic energy of the electron. Due to its negative potential, the outer cylinder deflects the electron towards the electron detector. Only electrons within the solid angle cone are detected. The resulting signal is proportional to the number of electrons detected as a function of kinetic energy.

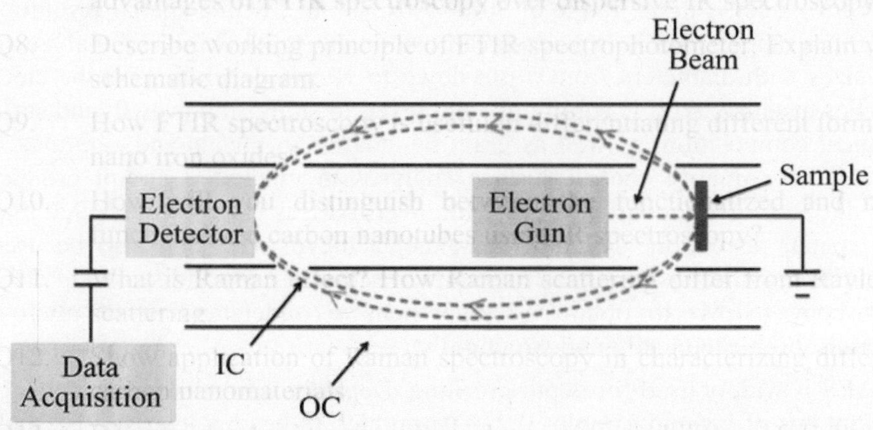

Fig. 4.34. Schematic of a cylindrical mirror analyzer

Concentric hemispherical analyzer (CHA): A CHA contains three parts *i.e.* a retarding and focusing input lens assembly, an inner and outer hemisphere (IH and OH) and an electron detector as illustrated in **Figure 4.35.** Electrons ejected from the surface enter the input lens, which focuses the electrons and retards their energy for better resolution. Electrons then enter the hemispheres through an entrance slit. A potential difference is applied on the hemispheres so that only electrons with a small range of energy differences reach the exit. Finally, an electron detector analyzes the electrons.

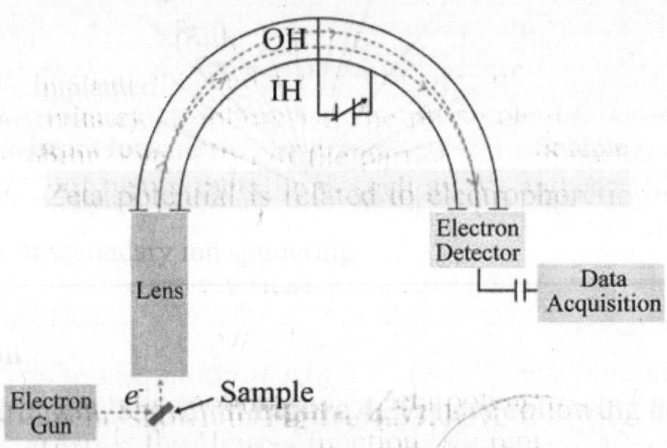

Fig. 4.35. Schematic of a concentric hemispherical analyzer

Applications

AES is an excellent complementary technique to SEM (Scanning Electron Microscopy) in that AES can provide elemental maps with significantly better spatial resolution and surface sensitivity than SEM/EDS. The typical information depth for Auger is in the range of 30-60 A° (3-6 nm), with a spatial resolution in the 10 nm range. AES is in widespread use owing to its ability to analyze small spot sizes with diameters from 5 μm down to 10 nm depending on the electron gun. For instance, AES is commonly employed to study film growth and surface-chemical composition, as well as grain boundaries in metals and ceramics. It is also used for quality control surface analyses in integrated circuit production lines due to short acquisition times. Moreover, AES is used for areas that require high spatial resolution, which XPS cannot achieve. AES can also be used in conjunction with transmission electron microscopy (TEM) and scanning electron microscopy (SEM) to obtain a comprehensive understanding of microscale materials, both chemically and structurally.

AES is widely used for depth profiling. Depth profiling allows the elemental distributions of layered samples 0.2 – 1 μm thick to be characterized beyond the escape depth limit of an electron. Varying the incident and collection angles, and the primary beam energy controls the analysis depth. In general, the depth resolution decreases with the square root of the sample thickness. Notably, in AES, it is possible to simultaneously sputter and collect Auger data for depth profiling. The sputtering time indicates the depth and the intensity indicates elemental concentrations. Since, the sputtering process does not affect the ejection of the Auger electron, helium or argon ions can be used to sputter the surface and create the trench, while collecting Auger data at the same time. The depth profile does not have the problem of diffusion of hydrocarbons into the trenches. Thus, AES is better for depth profiles of reactive metals (*e.g.*, gold or any metal or semiconductor). Yet, care should be taken because sputtering can mix up different elements, changing the sample composition.

Limitations

Since AES is a three-electron process, elements with less than three electrons cannot be analyzed, hydrogen and helium cannot be detected. Nonetheless, detection is better for lighter elements with fewer transitions. Numerous transition peaks in heavier elements can cause peak overlap, as can the increased peak width of higher energy transitions. Detection limits of AES include 0.1 – 1% of a monolayer, $10^{-16} - 10^{-15}$ g of material, and $10^{12} - 10^{13}$ atoms/cm^2. Another limitation is sample destruction. Although focusing of the electron beam can improve resolution; the high-energy electrons can destroy the sample. To limit destruction, beam current densities of greater than 1 mA/cm^2 should be used. Furthermore, charging of the electron beam on insulating samples can deteriorate the sample and result in high-energy peak shifts or the appearance of a large peak.

4.6.3 Secondary-Ion Mass Spectrometry (SIMS)

The Ion-Beam detecting techniques are classified as: secondary-ion mass spectrometry (SIMS), ion-scattering spectroscopy, Rutherford backscattering spectroscopy, and laser–microprobe mass spectrometry. The origins of SIMS can be traced back to the early part of the last century and the discharge tube experiments of J.J. Thomson, who was the first person to investigate the effects of positive ions on a metal plate. The first SIMS analytical instruments were built in the late 1940s by Herzog and co-workers, and SIMS was used in the 1960s to analyze the spatial and isotopic distribution of all the elements from hydrogen to uranium present in samples of moon rock brought back to earth.

In a SIMS technique, the sample is bombarded with a beam of charged particles with energies in the 1-25 keV range. These incoming particles are called primary ions. The primary ions deposit energy into the surface layers. Around the impact site and to the depth of ~3 nm, many bonds are broken and there is much random displacement and movement of atoms. This region is called the collision cascade. Collisions which result in translational energy being directed back to the surface layer lead to ejection of material. This process is termed sputtering of the surface (**Fig. 4.36**). Thus SIMS is based upon the sputtering of atomic layers from the surface of a sample by a primary ion beam.

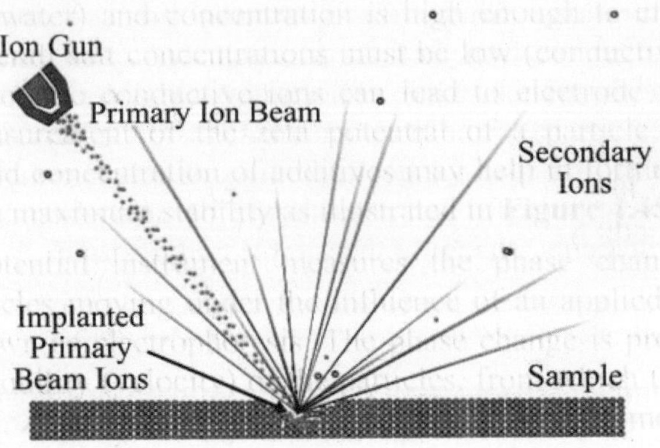

Fig. 4.36. Process of secondary ion sputtering

Instrumentation

Basic SIMS instrument, shown in **Figure 4.37,** have following components:

1. A primary beam source (usually $O2^+$, O^-, Cs^+, Ar^+, Ga^+ or neutrals) to supply the bombarding species.
2. A target or sample that must be solid and stable in a vacuum.
3. A method of collecting the ejected secondary ions.
4. A mass analyzer to isolate the ion of interest (quadrupole, magnetic sector, double focusing magnetic sector or time of flight).
5. An ion detection system to record the magnitude of the secondary ion signal (photographic plate, Faraday cup, electron multiplier or a CCD camera, and image plate).

The primary ion impact triggers a cascade of atomic collisions leading to atom, molecule fragments and ion ejection. The secondary ions are characteristic of the composition of the analysed area. They are separated according to their mass, and an image containing quantitative information is formed for a selected mass. SIMS reveals elemental and isotopic surface composition. It is one of the most powerful and versatile analytical techniques for semiconductor characterization with detection limits for some elements in the 10^{14} to 10^{15} cm^{-3} range if there is very little background interference signal. Lateral resolution of conventional SIMS is typically 100 μm but can be as small as 0.5 μm with depth resolution of 5 to 10 nm. Cesium (for negatively charged secondary ions) and oxygen (for positively charged secondary ions) are commonly used excitation ions. Other non-reactive primary ions like Ga or Bi, widely used in liquid-metal ion sources in focused ion beam systems with lateral resolutions below 100 nm, yield 100-1000 times lower signal intensities for elements.

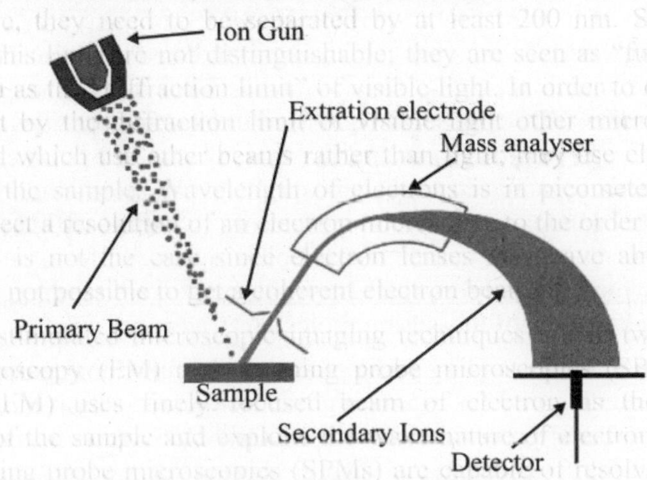

Fig. 4.37. Basic components of secondary ion mass spectrometer

Secondary-ion mass spectrometry (SIMS) is the highly developed of the mass spectrometric surface methods, useful for determining both the atomic and the molecular composition of solid surfaces. There are basically three variations of the SIMS experiment: **Static SIMS** used for elemental analysis of sub–monolayers on surfaces. In this mode the surface integrity is maintained during the time scale of the experiment; **Dynamic SIMS** used to obtain compositional information as a function of depth below the surface which takes advantage of the destructive nature of SIMS to obtain information on various layers of materials, and **Imaging SIMS**, also called scanning SIMS, is used to provide spatial images of surfaces.

Applications

Ion beams can be used in a variety of ways to obtain information about the nature of nanoparticles. One of the primary uses of TOF-SIMS is to extract molecular information about the functional groups and, possibly, molecular orientation of molecular coatings on particles surfaces. Surface selectivity arises from the depth of the region where the sputtered ions arise. Use of cluster ion beams has enhanced the surface sensitivity and, for some materials, minimized the region of damage, allowing sputter depth profiles of organic layers. To extract surface molecular information, SIMS is used in a "static" mode that involves a low density and low total dose of ions, so the surface damage and alteration are minimized. Both atomic and molecular secondary ions are used to extract the surface information.

Like the electron spectroscopy, SIMS is useful for obtaining information about surface layers, functional groups added to the surface, and contamination. Two differe-nces between the electron spectroscopy and SIMS are: (1) the high sensitivity (HS) of time-of-flight SIMS (TOF-SIMS) to many trace elements and functional groups and (2) the changes (damage) induced to the surface due to ion sputtering. The functional group sensitivity has been usefully applied in many ways. For example, TOF-SIMS has been used to examine peptides conjugated to Au nanoparticles as part of a protein kinase assay and to examine a multilayer, plasma-produced organic coating deposited on alumina nanoparticles. For relatively large nanoparticles produced during welding, SIMS with sputter profiling has been used to examine the complex layers that form on the particles. Unlike XPS, which often infers chemistry indirectly from binding energy shifts, SIMS has the advantage of measuring molecular fragments directly.

In addition to understanding surface coatings or functional groups, SIMS has proven to be equally good for examining the basic composition of nanoparticles, both as they are being processed inside the TOF-SIMS vacuum system and after they have been introduced into the vacuum system for analysis. *In situ* thermo-TOF-SIMS was used to examine the thermal decomposition of zinc acetate dehydrate during nanoparticle formation within a SIMS system. SIMS, in combination with TEM and *in-situ* optical transmission spectroscopy, has been used to study the composition and plasmon resonance of unique ZrN nanopart-icles produced by laser ablation/evaporation and adiabatic expansion from zirconium nitride powder targets. TOF-SIMS has also been used to characterize the composition and oxidation state of spark-generated nanoparticles made from pairs of Ir-Ir and Ir-C electrodes. From detailed analysis of the SIMS data, it was also possible to obtain information about particle size.

Because SIMS depends on the sputtering process, it inherently damages a nanostructured material under study, and this must be taken into consideration. Although many nanomaterials and nanoparticles have been successfully examined using SIMS, it is now established that nanoparticles sputter at different rates than flat surfaces or films. Recent work shows that in several circumstances, nanoparticles melt or evaporate after direct or grazing impact from many types of primary ions. Thus, it has been demonstrated that SIMS can be a useful and important tool for characterization of nanomaterials, but appropriate conside-ration should be given to potential implications of enhanced sputter rates or sample melting.

4.7 SCATTERING TECHNIQUES

Scattering is a general physical process where some forms of radiation, such as electro magnetic radiation, sound, or moving particles, are forced to deviate from a straight trajectory by one or more paths due to localized non-uniformities in the medium through which they pass. Scattering results from the interaction of the electrons in the molecules with oscillating electric field of radiation. Thus, a dipole is induced in the molecules, which oscillate with the electric field. Since an oscillating dipole is a source of electromagnetic radiation, the molecules emit light, called scattered light. The difference in energy between the incident and scattered radiation determines the nature of scattering whether the phenomena is elastic (Rayleigh scattering), quasi-elastic scattering (Rayleigh-Brillouin scattering) or inelastic scattering (Raman scattering).

Table 4.2 Radiations used for Scattering based techniques

Type of radiation (wavelength range)	Interaction	Scattering methods	Sample thickness	Limitations
X-rays (0.1-5 Å)	Electron density	Small angle X-ray scattering and Wide angle X-ray scattering	< 1 mm	Absorption
Neutrons (1 Å -15 Å)	Nuclei	Small-angle neutron scattering and wide-angle neutron scattering	1-2 mm	Low fluxes
Electrons (0.1 Å)	Electron cloud	Low-energy electron diffraction	100 μm	Low penetration
Laser light (1μm)	Polarizability (Refractive Index)	Dynamic light scattering	1-5 mm	Scattering from dust

The technique of light scattering has an advantage of being non-invasive probe method provided the intensity is not high enough to ionize the system. To probe a wide range of length scale, different scattering mechanisms are used, *i.e.* typical length scale accessible in light scattering is up to 1000 nm whereas in the case of X-ray scattering it is up to 100 nm. For probe length scale smaller than this, one has to go for electron scattering. As such, the scattering techniques are powerful characterization tools in biology, polymer and materials science, biochemical sciences and in colloidal domain. Scattering techniques provide the most obvious methods for obtaining quantitative information on size, shape and structure of smaller particles, since they are based on interactions between incident radiations (*e.g.*, light, X-ray or neutrons) and particles. **Table 4.2** list out different kind of radiation used in various scattering techniques.

4.7.1 X-ray diffraction technique

X-rays are short wavelength (~10^{-5} A° to 100 A°) electromagnetic radiations produced by the deceleration of high energy electrons of electronic transitions involving electrons in the inner orbitals of atoms. The energy of emitted X-rays is equal to the difference between the binding energies of the electrons involved in the transition. Since X-ray wave-lengths are in the same order as the sizes of nano-structures, X-ray diffraction is an important characterization tool. X-ray diffractograms hold a wealth of structural information about the macro-micro and nano-materials such as: (i) the crystallite size, (ii) size distribution, (iii) the defect structure/distribution, (iv) texture, (v) micro and macro strain, etc. X-ray diffraction (XRD) provides complimentary information about the nano-structure by showing the average coherence length as a function of direction. X-ray diffraction results are averaged over a large volume giving an overall view of the nano-structure. X-ray crystallography, X-ray powder diffraction (XRD), Small angle X-ray scattering are all based on elastic scattering. The X-ray powder diffraction is most widely used for the identification of unknown crystalline materials (*e.g.* minerals, inorganic compounds). Determination of unknown solids is critical to studies in geology, environmental science, material science, engineering and biology.

Wide angle X-ray diffraction (XRD)

The fundamental equation on which XRD is based is Bragg's law, which links lattice spacing with the observed angles of constructive interferences when X-ray scatter from a crystalline material. For nanomaterials, measurements are made on powders containing a very large number of randomly oriented particles. Thus diffraction rings or halos are seen at various angles with respect to the incoming beam, each originating from a subset of particles oriented correctly to satisfy the Bragg equation, and the corresponding technique is known as the powder method, or Debye-Scherrer method. Analysis of the position and intensity of the peaks in the powder diffraction pattern allows the identification and quantification of the phase composition of the sample. According to Bragg's Law, when a narrow beam of radiation of wavelength λ, strikes the crystal surface at incident angle θ; scattering occurs as a consequence of interaction of the radiation with regular array of atoms (scatterers). When distance, 2d $\sin\theta$ is equal to $n\lambda$, where n is an integer, and d is interplanar distance of crystal, crystal will appear to reflect the X-radiation. The conditions for constructive interference of the beam at angle θ are (Eq. 4.8)

$$n\lambda = 2d\sin\theta \tag{4.8}$$

Thus among scattering, diffraction takes place when the distance between the scattering centers are of the same order of magnitude as the wavelength ($n\ \lambda$) of the radiation. Bragg's equation is of fundamental importance. The X-rays appear to be reflected from the crystal only if the angle of incidence satisfies the condition that $\sin\theta = n\lambda / 2d$. In that case, part of the incoming beam is deflected by an angle 2θ, producing a *reflection* spot in the diffraction pattern. At all other angles, destructive interferences occur. X-rays are used to produce the diffraction pattern because their wavelength λ is typically the same order of magnitude (1–100 angstroms) as the spacing d between planes in the crystal.

Another fundamental equation applied to XRD analysis is the Debye-Scherrer relation, which relates the broadening of XRD peaks to the (average) finite size of the individual diffracting domains. This finite size can correspond with the crystals within a solid, but it can also corresponds with the individual (monocrystalline) particles in a agglomerated or aggregated nanomaterials. The Scherrer equation can be written as (Eqn. 4.9):

$$\tau = \frac{K\lambda}{\beta \cos \theta} \tag{4.9}$$

where, τ is the mean size of the ordered (crystalline) domains, which may be smaller or equal to the grain size; K is a dimensionless shape factor, with a value close to unity. The shape factor has a typical value of about 0.9, but varies with the actual shape of the crystallite; λ is the X-ray wavelength; β is the line broadening at half the maximum intensity (FWHM), θ is the Bragg angle.

The Scherrer equation provides a powerful and relatively simple way of determining the average particle size of a nanomaterial, but not the size distribution. The technique is more reliable for the particles in the 1-100 nm size range since the peak broadening is much more pronounced for smaller particles. The method is limited to crystalline particles. For amorphous or very poorly crystalline particles, Fourier Transform technique can be used in principle to determine the average particle size, however they may not be very accurate.

Instrumentation

Many XRD systems are available commercially, most offering the powder diffraction technique and being provided with software for peak shape analysis and deduction of average crystallite size. X-ray diffractometer consist of three basic elements: an X-ray tube, a sample holder, and an X-ray detector. X-rays are generated in a cathode ray tube by heating a filament to produce electrons, accelerating the electrons toward a target by applying a voltage, and bombarding the target material with electrons. When electrons have sufficient energy to dislodge inner shell electrons of the target material, characteristic X-ray spectra are produced. These spectra consist of several components, the most common being K_{α} and K_{β}. The specific wavelengths are characteristic of the target material (Cu, Fe, Mo, Cr). Filtering, by foils or crystal monochromator, is required to produce monochromatic X-rays needed for diffraction. Copper is the most common target material for single-crystal diffraction, with CuK_{α} radiation = 1.5418Å. These X-rays are collimated and directed onto the sample. As the sample and detector are rotated, the intensity of the reflected X-rays is recorded. When the geometry of the incident X-rays impinging the sample satisfies the Bragg Equation, constructive interference occurs and a peak in intensity occurs. A detector records and processes this X-ray signal and converts the signal to a count rate which is then output to a device such as a printer or computer monitor.

The geometry of an X-ray diffractometer is such that the sample rotates in the path of the collimated X-ray beam at an angle θ while the X-ray detector is mounted on an arm to collect the diffracted X-rays and rotates at an angle of 2θ.

The instrument used to maintain the angle and rotate the sample is termed a goniometer. For typical powder patterns, data is collected at 2θ from ~5° to 70°, angles that are preset in the X-ray scan. **Figure 4.38** shows typical features of a basic XRD Spectrometer.

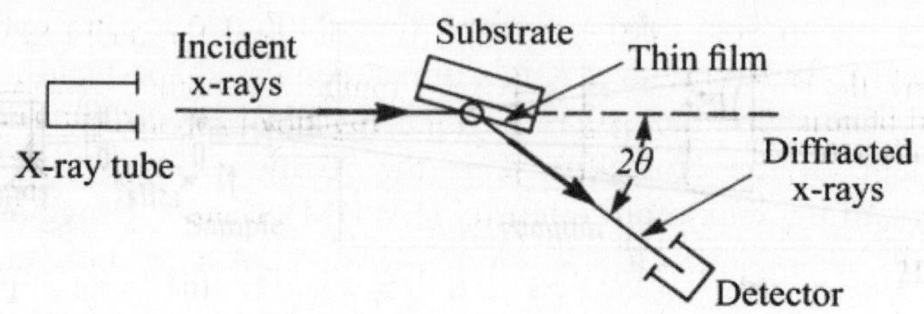

Fig. 4.38. Typical features of a basic XRD spectrometer

Figure 4.39 shows a typical powder diffractogram of NaCl, peak positions occur where the X-ray beam has been diffracted by the crystal lattice. The unique set of d-spacing derived from this pattern can be used to 'fingerprint' the mineral. Lattice strain on diffraction peak position and width

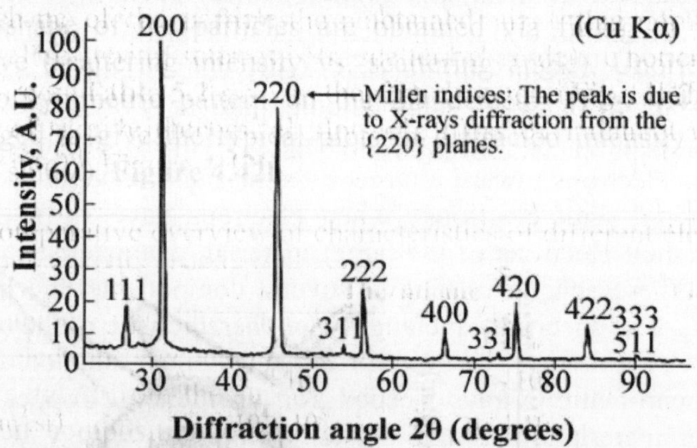

Fig. 4.39. XRD pattern of NaCl powder; b. crystalline and amorphous phases of poly (ethylene oxide)

X-ray powder diffraction method is unique in that it is the only analytical method that is capable of providing qualitative and quantitative information about the compounds presents in a solid sample. X-ray diffraction pattern is unique for each crystalline substance and chemical identity is assumed by matching the diffraction pattern of an authentic sample. **Figure 4.40** shows X-ray diffraction patterns of nanostructured metal oxide film and carbon nano materials.

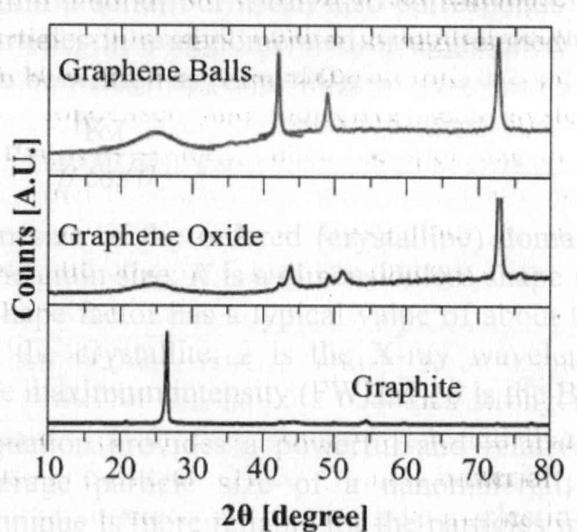

Fig. 4.40. X-ray diffraction patterns of different carbon nanomaterials

Small-angle X-ray scattering (SAXS)

Small-angle X-ray scattering (SAXS) is a technique where the elastic scattering of X-rays (wavelength 0.1 ... 0.2 nm) by a sample which has in homogeneities in the nm-range, is recorded at very low angles (typically 0.1 - 10°). This angular range contains information about the shape and size of macromolecules, characteristic distances of partially ordered materials, pore sizes, and other data. To satisfy Bragg's Law, θ must change as *d* changes *e.g.*, θ decreases as *d* increases thus making SAXS to be capable of delivering structural information of macromolecules between 5 and 25 nm, of repeat distances in partially ordered systems of up to 150 nm. **SAXS** is ideally suited for the study of ordered systems with large repeat distances, such as liquid crystalline mesophases, complex fluids and nano- to mesoporous materials. In addition, the elastic scattering of X-rays that can be observed at small and very small angles gives information about the shape and size of macromolecules.

Instrumentation

The SAXS instrument consists of an X-ray source, a pinhole or a slit collimation system, a sample environment, a scattering vessel and a 2D detector. The X-ray source is either a laboratory source like a Kratky or a rotating anode (whereby X-rays are produced when charged particles hit a Cu or an Mo target), or a synchrotron facility (whereby X-rays are produced by accelerated particles moving in a magnetic field). Synchrotron light provides a higher X-ray flux. The pinhole collimation is preferred but is characterized by low flux-on-sample. Slit collimation yields higher fluxes but produces slit-smeared data. For example, most laboratory X-ray sources use slit collimation whereas most synchrotron sources use pinhole collimation. The Cu-K α line is characterized by a source wavelength of 1.54 Å. Schematic of a typical SAXS setup is shown in **Figure 4.41**. SAXS measures the size of particle based on scattering of X-rays at the particle surface.

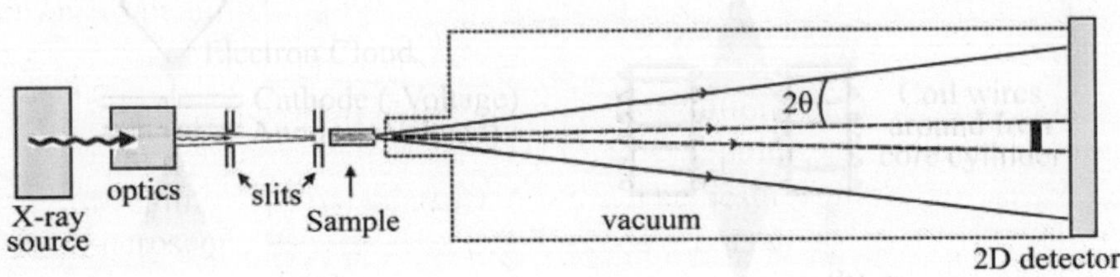

Fig. 4.41. Schematic of a typical SAXS setup

Light passing through a medium that is not completely homogenous is forced to deviate from the straight line. Scattering angle, *i.e.* the degree of deviation from straight line, depends on the wave length of light and the size of the particle on which it is scattered. X-rays which have wavelengths in the range of 0.1 nm and 1 nm are scattered by particles in the range between 1 and 100nm. The intensity of scattered radiation is measured in dependence of the scattering angle.

Size and shape of the particles are obtained via fitting of the measured scattering curve (scattering intensity vs. scattering angle). Unoriented samples yield a centro symmetric pattern on the 2D detector (**Fig. 4.42a**), which is radially averaged to give the typical plots of diffracted intensity vs. scattering vector q (= $4\pi \sin\theta/\lambda$), **Figure 4.42b**.

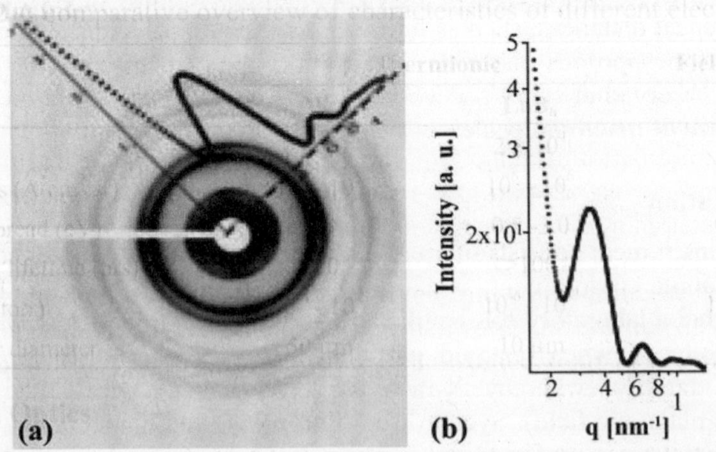

Fig. 4.42. (a) Centrosymmetric pattern on the 2D detector, (b) typical plots of diffracted intensity vs. scattering vector

Different algorithms are used for calculations of particle size distribution from raw SAXS data. For relatively monodispersed samples, oscillations are observed in the scattering curve. The periodicity is directly related to the mean particle diameter. By fitting the curve, number, volume and intensity, weighted mean diameter can be obtained.

Application of SAXS

SAXS is used for the determination of the microscale or nanoscale structure of particle systems in terms of such parameters as averaged particle sizes, shapes, distribution, and surface-to-volume ratio. SAXS can be applied to any two phase system where the phases have different density. In such a system, SAXS measures a phase with the smaller volume fraction. This means that in suspension but not to powders, as the volume fraction has to be below a few percent to avoid particle-particle interactions.

Not only particles, but also the structures of ordered systems like lamellae, and fractal-like materials can be studied. The method is accurate, non-destructive and usually requires only a minimum of sample preparation. Applications are very broad and include colloids of all types, metals, cement, oil, polymers, plastics, proteins, foods and pharmaceuticals and can be found in research as well as in quality control. In the case of biological macromolecules such as proteins, the advantage of SAXS versus crystallography is that a crystalline sample is not needed.

4.7.2 Dynamic Light Scattering

Dynamic Light Scattering (DLS) measures the hydrodynamic radius (D_h) of a particle which determines how fast a particle moves in a suspension through Brownian motion. To measure D_h, DLS uses a laser beam, which, when passes through a suspension, is scattered by the nanoparticles. The Brownian motion of the particles causes fluctuations in the intensity of scattered light around a mean value. When laser light hits small particles (below 250 nm), it scatters in all directions (Rayleigh scattering). In case of a laser light source the time dependent fluctuations in the scattering intensity is observed. This is due to the fact that the small molecules undergo a diffusive Brownian motion when in solution and the distance between the nanoparticles in the solution is constantly changing with time. This scattered light then undergoes either constructive or destructive interference by the surrounding particles, and within this intensity fluctuation, information is stored about the time scale of movement of the particles.

In most DLS instruments, the autocorrelation function of the fluctuations is recorded as a function of time. Size of the particles can be obtained, for example, by fitting the autocorrelation function to an exponential decay function that delivers an average diffusion coefficient for the particles. This diffusion coefficient (D_t) can be used to calculate an average radii of the particles via the

Stoke-Einstein relationship ($D_h = k_B T / 6 \pi\eta D_t$) where, T is the sample temperature in absolute units, k_B is the Boltzmann constant, η is the sample viscosity. DLS can thus be used to determine the size distribution profile of small particles in suspension including nanoparticles.

Dynamic Light Scattering (DLS) is an important tool for characterizing the size of nanoparticles in solution. Basic components and optics for a DLS experiment are shown schematically in **Fig. 4.43.** DLS measures the light scattered from a laser that passes through a colloidal solution and by analyzing the modulation of the scattered light intensity as a function of time, the hydrodynamic size of particles and particle agglomerates can be determined. Larger particles will diffuse slower than smaller particles and the DLS instrument measures the time dependence of the scattered light to generate a correlation function that can be mathematically linked to a particle size. DLS is a valuable tool for determining and measuring the agglomeration state of nanoparticles as a function of time or suspending solution. When DLS sizing data is compared to Transmission Electron Microscopy images, the aggregation state of the particles can be determined. In a non-agglomerated suspension, the DLS measured diameter will be similar or slightly larger than the TEM size. If the particles are agglomerated, the DLS measurement is often much larger than the TEM size and can have a high polydispersity index (large variability in the particle size).

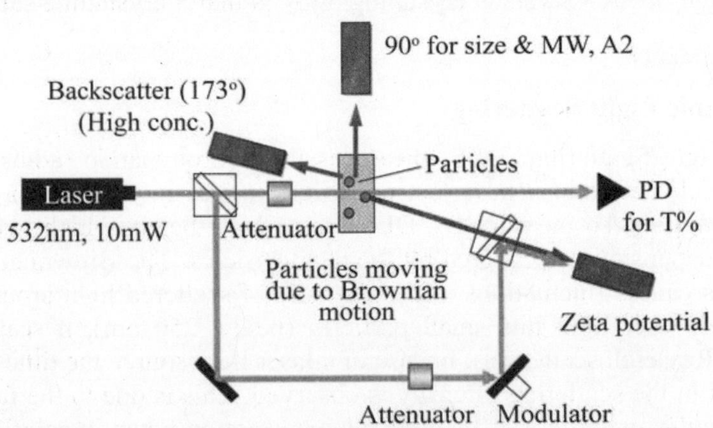

Fig. 4.43. Schematic representation of DLS optics

Applications

DLS is used to characterize size of various particles including proteins, polymers, micelles, carbohydrates, and nanoparticles. If the system is monodisperse, the mean effective diameter of the particles can be determined. This measurement depends on the size of the particle core, the size of surface structures, particle concentration, and the type of ions in the medium. Sample preparation is critical in DLS experiment. Sample is prepared either by filtration or centrifugation to remove dust and artifacts from the solution.

Since DLS essentially measures fluctuations in scattered light intensity due to diffusing particles, the diffusion coefficient of the particles can be determined. DLS software of commercial instruments typically displays the particle population at different diameters. If the system is monodisperse, there should only be one population, whereas a polydisperse system would show multiple particle populations. If there is more than one size population present in a sample then CONTIN analysis must be applied. For more than two populations CONTIN analysis at several scattering angles is required.

Stability studies can be done conveniently using DLS. Periodical DLS measurements of a sample can show whether the particles aggregate over time by seeing whether the hydrodynamic radius of the particle increases. If particles aggregate, there will be a larger population of particles with a larger radius. Additionally, in certain DLS machines, stability depending on temperature can be analyzed by controlling the temperature *in situ*. DLS is mostly used for nanoparticle size and its distribution.

Limitations of DLS

DLS, in its standardized form, does not produce particle size distribution, but a light scattering intensity-weighted average value. In addition, a well-known difficulty of DLS measurement is that the results are strongly biased in the presence of a small fraction of large particles. This is due to the fact that the intensity of the scattered light is inversely proportional to the sixth power of the radius of the nanoparticle. Thus a 50 nm particle will scatter 10^6 as much light as a 5nm particle as a result average particle size value determined by DLS are misleading. Hence, DLS does not distinguish between constituent particles and aggregates/ agglomerates. Moreover DLS is performed in liquids, which further limits its use for analysis of particles not dispersed in liquids.

4.8 ZETA POTENTIAL ANALYSIS

Zeta potential is a scientific term for electro-kinetic potential in colloidal systems. In the colloidal chemistry literature, it is usually denoted using the Greek letter zeta (ζ), hence Zeta potential. From a theoretical viewpoint, the zeta potential is the electric potential in the interfacial double layer (DL) at the location of the slipping plane versus a point in the bulk fluid away from the interface (**Fig. 4.44**). In other words, zeta potential is the potential difference between the dispersion medium and the stationary layer of fluid attached to the dispersed particle. A value of 25 mV (positive or negative) can be taken as the arbitrary value that separates low-charged surfaces from highly charged surfaces. The significance of zeta potential is that its value can be related to the stability of colloidal dispersions. The zeta potential indicates the degree of repulsion between adjacent, similarly charged particles in dispersion. For molecules and particles that are small enough, a high zeta potential will confer stability, *i.e.*, the solution or dispersion will resist aggregation. When the potential is low, attraction exceeds repulsion and the dispersion will break and flocculate. So, colloids with high zeta potential (negative or positive) are electrically stabilized while colloids with low zeta potentials tend to coagulate or flocculate as outlined in the **Table 4.3**.

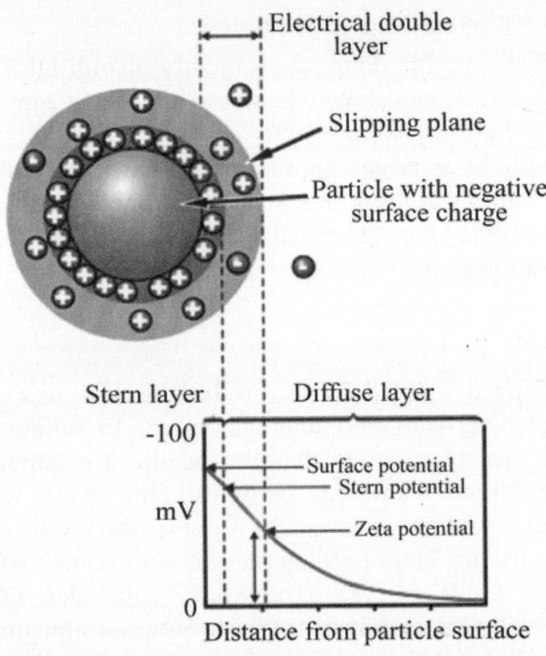

Fig. 4.44. Electric double layer around a particle showing Stern Layer

Table 4.3. Zeta potential vs. stability of colloids

Zeta potential (mV)	Stability behavior of the colloid
0 to ± 5	Rapid coagulation or flocculation
±10 to ± 30	Incipient instability
±30 to ± 40	Moderate stability
±40 to ± 60	Good stability
more than ± 61	Excellent stability

Zeta potential is widely used for quantification of the magnitude of the electrical charge at the double layer. However, zeta potential is not equal to the

Stern potential or electric surface potential in the double layer. Nevertheless, zeta potential is often the only available path for characterization of double-layer properties. Zeta potential should not be confused with electrode potential or electrochemical potential. Electrochemical reactions are generally not involved in the development of zeta potential. In an ionic solution, the surface of a charged particle is firmly bound to opposite charged ions, forming a thin liquid layer named the Stern layer, which is encompassed by an outer diffuse layer consisting of loosely associated ions. These two layers compose the so-called electrical double layer. The tangential motion is driven by an external force or Brownian motion of the charged particle. The movement of the charged particle shears ions migrating with the charge particle in the diffuse layer from ions staying with the bulk dispersant outside the layer. The electric potential on the shear surface is the zeta potential.

Measurement of Zeta potential

Zeta potential, the electric potential at the boundary of the double layer is an important tool for understanding the state of the nanoparticle surface and predicting the long term stability of the nanoparticle. Zeta potential analysis is quite useful in understanding the surface modification of the nanoparticles.

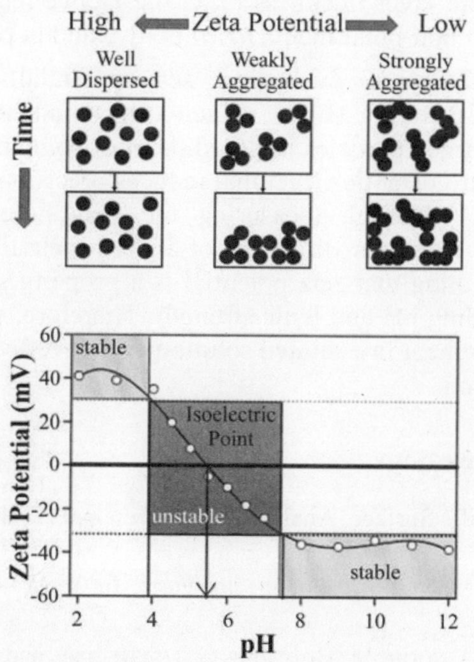

Fig. 4.45. Effect of zeta potential on suspension properties

Zeta potential is sensitive to environmental changes including pH, ionic strength. Therefore, care must be exercised in preparing sample for Zeta potential measurement. Sample must be monodisperse in particulate free, polar solvent (*e.g.* high purity water) and concentration is high enough to effectively scatter 633 nm light. Overall salt concentrations must be low (conductivities <1mS/cm) since movement of the conductive ions can lead to electrode polarization and degradation. Measurement of the zeta potential of a particle dispersion as a function of pH and concentration of additives may help in formulating condition for a product with maximum stability as illustrated in **Figure 4.45**.

The Zeta potential instrument measures the phase change in the light scattered by particles moving under the influence of an applied electric field, a phenomenon known as electrophoresis. The phase change is proportional to the electrophoretic mobility (velocity) of the particles, from which the zeta potential can be calculated. Zeta potential is related to electrophoretic mobility by Henry Equation (4.10):

$$U_E = \frac{2\varepsilon Z f(Ka)}{3\eta} \tag{4.10}$$

where, U_E = electrophoretic mobility; Z = Zeta potential, ε = dielectric constant, η = viscosity and $f(Ka)$ is the Henrys function; Ka represents the ratio of particle radius (a) to the double layer thickness ($1/K$, the Debye length). Unit of K are reciprocal of length. In non-polar media $f(Ka)$ is ~1.0 and in polar media ~ 1.5.

Among the methods of evaluating zeta potential, the technique of electrophoretic light scattering (ELS), which can simultaneously measure the velocities of many charged particles in liquid, is most commonly used. However, it still suffers the electro osmotic effect that reduces precision and reproducibility of the measurement. Although measuring the Zeta potential of suspended particles after dilution reduces difficulty of light penetration into the sample solution, it is worth noting that zeta potential is a property sensitive to environmental changes including pH and ionic strength. Therefore, a precise, repeatable Zeta potential measurement in a diluted solution cannot reflect the true value in a concentrated suspension.

Suggested further readings

1. J.C. Vickerman Ed., Surface Analysis: The Principal Techniques, 2nd Edition, Wiley-VCH, Weinheim, 2011.
2. Jamal Uddin Ed., Macro to Nano Spectroscopy, In Tech Janeza Trdine, Croatia, 2012.
3. M.S. Dresselhaus, A. Jorio, M. Hofmann, G. Dresselhaus and R. Saito, Perspectives on carbon nanotubes and graphene Raman Spectroscopy, *Nano Lett.*, 10 (2010) 751–758.

4. I. T. Lucas, A. S. McLeod, J. S. Syzdek et al., Near-Field Spectroscopy and Imaging of Single Li_xFePO_4 microcrystals, *Nano Lett.*, 15 (2015) 1–7.
5. R. D. Rodriguez, E. Sheremet, T. Deckert-Gaudig, et al. Surface- and tip-enhanced Raman Spectroscopy reveals spin-waves in iron oxide nanoparticles, *Nanoscale*, **7** (2015) 9545-9551.
6. P. Lin et al. Techniques for Physicochemical Characterization of nanomaterials, Biotechnol. Adv., 32 (2014) 711-726.
7. P. Whitefield and L. Mitchell, X-Ray Diffraction Analysis of Nanoparticles: Recent Developments, Potential problems and Some Solutions, Int. J. Nanoscience, 3 (2004) 757- 763.

REVIEW QUESTIONS

Q1. Give range of electromagnetic radiations useful for various spectroscopic techniques.

Q2. What is Beer-Lamberts law and how it is applied to estimate the concentration.

Q3. What is Mie theory and how it is used in characterization of nano particles?

Q4. What do understand by surface plasmon resonance (SPR)? How SPR affects the absorption spectra of nano particles?

Q5. How UV-Vis absorption spectra is useful in finding the size and size distribution of semiconducting nanoparticles?

Q6. Enumerate basic difference between UV-Vis absorption spectroscopy and Infrared Absorption Spectroscopy.

Q7. What is Fourier Transform Infrared (FTIR) Spectroscopy? Give advantages of FTIR spectroscopy over dispersive IR spectroscopy.

Q8. Describe working principle of FTIR spectrophotometer. Explain with schematic diagram.

Q9. How FTIR spectroscopy is useful in differentiating different forms of nano iron oxides?

Q10. How will you distinguish between the functionalized and non-functionalized carbon nanotubes using IR spectroscopy?

Q11. What is Raman effect? How Raman scattering differ from Rayleigh scattering.

Q12. Show application of Raman spectroscopy in characterizing different carbon nanomaterials.

Q13. Discuss the basic principle and unique capabilities of Enhanced Raman spectroscopy?

Q14. Enumerate the distinguishing feature of Surface enhanced Raman spectroscopy (SERS) and Tip enhanced Raman Spectroscopy (TERS).

Q15. What do you understand by photoelectric effect and explain the spectroscopic technique based on this effect.

Q16. How XPS is useful is of qualitative surface analysis? Give a brief account of Instrumentation of XPS with the details of source and detectors.

Q17. How XPS is useful in characterization of nanomaterials? Support your answer with suitable examples.

Q18. What is Auger process? How does, Auger spectroscopy differs from X-ray photo electron spectroscopy?

Q19. Give a brief account of applications and limitations of AES technique.

Q20. Discuss principle and applications of SIMS for characterization of nanomaterials.

Q21. Mention important features of different scattering techniques useful for characterization of nanomaterials.

Q22. Give a brief account of basic components of X-ray diffraction spectrometer. Differentiate between WAXS and SAXS in terms of principle and applications?

Q23. How XRD is used in estimating (a). size of nanoparticles; (b.) identification of phase of nano crystalline materials?

Q24. What do you understand by hydrodynamic radius? How Dynamic Light Scattering technique is useful in estimating the diameter of spherical nanoparticles?

Q25. What is Stern potential? Show usefulness of Zeta potential measurement in characterization of nanomaterials

■■■■

Characterization of Nanomaterials Part II: Microscopic Imaging Techniques

5

5.1 INTRODUCTION

Nanoparticles, one of the "building blocks" of nanoscience and nanotechnology, have been all around us, without our knowledge, since the origin of human history. The fundamentals of nanoscience and nanotechnology lies in the fact that properties of material changes drastically when their size is reduced to the nanoscale range. Thus, there is a significant focus on the tools and techniques with high spatial resolution and which provides atomic level manipulation of materials to give an understanding on the noval properties of nanomaterials. In fact, the credit to the success of nanoscience and nanotechnology largely goes to the development of a variety of advanced microscopy instruments that has provided the opportunity to see the objects at atomic scale, generate an understanding about their behavior and imagined the novel ways to manipulate nanomaterials. Nanotechnology really took off in mid 1980s with the invention of revolutionary instruments *viz.* Scanning Tunneling Microscope and Atomic Force Microscope with literally opened the doors to nano world. Compared to bulk materials, nanomaterials may exhibit changed or additional properties based on their increased surface to volume ratio and/or quantum physical effects. Characterization of physical, chemical and biological properties on a nanoscale level are essential prerequisite for commercialization, risk assessment of nanomaterials and to investigate how nanomaterials behave under different chemical and physical conditions, how they move and distribute in different environmental compartments like water, soil and air and how they interact with the biosphere and the human organism.

Nanotechnology has motivated an upsurge in research and development activities in this due to invention and innovation of sophisticated nano characterization techniques to allow a better control of morphology, size and dimensions of material in nano range. Nanostructured materials are subjected to nearly every type of chemical and physical analysis possible; of these, spectroscopic and scattering techniques have been discussed in Chapter 4. Present chapter gives a concise account of essential microscopic imaging techniques in characterizing the morphological properties of the nanostructured materials.

Imaging through Microscopy

Microscopy is the technical field of using a variety of microscopes to view very small samples and objects which otherwise cannot be seen with the unaided eye. The well-known branches of microscopy are optical, electron, and scanning probe microscopy. Optical and electron microscopy, known for long, involve the diffraction, reflection, or refraction of electromagnetic radiation/electron beams interacting with the specimen, and the subsequent collection of this scattered radiation or another signal in order to create an image. This process may be carried out by wide-field irradiation of the sample (*e.g.* standard light microscopy and transmission electron microscopy) or by scanning of a fine beam over the sample (*e.g.* confocal laser scanning microscopy and scanning electron microscopy). The development of electron microscopy revolutionized the field of biology and continue as an essential technique in the life and physical sciences. Scanning probe microscopy, on the other hand, involves the interaction of a scanning probe with the surface of an object of interest. The invention of scanning probe microscope is believed to be the main equipment facilitating the development of hundreds of new nanomaterials; in other words the opening for the field of nanoscience and nanotechnology.

Optical microscopy is the classical method for obtaining detailed information about physical nature of surfaces. The resolution of optical microscopy is limited by diffraction effects to the wavelength of light. An optical microscope uses visible light (*i.e.* an electromagnetic radiation) and a system of lenses to magnify the images of small samples. For this reason it is also called a light microscope. Optical microscopes are the oldest and simplest of the microscopes. 'Resolution', the ability to discriminate between two closely spaced objects, is an important aspect of any microscope. The resolution limit of an optical microscope is imposed by the wavelength of the visible light. As a general rule of thumb, the resolution is about half the wavelength used in the measurement. Visible light has wavelengths between 400 and 700 nm. The resolving power of an optical microscope is around 0.2 μm or 200 nm. Thus for two objects to be distinguishable, they need to be separated by at least 200 nm. Single objects smaller than this limit are not distinguishable; they are seen as "fuzzy objects". This is known as the "diffraction limit" of visible light. In order to overcome the limitations set by the diffraction limit of visible light other microscopes have been designed which use other beams rather than light; they use electron beams to illuminate the sample. Wavelength of electrons is in picometer range, it is natural to expect a resolution of an electron microscope to the order of picometer. However this is not the case since electron lenses also have aberrations and moreover it is not possible to get a coherent electron beam.

Electron stimulated microscopic imaging techniques fall in two categories: electron microscopy (EM) and scanning probe microscopies (SPM). Electron microscope (EM) uses finely focused beam of electron as the source for illumination of the sample and exploits the wave nature of electron to obtain an image. Scanning probe microscopies (SPMs) are capable of resolving details of surface down to atomic level and reveal 3-dimensional details unlike optical and electron microscopes

5.2 ELECTRON MICROSCOPY

Essentially a microscope consist of an illuminating source, a condenser lens to converge the beam on to the sample, an objective lens to magnify the image and a projector lens to project the image to be photographed or stored. Electron microscope, discovered by Knoll and Ruska in 1926, uses finely focused beam of electron as the source for illumination of sample and exploits the wave nature of electron to obtain an image. Electron microscopes have much greater resolving power than light microscopes that use electromagnetic radiation and can obtain much higher magnifications of up to 2 million times, while the best light microscopes are limited to magnifications of 2000 times. The maximum usable magnification is equal to the resolving power of the eye divided by the resolving power of the instrument. The resolving power of human naked eye is ~250 μm and therefore the meaningful minimum magnification with light microscope with resolving power as 250 nm is to the order of 1000. Theoretically resolving power of an electron microscope would be 0.0027 nm for electron energy of 50,000 V; reaching magnification of the order of 10^9. However this is not achieved due to aberration of the electromagnetic lenses and complex nature of electron-sample interaction.

Electron and light microscopes both have resolution limitations, imposed by the wavelength of the radiation they use. The greater resolution and higher magnification of the electron microscope is due to fact that wavelength of an electron (de Broglie wavelength), is much smaller than the wavelength of photon of visible light.

The de Broglie wave equation (Eq. 5.1) relates the velocity of electron (v) with its wavelength (λ)

$$\lambda = h / mv \qquad 5.1$$

Where. h is Plank's constant (6.625×10^{-34} N m s), and m is mass of an electron (9.109×10^{-31} Kg). Thus when an electron of charge, e (-1.602×10^{-19} C) and of mass m, passes through a potential difference of V volts, it acquires a kinetic energy as given by eq. 5.2 and thus velocity and wavelength of electron can be expressed as (Eqs. 5.3 and 5.4).

$$\tfrac{1}{2} mv^2 = e V \qquad 5.2$$

$$v = (2 eV / m)^{1/2} \qquad 5.3$$

$$\lambda = h/(2 m eV)^{1/2} \qquad 5.4$$

At high potential, when electron velocity approaches the speed of light, c (2.998×10^8 m/s), the relativistic corrections to mass is applied. Accordingly the estimated velocity (v) and the wavelengths ((λ) of electrons achieved at different acceleration voltage are included in **Table 5.1.**

Table 5.1. Electron wavelength and velocity at different acceleration voltage

V, volts	**λ (nm)**	**v ($\times 10^{10}$ cm/s)**
50,000	0.0055	1.283
100,000	0.0039	1.699
1,000,000	0.0012	2.917

The resolution (d) and the resolving power (R) of an instrument is expressed as (Eqs. 5.5, 5.6)

$$d = 0.61\, \lambda / n.\sin \theta \qquad 5.5$$

$$\mathrm{R} = \lambda / 2\, \mathrm{NA} \qquad 5.6$$

Where, n is the refractive index of the medium and θ is the semi-angular aperture of the lens. The quantity ($n.\sin \theta$) is the numerical aperture (NA) of the lens. The theoretical limit of resolving power of an instrument is determined by the wavelengths of the electron beam used and the numerical aperture of the system. The numerical aperture, NA is engraved on each objective and condenser lens system and is a measure of electron gathering ability of objective and electron providing ability of the condenser.

Conceptually, electron microscopes are similar to an optical microscope in the sense that both microscopes use a radiation to visualize a sample: photons in the case of an optical microscope, and electrons (*i.e.* particles) in the case of electron microscopes. Electrons are negatively charged particles within the atom. Unlike light photons, electrons cannot be focused by glass lenses, but electromagnets are used as lenses for guiding and concentrating electrons on to the specimen. Electrons are one type of ionizing radiation, capable of removing one of the tightly bound inner-shell electrons from the attractive field of the nucleus and produces many of the secondary signals from the specimen as some of the electrons are transmitted through the sample, get backscattered or emits secondary electrons which are used in electron microscopy as schematically shown in **Fig. 5.1.**

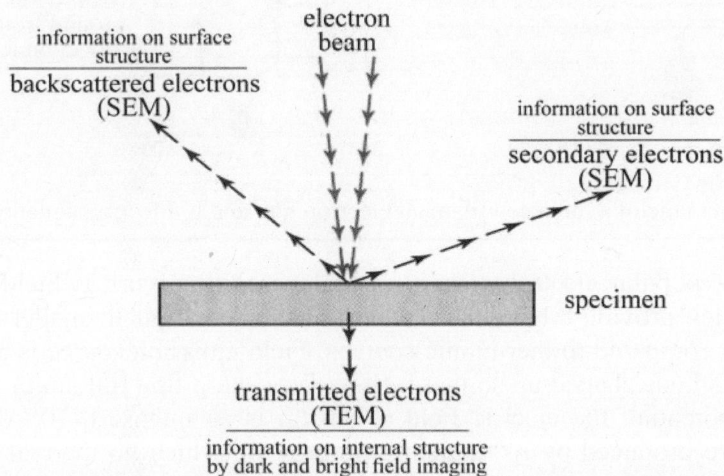

Fig. 5.1. Electron-sample interaction and resulting electron microscopic techniques

Electron Gun: source of electrons

Electron gun provides a stable beam of electrons used in electron microscopy. The most common form of an electron gun is a thermionic emitter, wherein the work function of the metal is overcome by the surface temperature of the filament. The typical thermionic electron gun consists of three parts, a tungsten filament, a grid cap (Wehnelt cap) surrounding the filament and the anode. The most common electron gun **(Fig. 5.2)** consist of a heated (~2000-2700ºK) tungsten filament of 100 μm diameter and bent into a shape of hairpin with a V-shaped tip. The cathodic filament is maintained at a potential of 1-50 kV with respect to anode contained in the gun. The grid cap is biased negatively with respect to the filament. The effect of electric field in the gun is to cause the emitted electrons to converge on a tiny spot called the crossover having a diameter d_0. The confined beam of electrons is accelerated to the anode held at ground potential, thus creating an electron beam that streams down the microscope column. The electrons are accelerated to energy between 1 and 30 keV. Cathodes constructed in the form of lanthanum hexaboride (LaB_6) rods are also used in electron guns when greater brightness is desired. LaB_6 based thermionic source is expensive and requires a better vacuum system to prevent oxide formation, which cause the efficiency of the source to deteriorate rapidly. The filament gradually evaporates at high operative temperature due to high operative current and reduces the filament life.

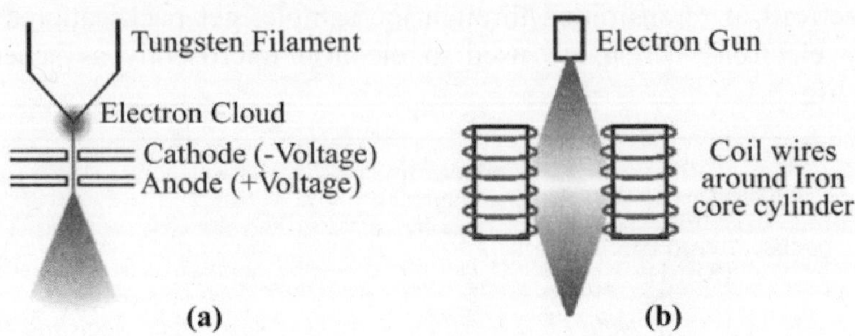

Fig. 5.2. Schematic of a. tungsten filament electron gun and b. Electromagnetic lens

Another popular electron source for high-resolution work is Field emission sources which provide a beam of electrons that have a much smaller crossover diameter as compared to thermionic sources. Field emission source is a tungsten or carbon cathode shaped up so that it has a very sharp tip (100 nm or less) held at a high potential, the electric field at the tip is so intense (>10^7 V/cm) that electrons are produced by a "tunneling process" in which no thermal energy is required to free the electrons from the potentials barrier that normally prevents their emission. This type of sources is fragile and also require better vacuum than thermionic sources. **Table 5.2** give a comparative overview of different electron guns in terms of size, brightness, energy, life, cross over diameter and vacuum required.

Table 5.2 A comparative overview of characteristics of different electron guns

Emission	**Thermionic**		**Field Emission**
	W	**LaB_6**	**FE**
Size (nm)	1×10^5	2×10^4	0.2
Brightness (A/cm^2.sr)	10^4 - 10^5	10^5 - 10^6	10^7 - 10^9
Energy Spread (eV)	1-5	0.5 -3.0	0.2 - 0.3
Operating lifetime (hrs)	>20	>100	>300
Vacuum (torr)	10^{-4} - 10^{-5}	10^{-6} - 10^{-7}	10^{-9} - 10^{-10}
Crossover diameter	50 μm	10 μm	10 nm

Electron Optics

In optical microscopy, optical lenses are used where focus of light is achieved through light refraction whereas in electron microscopy electromagnetic lenses are used where focus is achieved through Lorentz forces **(Fig. 5.2b)**.

Electromagnetic lens a consist of coil of wires enclosed in a cylindrically symmetrical iron casing. A magnetic field is generated between the pole pieces by applying a current through the coils. Electron microscope contains two types of magnetic lenses; condenser and objective lens system. The magnetic condenser and objective lens system reduce the spot size of electron beam to a diameter of 2-10 nm when it reaches the specimen. The condenser lens system, which consists of one of more lenses, is responsible for the throughput of the electron beam reaching the objective lens; the objective lens determines the size of the beam hitting the sample surface.

The rotationally symmetric magnetic field is inhomogeneous in such a way that it is weak in the center of the gap and becomes stronger close to the bore. Electrons close to the center are less strongly deflected than those passing the lens far from the axis. The overall effect is that a beam of parallel electrons is focused into a spot (so-called cross-over). In a magnetic field, an electron experiences the Lorentz force F (Eqs. 5.7, 5.8):

$$F = \mathrm{e}\,(E + v \times B) \quad 5.7$$

$$|F| = \mathrm{e}\,\mathrm{v}\,B \sin(v, B) \quad 5.8$$

Where, *E* and *B* represent strength of electric field and magnetic field; *e* and *v* represent the charge and velocity of electrons respectively. The focusing effect of a magnetic lens increases with the magnetic field B, which can be controlled via the current flowing through the coils. As described by the vector product, the total resulting force, F is perpendicular to *v* and B. This leads to a helical trajectory of the electrons and to the magnetic rotation (image is rotated in respect of the object). The degree of rotation depends on the strength of magnetic lens and the focal length can be altered by changing the strength of the current.

5.2.1 Scanning Electron Microscope (SEM)

Modes of Electron Microscopy: SEM and TEM

Electron Microscopy is a tool for visualizing otherwise invisible world of micro space (10^{-6} m) and nano space (10^{-9}m). There are two modes of a electron microscope, scanning mode, conceptualized by Manfred von Ardenne in 1937, where a probe beam is scanned over the surface and image is created point by point and in transmission mode, the probe beam passes through the specimen and differentially refracted and absorbed and image is created. Accordingly there are two important types of electron microscopes: scanning electron microscope (SEM) and the transmission electron microscope (TEM). Although scanning and transmission electron microscopy methods have many similarities, both uses electron as the source for illumination, however differ in many other aspects. In TEM, the electron beam passes through the sample specimen where as in SEM, electron beam scans over the surface. Scanning electron microscopy can be thought of as providing images of external morphology, similar to those accessed by the human eye. In contract, transmission electron microscopy probes the internal structure of solids and gives us information about microstructural detail that is not familiar to the eye.

A Scanning Electron Microscope reveals details and complexities that are inaccessible by light microscopy. SEM uses a focused high energy beam of electrons to magnify an object. The key features of scanning electron microscope are:

1. *Resolution at high magnification*: Resolution can be defined as the least distance between two closely opposed points, at which they may be recognized as two separate entities. The best resolution possible in a light microscope (LM) is about 200 nm whereas a typical SEM has a resolution of about 5 nm. SEMs can magnify an object from about 10 times up to 300,000 times whereas light microscope which can magnify up to 10000 times.
2. *Depth of field:* Topographical detail *i.e.* the height of a specimen that appears in focus in an image of SEM is >300 times the depth of field as compared to the LM. Moreover the three dimensional (3D) appearance of the specimen image, is the most valuable feature of the SEM. This is because such images, even at low magnifications, much more information can be obtained about a specimen than is available using the LM. The use of "stereo pair" SEM images can give even greater information about the sample.
3. *Microanalysis:* Analysis of sample composition including information about chemical composition, as well as crystallographic, magnetic and electrical characteristics is also possible with modern instrumentation.

Basic Principle

When a finely focused beam of electrons impinges on the surface of the solid sample and interacts with the atom of sample, several signals in the form of backscattered electrons, secondary electrons, Auger electrons and characteristic X-rays are generated from the surface which contain information about surface topography and composition of specimen. **Figure 5.3** shows signals used in SEM, generated by electron beam-sample interactions. The interactions of a solid with an electron beam can be divided into two categories, *elastic* interactions that affect the trajectories of the electrons in the beam without altering their energies significantly and *inelastic* interactions, which result in transfer of part of all of the energy of the electrons to the solid. The excited solid then emits secondary electrons, Auger Electrons, X-ray and sometimes longer-wavelength photons.

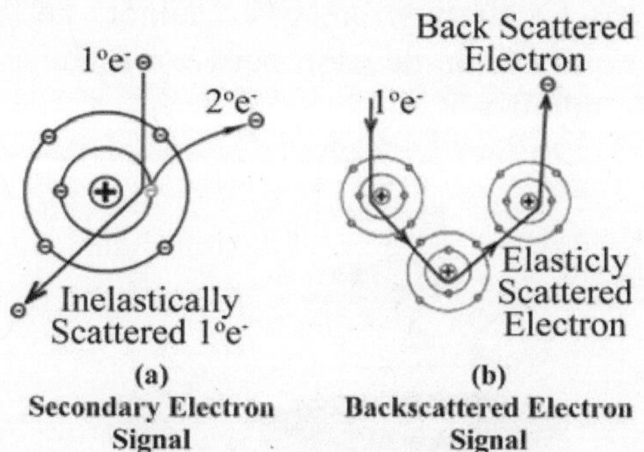

Fig. 5.3. Signals used in SEM: Secondary electrons and backscattered electrons

In a typical SEM, a source of electrons is focused into a beam, with a fine spot size of ~5 nm, having energy ranging from a few hundred eV to 50 KeV, that is rastered over the surface of the specimen by deflection coils. As the electron strikes and penetrate the surface, a number of interactions occur, which result in the emission of electrons and photons from the sample, and the SEM images are produced by collecting the emitted electrons on a cathode ray tube (CRT). Basic components of scanning electron microscope are shown in **Fig. 5.4.** Various SEM techniques are differentiated on the basis of what is subsequently detected and imaged. The principle images produced in the SEM are of three types: secondary electron images back scattered electron images and elemental X-ray maps.

A high energy primary electron interacting with an atom, results in either inelastic scattering (deflecting the electrons with loss of energy) with atomic electrons or elastic scattering (deflecting the electrons with no loss of energy) with the atomic nucleus. In an inelastic collision with an electron, the primary electron transfers part of its energy to the other electron. When the energy transferred is large enough, the other electron will emit from the sample. If the emitted electron has the energy of less than 50 eV, it is referred to as secondary electron. Back scattered electrons are the high energy electrons that are elastically scattered and essentially possess the same energy as the incident or primary electrons. The number of secondary electrons is generally one half to one fifth or less the number of backscattered electrons. Secondary electrons are produced as a result of interactions between energetic beam electrons and weakly bound conduction electrons in the solid, which leads to ejection of the conduction band electrons with a few electron volts of energy. Secondary electron are produced from a depth of only 50-500 A^{o} and exit in a beam that is slightly larger in diameter than the incident beam. Beam of backscattered electrons has a much larger diameter than the incident beam. The diameter of backscattered beam is one of the factors limiting the resolution of an electron microscope. The probability of backscattering increases with the atomic number of the sample material. Although backscattering images cannot be used for elemental identification, useful contrast can develop between regions of the specimen that differ widely in atomic number, Z.

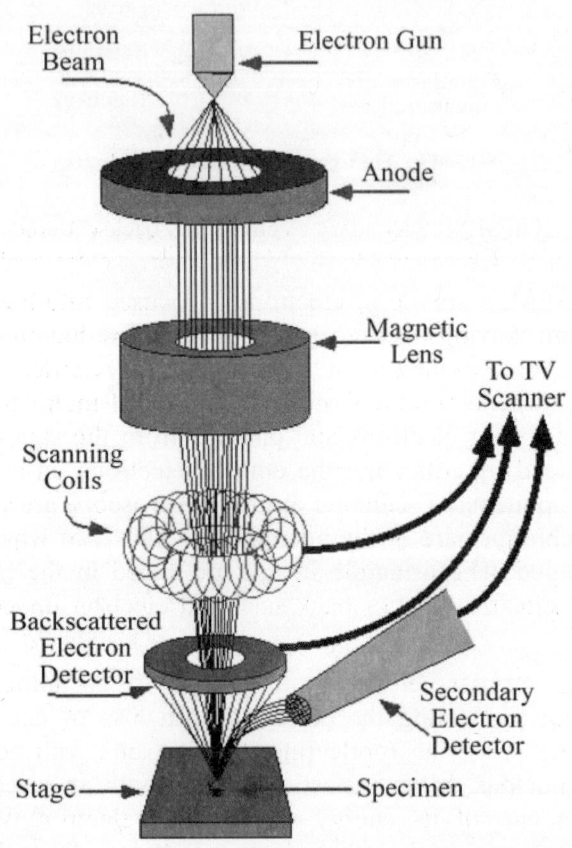

Fig. 5.4. Basic components of scanning electron microscope

Instrumentation

A typical laboratory bench top model of scanning electron microscope (SEM) is shown in **Figure 5.5**. Its essential components include the following:

- Electron Source ("Gun")
- Electron Lenses
- Sample Stage
- Detectors for all signals of interest
- Display / Data output devices

- Infrastructure Requirements: power supply, vacuum system, cooling system, vibration free floor, room free of ambient magnetic and electric fields.

SEMs always have at least one detector, usually a secondary electron detector. Mostly, additional detectors are also there in SEMs. The specific capabilities of a particular instrument are critically dependent on which detectors it accommodates.

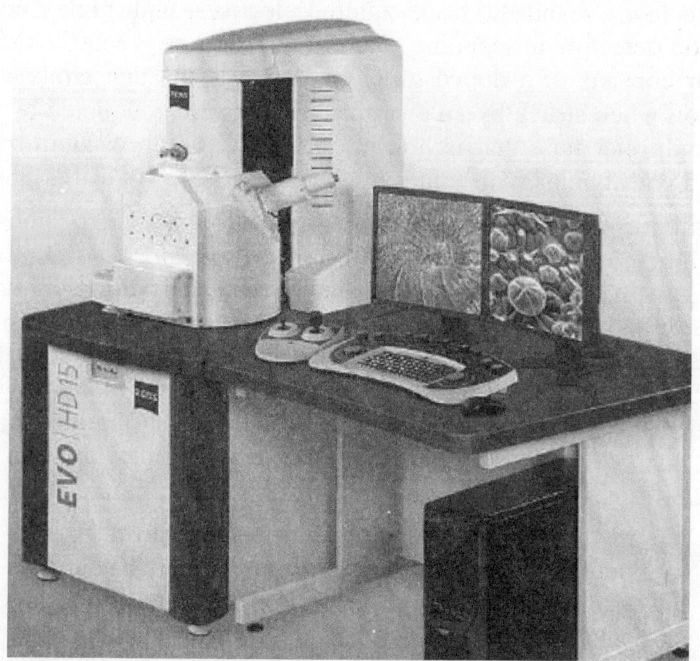

Fig. 5.5. SEM Instrument (EVO® HD Scanning Electron Microscope, Carl Zeiss)

Samples and Sample Holder

The specimen for SEM should be conducting for the electron beam to scan the surface and that the electrons have a path to ground for conventional imaging. Examining non-conducting materials particularly polymers and biological material may present other difficulties, such as thermal degradation, radiation damage, and sample volatility in the high vacuum. Non-conductive solid specimen is generally coated with a layer of conductive material by low vacuum sputter coating or high vacuum evaporation.

Sample chambers are designed for rapid changing of samples. Large-capacity vacuum pumps are used to hasten the switch from ambient pressure to ~10^{-4} torr or less for conventional SEMs. The sample holder, or stage, in most instruments is capable of holding samples many centimeters on an edge. Furthermore, the stage can be moved in the *x, y* and *z* directions, and it can be rotated about each axis. As a consequence, the surfaces of most samples can be viewed from almost any perspective. Environmental SEMs use much higher pressures in the sample chamber and allow for variations in temperature and gas composition.

Detector

In SEM, two types of electrons come out of sample; secondary electrons having energies of a few eV and the backscattered electrons with higher energies. The most common detectors in scanning electron microscope is scintillation detectors. The detector consists of a doped glass or plastic target that emits a cascade of visible photon when struck by an electron. Photons are conducted by a light pipe to a photomultiplier tube that is housed outside the high-vacuum region of the instrument. Typical gains with scintillation detectors are 10^5 to 10^6 lines.

A scintillator-photomultiplier detector based on Everhart-Thornley design, shown in **Figure 5.6,** is used to detect the secondary electrons emitted from the specimen. Low-energy secondary electrons (trajectories shown by dashed lines) are collected by applying a suitable bias to the Faraday grid. These electrons are further accelerated in order to give them sufficient energy to scintillate the phosphor that is highly efficient at converting the energy contained in the electrons into ultraviolet light photons (4000 Å). The response time of the phosphor is fast and permits high resolution scanning.

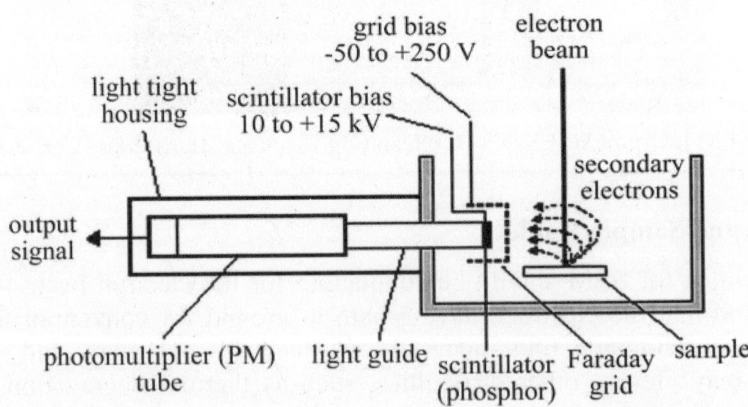

Fig. 5.6. A secondary electron scintillator-photomultiplier detector based on Everhart and Thornley design

Semiconductor detector, consist of flat wafer of a semiconductor are also used in electron microscopy. When high-energy electrons strikes the detector, electron–hole pairs are produced that result in an increase in conductivity. Current gains with typical semiconductor detector are 10^3 to 10^4. The X-rays produced in the scanning electron microscope are usually detected and measured with energy dispersive system. The characteristic X-rays are used for qualitative and quantitative information about the sample.

Applications

SEM is one of the most widely used techniques used in characterization of nanomaterials and nanostructures. SEM shows detailed three dimensional images with a resolution of few nanometers and at much higher magnification (up to x 300,000) as compared to light microscope (up to x10000). Although images are black and white since images are created without light waves. SEM not only produces topographical information, it also provides the chemical composition information near the surface. The surface structure of polymers, nanocomposites, fracture surfaces, nanofibers, nanoparticles and nanocoatings can be imaged through SEM with great clarity. The SEM technique is also used to view dispersion of nanoparticles such as carbon nano tubes, nanoclays and hybrid poly hederal oligomeric silserquionzone (POSS) nanofillers in the bulk and on the surface of nanocomposites fibers. SEM imaging is a prime characterization technique in biomedical, environmental, textile, tissue engineering, cell culture & cell development, scaffold construction etc. **Figure 5.7** illustrate some of the SEM images of nanomaterials the main limitations of SEM is that sample must have surface electrical conductivity. Specialized SEM has been developed to cater for non-conductive samples in their natural states.

Fig. 5.7. SEM images of (a) nylon 6 nanofibers impregnated with Ag nanoparticles (b) ZnO nanoflowers (c) SWCNT

The versatility of the SEM and electron microprobe to study the minute details of a sample lies in a wide variety of signals generated when the electron beam interacts with the solid.

In basic scanning electron microscopy instruments, the primary detection mode is imaging of secondary electrons. X-rays the second most common imaging mode for SEM used to identify the elemental composition of the sample

(X-rays being characteristic). The technique is called as energy dispersive X-ray (EDX).

Energy Dispersive X-ray Analysis (EDX)

Energy Dispersive X-ray Analysis is a technique to analyze near surface elements and their composition at different position, thus obtaining an overall mapping of the sample. EDX is used in conjunction with SEM. Under SEM conditions when electron beam strikes the surface of a conducting sample, the energy of beam (typically in the range 10-20 keV), causes emission of characteristic X-rays (see **Figure 5.1**). The energy of X-rays emitted depends on the nature of material under examination. The X-rays are generated in a region about 2 μ in depth and thus EDX is not truly a surface technique. By moving the electron beam across the material an image of each element in the sample can be obtained. Presence of heavy metal ions such as Au and Ag nanoparticles on the surface can easily be identified using EDX techniques (**Fig. 5.8**).

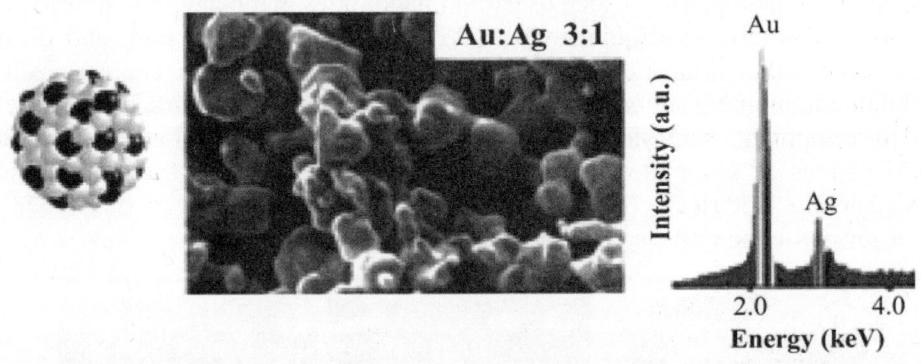

Fig. 5.8. SEM image and EDX spectra of nanoporous material made up of 3:1 Au:Ag

Elements of low atomic number are difficult to detect by EDX; however, Field Emission scanning electron microscopy (FESEM) and corresponding EDX spectrum may overcome this limitation. Field Emission SEM (FESEM) operates at low voltage-high vacuum and at low vacuum-high voltage, respectively. Wet, oily, volatile and non-conductive specimen in their natural state can be examined using this technique. Conductive coating that mask valuable information are not required. Both secondary and backscattered electrons may be detected with an FESEM. Everhart-Thornley detector cannot be used because the high bias voltage of scintillator would cause electrical breakdown at high pressure. Instead, gas-phase secondary electron detectors, which make use of cascade amplification, are employed. **Figure 5.9** depicts FESEM image and EDX spectrum of Carbon nano onions.

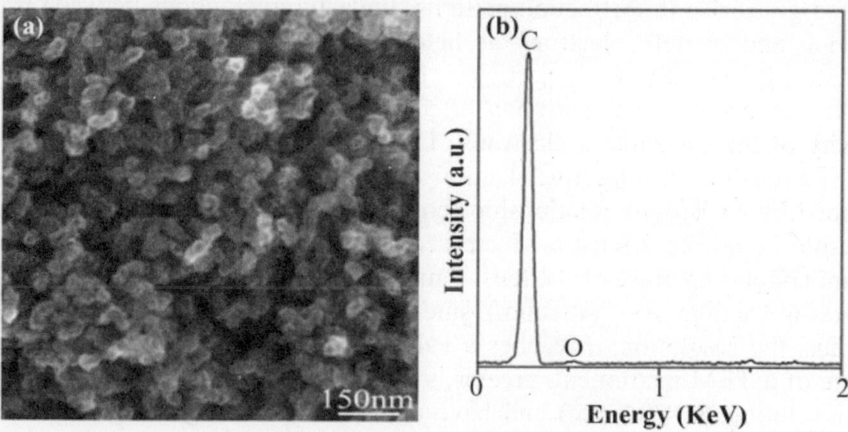

Fig. 5.9. FESEM image and EDX spectrum of Carbon nano onions

5.2.2 Transmission electron microscope (TEM)

Transmission electron microscopy (TEM) is an electron microscopy which utilizes electrons rather than photons to image the specimens. A beam of electrons is transmitted through an ultra-thin specimen (<200 nm); upon interaction of transmitted electrons through the specimen, image is formed, magnified and focused onto an image device, such as fluorescent screen or a layer of photographic film or to be detected by a camera. The microscope is kept under high vacuum to insure unimpeded travel of electrons to and by the specimen. Electron beams is focused through a series of electromagnetic lenses on to the specimen as shown schematically in **Figure 5.10.** The big advantages of TEM are the high magnification (X1000 000) and its ability to provide both image and diffraction information from a single sample. Interaction of electrons transmitted through the specimen results in scattering which can be elastic (no energy change) or inelastic (energy change). Elastic scattering can be coherent or incoherent. Well orderly arranged atoms in a crystalline substance result in coherent scattering, giving spot patterns. Inelastic processes give characteristic absorption or emission, specific to the compound/elements/ chemical structure. In TEM one can switch between imaging the sample and viewing its diffraction pattern by changing the strength of the intermediate lens.

Basic principle

In transmission electron microscope, electrons are accelerated to 100 KeV to 1 MeV, projected onto a thin specimen by means of the condenser lens system, and penetrate the sample thickness either non-scattered or scattered. The scattering processes experienced by electrons during their passage through the specimen determine the kind of information obtained. Elastic scattering involves no energy loss and gives rise to diffraction patterns. Inelastic interactions between primary electrons and sample electrons at heterogeneities such as grain boundaries, dislocations, second phase particles, defects, density variations, etc., cause complex absorption and scattering effects, leading to a spatial variation in the intensity of the transmitted electrons. The high magnification and resolution of TEM is a result of the effective electron wavelengths (λ), and the resolution (d) as governed by de Broglie relationship (eq. 5.1) and eq. 5.5. In order to maximize the resolving power, λ must be decreased, and refractive index of the medium, n or semi angular aperture of the lens, θ must be increased. In case of TEM, which operates in vacuum, $n = 1$ (vacuum), sine $\theta = 10^{-2}$ and λ is of the order of 0.005 nm and thus, the resolution, d reaches a value of 0.3 nm. The higher the operating voltage of a TEM instrument, greater is the lateral spatial resolution (*e.g.* at 400 KV, resolution is < 0.2 nm) and have additional advantage of greater electron penetration, allowing to work with thicker samples.

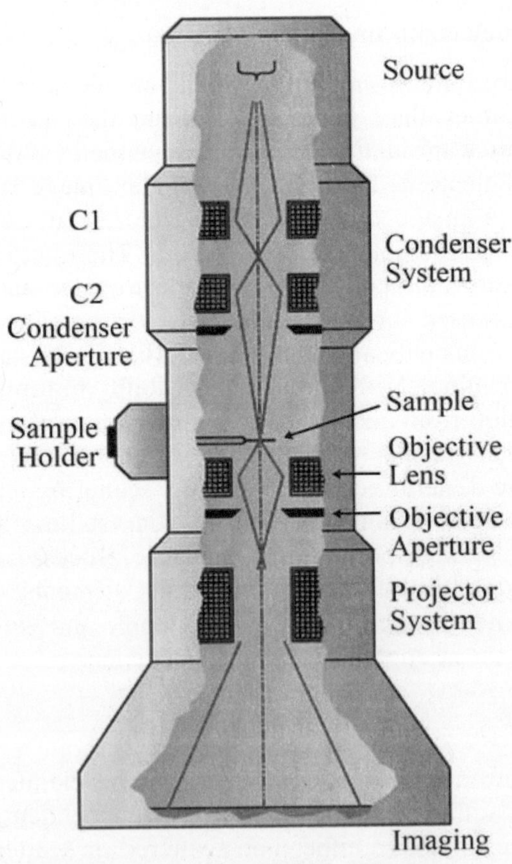

Fig. 5.10. Laboratory Model of Transmission Electron Microscope

TEM images: Bright field image and Dark field image

The contrast in a TEM image is not like that in a light microscope. In TEM the crystalline sample interacts with the electron beam mostly by diffraction. The intensity of the diffraction depends on the orientation of the plane of the atoms in a crystal relative to the electron beam; at certain angles the electron beam is diffracted strongly from the axis of the incoming beam, while at other angles the beam is largely transmitted. There are two main mechanism of contrast in an image, the transmitted and scattered beams which can be recombined at the image plane, thus preserving their amplitude and phases. This results in phase contrast image of the object. The amplitude contrast image can be obtained by eliminating the diffracted beams. This is achieved by placing suitable apertures below the back focal plane of the objective lens. This image, using transmitted beam is called the 'bright field image'. In 'dark field image', the diffracted beam is used for imaging either by using a movable aperture or by shifting the incident beam, keeping the aperture constant. **Figure 5.11** shows arrangement of optics in two modes of TEM along with typical images of spherical δ' precipitates in an Al-Li alloy in bright field and dark field modes, respectively.

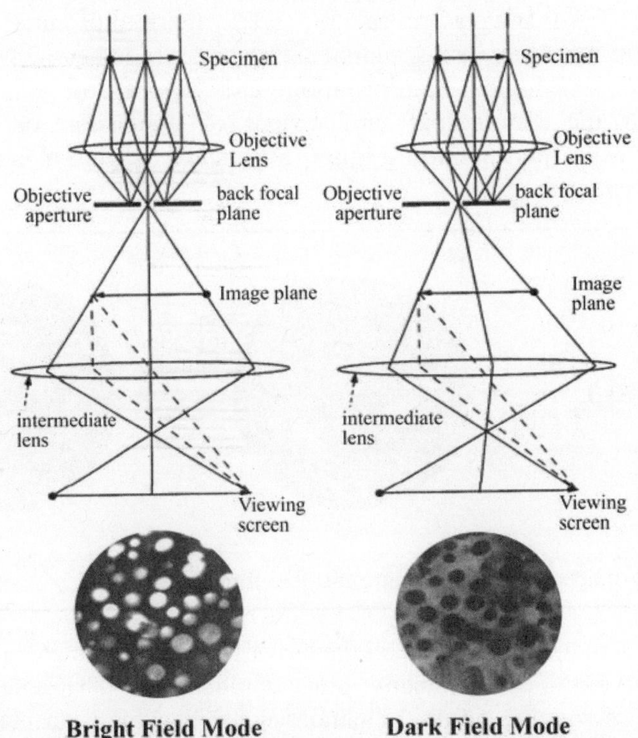

Fig. 5.11. Bright field and dark field modes and typical TEM images of spherical δ precipitate in an Al-Li alloy in bright field mode and dark field mode

Modern TEMs are equipped with specimen holders that allow tilting the specimen to a range of angles in order to obtain specific diffraction conditions. This produces a variation in the electron intensity that reveals information on the crystal structure. The technique of bright field imaging is particularly sensitive to the extended crystal lattice defects.

Sample preparation: The most difficult aspect of the TEM technique is in preparation of samples. The specimen for TEM must be prepared as thin film/foil so that electron beam can penetrate and able to withstand the vacuum present inside the instrument. Nanopowders and nanotubes can be quickly produced by the deposition of a dilute sample solution containing the specimen onto the support grid. The Biological specimens are, however, inherently unstable under vacuum and thus have to be fixed usually in glutaraldehyde and paraformaldehyde, dehydrated and stabilized in resins before they can be placed in a vacuum. Polymeric nanocomposites and biological samples are cut into thin films/slices (< 100 nm) using ultra-microtome with diamond knife at cryogenic condition (in liquid nitrogen). These sections are electron transparent and specimen contrast is enhanced using metal stains, such as lead citrate, uranyl acetate and phospho-tungstic acid.

Applications: TEM images reveal the size, distribution and dispersion of nanoparticles in nanocomposites, nanocoatings etc.; together with, it also reveal the extent of exfoliation, intercalation and orientation of nanoparticles. Monitoring of the morphology and extent of dispersion or inter-calation/ exfoliation of nano particulates. **Figure 5.12** shows different stages of Fe_3O_4 nanocrystals synthesis.

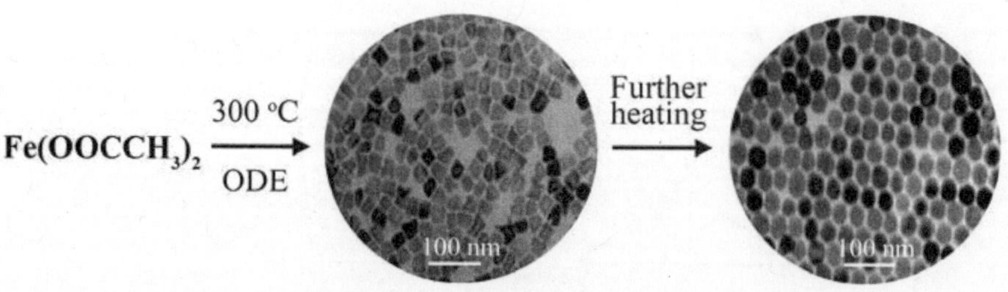

Fig. 5.12. TEM images showing formation of Fe_3O_4 nanocrystals

In addition to the structural characterization and chemical analyses, TEM also has been explored for many other applications. Examples include the determination of melting points of nanocrystals, in which, an electron beam is used to heat up the nanocrystals and the melting points are determined by the disappearance of electron diffraction. Another example is the measurement of mechanical and electrical properties of individual nanowires and nanotubes. This technique allows a one-to-one correlation between the structure and properties of the nanowires.

Laser Scanning Confocal Micro-scope (LSCM) provides non-invasive investigation of migration of nanoparticles across the bio-barrier and reveals 3D morphology of nanoparticles.

High Resolution Transmission Electron Microscopy (HRTEM)

TEM with resolving powers ~1A^o are available now. At these high resolutions, individual atoms and crystalline defects can be imaged. As a result high resolution TEM (HRTEM) is one of the most essential tools of nanoscience. The microscope is kept under high vacuum to ensure unimpeded travel of electrons to and by the specimen. The electrons is transmitted by the specimen on to a phosphor screen for visualization. External structures of microbes can be viewed at high magnification using the method of scanning electron microscopy. The specimen is stabilized through either chemical or cryo fixation, covered through a layer of conductive material carbon or gold/platinum and the specimen is scanned with an electron beam. The Electrons are either reflected or passed by the specimen and a digital image is creating from the composite data, creating 3-dimensional structures. This technique has been further establishment for examination of specimens at low vacuum and ambient conditions. **Figure 5.13** shows HRTEM images of carbon nano-onions.

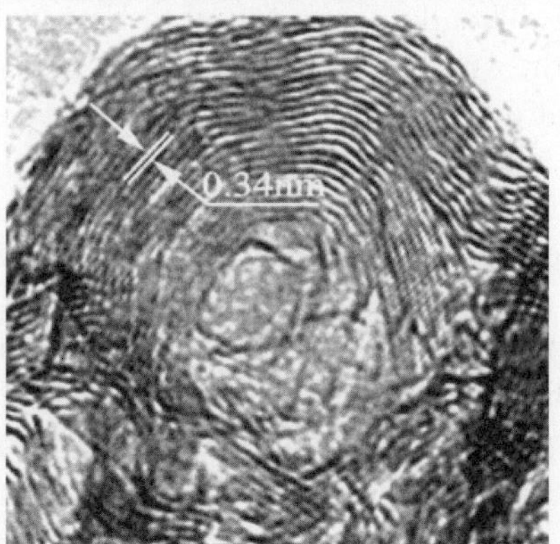

Fig. 5.13. HRTEM images of carbon nano-onions

5.3 SCANNING PROBE MICROSCOPY

Scanning probe microscopies (SPMs) are capable of resolving details of surfaces down to the atomic level. The first example of this type of microscope, the scanning tunneling microscope was described in 1982. Unlike optical and electron microscopes, SPMs reveal details not only on the lateral x- and y- axes of a sample but also on the z-axis, which is perpendicular to the surface. The resolution of SPMs can be as low as 0.1 nm (1 A°) in the x- and y- axes (resolution in the z-dimension is even better than 0.1 nm) as compared to resolution of a typical electron microscope. SPM provides images of surfaces by scanning the surface line by line with a probe. Scanning works very similarly to the way the blind read Braille, line by line, by moving a finger over buds on the paper. Both are based on scanning the surface of the sample in a raster pattern with a very sharp tip that moves up and down along the z-axis along the surface topography. This movement is measured and translated into an image with details on an atomic- size scale. Scanning tunneling microscopy, atomic force microscopy, scanning electrochemical microscope and numerous variations of these techniques fall under the category of scanning probe microscopy (SPM). All these techniques work on different principle but the most significant common feature is to produce a highly magnified image of the surface or bulk of the sample. Two types of SPMs that are most widely used in nanoscience and nanotechnology: the scanning tunneling microscope (STM) and atomic force microscope (AFM). Basic principle, methodology and applications of both of these techniques are described in the following subsequent sections.

5.3.1 Scanning Tunneling Microscope (STM)

In 1981, Binning and his co-workers introduced a totally new concept of imaging, the scanning tunneling microscope (STM). The invention of the STM earned Binning and his co-worker Heinrich Rohrer (at IBM Zürich) the Nobel Prize in Physics in 1986. This device is found to be capable of resolving features on an atomic-scale on the surface of a conducting solid specimen. They used a small metal tip placed at a very close distance from a conducting surface: when the two are placed very close together, but not actually touching, a bias between the two can allow electrons to tunnel through the vacuum between them. This creates a tunneling current which can be measured and which is a function of the electron density on the surface. Electron density is the probability of finding an electron in a particular place. There is high electron density around the atoms and bonds in molecules. This type of microscope is called the scanning tunneling microscope (STM). Variations in current as the probe moved over the surface are translated into an image.

STM instrument is found to be capable of resolving features on an atomic scale on the surface of a conducting solid. The resolution of STM is of the order of 0.1nm lateral resolution and 0.01 nm depth resolution. With these high resolution, individual atoms within materials can be routinely imaged and manipulated. The STM can create detailed 3D images of a sample with atomic resolution. This means that the resolution is actually so high that it is possible to see and distinguish the individual atoms (~ 0.2 nm) on the surface. The STM has proven to be a powerful tool in nanoscience and nanotechnologies. It is used in both applied and fundamental research to obtain atomic-scale images of metal and semiconducting surfaces. It provides a three-dimensional profile of the surface roughness; it allows observing surface defects, and determining the size and conformation of molecules and aggregates. Another outstanding application of the STM is that it can be used to manipulate individual atoms, trigger chemical reactions, as well as perform electronic spectroscopy. The STM can be used not only in ultra-high vacuum but also in air, water, and various other liquid or gas ambient, and at temperatures ranging from near zero Kelvin to a few hundred degrees Celsius.

Working principle: The STM is based on the concept of quantum mechanical tunneling, which occurs when a small bias voltage (mV to 3V) is applied between a sharp tip and conducting sample when the tip is within a few nanometers of the surface. The resulting *tunneling current* is a function of tip position, applied voltage, and the local density of states (LDOS) of the sample. The magnitude of tunneling current I_t is given by Eq. 5.9.

$$I_t = V e^{-Cd} \tag{5.9}$$

Where V is the bias voltage, C is a constant, and d is the distance between the probe and the surface. Information is acquired by monitoring the current as the tip's position scans across the surface, and is usually displayed in an image form.

At close range, fine control of the tip in all three dimensions when near the sample is typically piezoelectric, maintaining tip-sample separation d typically in the 4-7 Å (0.4-0.7 nm) range, which is the equilibrium position between attractive ($3< d <10$Å) and repulsive ($d <3$Å) interactions. In this situation, the voltage bias will cause electrons to tunnel between the tip and sample, creating a current that can be measured. Once tunneling is established, the tip's bias and position with respect to the sample can be varied and data are obtained from the resulting changes in current.

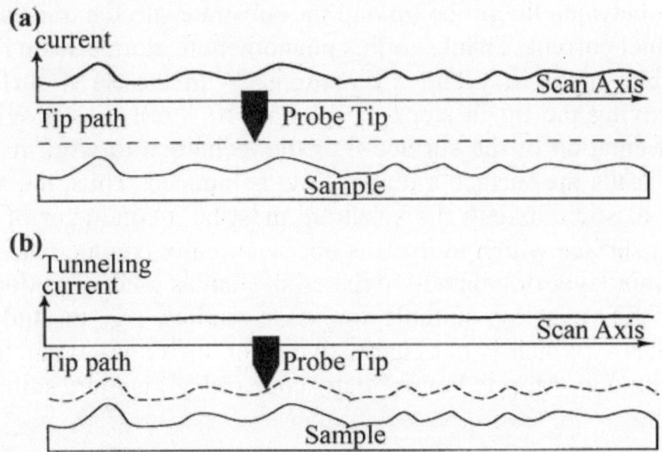

Fig. 5.14. Imaging mode of STM; Constant Height (a); and Constant Current (b)

In principle, the STM can be operated in two major modes *i.e.* Constant height and Constant Current. In constant height mode, the tip position is kept constant in the z direction while the tunneling current is monitored. In constant current mode, the tunneling current is kept constant while the z position of the tip changes to keep *d* constant as shown in **Fig. 5.14,** and the z position of the tip is monitored. The up-and-down motion of the tip reflects the topography of the surface. Most STM operates at constant current mode. In this mode, the contrast on the image is due to variations in charge density. In constant height mode, the voltage and height are both held constant while the current changes to keep the voltage from changing; this leads to an image made of current changes over the surface, which can be related to charge density. The benefit to using a constant height mode is that it is faster, as the piezoelectric movements require more time to register the height change in constant current mode. All images produced by STM are grayscale, with color optionally added in post-processing in order to visually emphasize important features.

STM images

When a conducting tip is brought very near to a metallic or semiconducting surface, at a distance of about 0.1 nm, it can induce the formation of a tunnel current between the tip and the surface: a bias between the two atoms (tip and surface) can allow electrons to tunnel through the vacuum between them and induce the formation of a current. Variations in current as the probe passes over the surface are translated into an image. As the conducting tip of an STM scans over a conducting or semiconducting surface, a "tunnel current" is formed, which arises from electrons jumping from the surface to tip of the STM probe. The probability of this happening depends largely on the distanced between the surface and tip, thus the size of the current depends on this distance. Small changes in the distance between the probe tip and the substrate surface translate into large changes in tunnel current. Thanks to this phenomenon, atomic scale resolution by STM is possible in the x, y and z directions. With the STM surfaces can be scanned by moving the tip in steps of 0.1 nm (10^{-10} m), thus providing a very accurate representation of the surface. For the technique to work it is necessary that as the tip scans the surface a tunnel flow is induced. Thus, the surface must be conductive to some extent; the substrate must be a conductor or semiconductor. To scan a surface which in itself is not electrically conductive, one can coat it with a very thin layer of conductive material, such as gold. This does, however, imply that the STM is less suitable for some studies, *e.g.* to study biological molecules as DNA (which is not conductive). For such non-conductive samples, other SPM technique, Atomic Force Microscope (AFM) is more suitable.

Instrumentation

The essential components of a STM are shown in **Fig. 5.15.** The components of an STM include scanning tip, piezoelectric controlled height and x,y scanner, coarse sample-to-tip control, vibration isolation system, and computer. The scanning waveforms, applying on the x and y piezos, make the tip raster scan on the sample surface. A bias voltage is applied between the sample and the tip to induce a tunneling current. The z piezo is controlled by a feedback system to maintain the tunneling current constant. The voltage on the z piezo represents the local height of the topography.

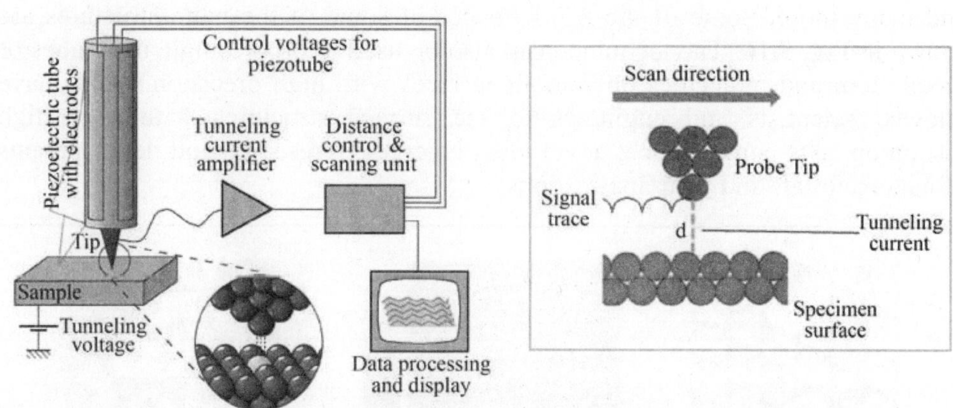

Fig. 5.15. Overview of STM with a close-up view of the tunneling effect between tip of the Probe, the surface atoms and a signal trace

A probe tip, is attached to a piezo-drive, which consists of three mutually perpendicular piezoelectric transducers: x piezo, y piezo, and z piezo. Upon applying a voltage, a piezoelectric transducer expands or contracts. By applying a saw tooth voltage on the x piezo and a voltage ramp on the y piezo, the tip scans on the x,y plane. Using the coarse positioner and the z piezo, the tip and the sample are brought to within a fraction of a nanometer each other. The electron wave functions in the tip overlap electron wave functions in the sample surface. A finite tunneling conductance is generated. By applying a bias voltage between the tip and the sample, a tunneling current is generated. To ensure stable operation, vibration isolation is essential.

Tips: The probe of an STM is a very thin needle called the "tip" that is so small that it ends in just a few atoms. The tip is made of conducting material (*e.g.* metal, typically tungsten or Pt–Ir alloy). The precise movement of the tip is controlled by a piezomotor. The tip of an STM is about 3 mm long and should be located very close to the surface to be scanned. In practice, the distance between the end of the tip and the surface must be less than 0.1 nm, without the tip actually hitting the surface. One of the fundamental elements of the STM is the tip of the probe that scans the surface, which must be sharpened to a very fine tip.

The fabrication of sharper probes allows better resolution of surface features. Ultimately, a probetip sharpened to one atom would provide the best resolution.

Applications

Scanning tunneling microscopy is a powerful tool in nanotechnology and nanoscience providing facilities for characterization and modification of variety of materials. STM is successfully used to detect and characterize materials like carbon nanotubes (single- and multi-walled), graphene, intriguing chalcogenides and many more. Some of the STM images of some of the nano structures are shown in **Fig. 5.16**. The technique can also be used to probe single nanotubes or metal atom and molecules on smooth surfaces with high precision. SPMs have allowed scientists and engineers to see surface structures with very high resolution, as a consequence, novel ideas emerges in research and developments of nanomaterials and their applications.

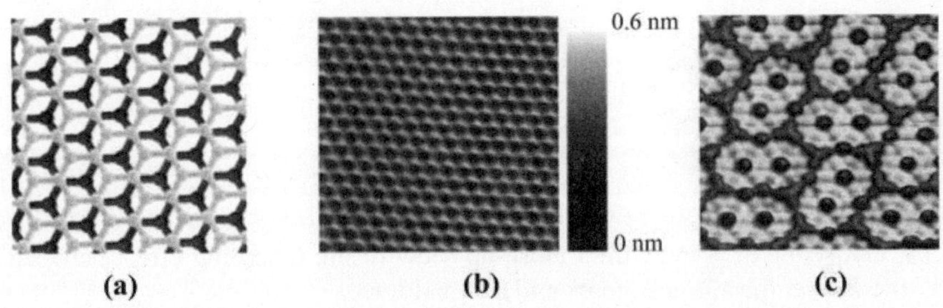

Fig. 5.16. Structure vs STM image (a) model of two layers of graphite (b) STM image of a graphite sheet with height scale (c) STM image of cobalt sulfide "nanoflower" structure synthesized on a Au (111) surface

5.3.2 Atomic Force Microscope

Atomic force microscope (AFM), another SPM technique, was developed in 1985, also by Binning and co-workers at IBM Zürich. An AFM measure the forces between tip and surface and therefore, does not require the surface to be conductive. The Atomic Force Microscope (AFM) is basically meant to overcome the intrinsic limitations of the STM, which was not suitable for imaging surfaces coated with biological entities such as DNA or proteins. The AFM operates in air, not under a vacuum and is non-destructive technique with a very high 3D spatial resolution. It offers the capability of 3D visualization and both qualitative and quantitative information on many physical properties including size, morphology, surface texture and roughness. Statistical information, including size, surface area, and volume distributions, can be determined as well. In general the sample size for AFM ranges up to 80 μm. In a single scan wide range of particle sizes can be characterized ranging from 1 nm - 10 μm. In addition, the AFM can characterize nanoparticles in multiple mediums including ambient air, controlled environments and even liquid dispersions.

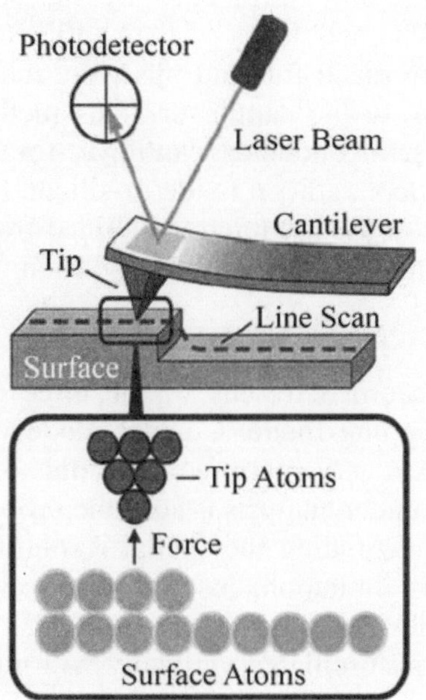

Fig. 5.17. Schematic representation of AFM

Basic Principle

In atomic force microscopy, a sharp tip which is mounted on a cantilever is moved along the surface to be probed and the changes of voltage in term of height are recorded. These height changes can be caused by particles lying on the surface, but can also be a measure of the roughness of the surface itself. In its most basic set-up, the instrument works similar to a record player, with the tip being the needle and the record the surface to be measured. In the simplest of set-ups the bend in the cantilever is kept constant and the instrument measures how much the cantilever needs to be moved up and down to achieve constant bending. The force acting between the cantilever and the sample surface causes minute deflections of the cantilever, which are detected by optical means. **Figure 5.17** shows schematically the most common method of detecting the deflections of the cantilever holding the tip. A laser beam is reflected off a spot on the cantilever to a segmented photodiode that detects the motion of the probe. The output from the photodiode then controls the force applied to the tip so that it remains constant. Optical control system in AFM is analogous to the tunneling-current system in the STM. The movement system is a tubular piezoelectric device that moves the sample in a x, y and z directions under the tip. The signal from laser beam detector is then fed back into the sample piezoelectric transducer, which causes the sample to move up and down, maintaining a constant force between the tip and the sample.

AFM images are not only influenced by the substrate surface, but also depend on the physical characteristics of the cantilever and tip. In early AFMs, cantilevers were cut from metal foil and tips were made from crushed diamond particles, glued manually to the cantilever. This method has been replaced by producing an integral semiconductor cantilever-tip assemblies, produced by etching single chip of silicon, silicon oxide or silicon nitride. The most common cantilever-tip assemblies in use are micro machined from monolithic silicon. The cantilever and tips are remarkably small, ideally, a single atom at the tip apex.

Modes of AFM: Contact, Noncontact and Tapping

AFMs images can be recorded by any of the three modes viz. contact mode, noncontact mode and tapping mode. Contact mode is most common however since the AFM probe tip is very sharp, with a radius of curvature of the order of 10 nm and if the surface under analysis is soft, the probe can penetrate it, with the risk of damaging it and degrading the spatial resolution of the resulting micrograph. To overcome this limitation, instruments working in non-contact mode (dynamic) have been developed. In these systems, the probe is not simply dragged on the surface but oscillated vertically with respect to the surface while it is scanned; thereby significantly reduces the damage that can be induced by the probe and allow imaging of soft, compressible samples, such as biomolecules and cells. On the other hand the tip of an AFM can be deliberately used to "scratch" and remove some molecules from a surface or to use the tip to write with an "ink". Both are "writing" methods in the sense that they allow creating nanostructures on a surface with any geometry. This technique is called dip pen nanolithography (DPN) and is discussed in Chapter 3.

Sample Preparation

Nanoparticles typically fall into one of two categories when it comes to sample preparation. The first category is where nanoparticles are rigidly attached to a solid structure. The second category is related to weakly adhered nanoparticles where to the substrate, *i.e.* dispersed in liquid or dry mediums. A good example of the first category is nanoparticles imbedded in a solid matrix, as in the case of nano-composites or nano precipitates. In such cases, typically a cross-section of the composite material is scanned to determine such properties as average particle size and spatial distribution. Examples of nanoparticles in the second category are quantum dots, diesel soot particles, carbon black, and colloidal suspensions. Sample preparation for the second category of nanoparticles involves the stable attachment of particles onto the substrate. Because the AFM works by scanning a mechanical probe across the sample surface, any structure being imaged must have greater affinity to the flat surface than to probe tip. When nanoparticles accidentally attach to the probe, the resulting images show reduced resolution. Streaking will occur in the images if nanoparticles are not rigidly attached to the flat surface while scanning in contact mode. To avoid such artifacts, close contact mode (near contact) or CFM (crystal sensing) is strongly recommended for such samples. In certain cases, it is necessary to affix nanoparticles to a sticky substrate within liquid or dry mediums. A cheap and easy way to do this is the use of double-sided sticky tape or other similar methods commonly used by microscopists. More refined techniques include the use of mica, "Tacky Dot" slides, and TempFix.

Applications

Qualitative Analysis

In AFM studies individual particles and groups of particles can be resolved. Microscope images can be critical in troubleshooting of quality control issues. The AFM offers visualization in three dimensions. Resolution in the vertical, or Z, axis is limited by the vibration environment of the instrument; whereas resolution in the horizontal, or X-Y, axis is limited by the diameter of tip utilized for scanning. Typically, AFM instruments have vertical resolutions of less than 0.1 nm and X-Y resolutions of around 1 nm. In material sensing mode, the AFM can distinguish between different materials, providing spatial distribution information on composite materials with otherwise uninformative topographies. In **Fig. 5.18** material in-homogeneity can be seen on a topographically flat organic film. Similarly, nanocomposites can be analyzed for dispersion of particulate matter.

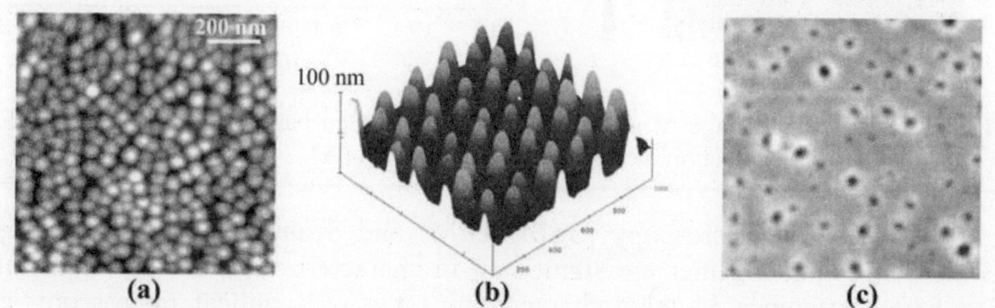

Fig. 5.18. AFM images showing: a. InAs/GaAs QDs grown on a Ge/Si substrate; b. Pt nanoparticles coated Si nanopillars; c. Cholesterol-BSA conjugate

Quantitative Analysis

Software-based image processing of AFM data can generate quantitative information from individual nanoparticles and between groups of nanoparticles. For individual particles, the size information (length, width, and height) and other physical properties (such as morphology and surface texture) can be measured. Statistics on groups of particles can also be measured through image analysis and data processing. Commonly desired ensemble statistics include particle counts, particle size distribution, surface area distribution and volume distribution. With knowledge of the material density, the mass distribution can be easily calculated. Image processing of an AFM image is shown in **Fig. 5.18**.

Statistics on groups of particles can also be measured through image analysis and data processing. Commonly desired ensemble statistics include particle counts, particle size distribution, surface area distribution and volume distribution. With the knowledge of the material density, mass distribution can be easily calculated. Image processing of an AFM image is shown in **Fig. 5.19** for graphene nano balls. Dry particles can be scanned in both ambient air and in controlled or inert environments, such as nitrogen or argon gas. Liquid dispersions of particles can also be scanned under similar conditions, provided the dispersant is not corrosive to the probe tip and can be anchored to the substrate. Particles dispersed in a solid matrix can also be analyzed by topographical or material sensing scans of cross-sections of the composite material. Such a technique is useful for investigating spatial nanocomposites.

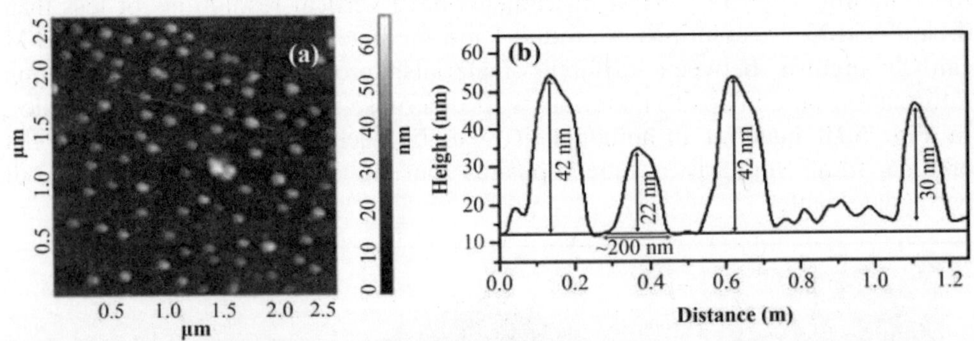

Fig. 5.19. (A) AFM image of graphene balls (the shaded bar is a measure of height); (B) A cross section along the line indicated in (A)

Both Electron microscopy (SEM, TEM) and Scanning probe microscopy (STM, AFM) put together are significant in characterization of nano materials. The basic difference in their characteristics vis-à-vis optical microscopy are summarized in **Table 5.3**.

Table 5.3. Characteristic features of Optical, Electron and Scanning probe microscope

Features	Optical microscope	Electron microscope	Scanning probe microscope
Operation	Air, Liquid	TEM: Vacuum SEM: Vacuum	STM: Vacuum AFM: Air, liquid dispersion, solid dispersion

Features	Optical microscope	Electron microscope	Scanning probe microscope
Imaging radiations	Visible light	SEM: Secondary electrons Backscattering electrons TEM: Transmitted electrons	STM: Quantum tunneling AFM: nN force between cantilever tip and surface
Wavelength of interacting radiations	400-700 nm	– 0.006 nm (vary with accelerating voltage)	N/A
Operating voltage	N/A	SEM: < 20-40 KV TEM: 40 KV~1MV	STM: < 1 V
Sample type	Not completely transparent	SEM: Conducting/ Non conducting TEM: Vacuum compatible thin film	STM: Conducting, Surface height < 10mm AFM: Conducting / Non conducting
Lateral resolution	1μm	SEM : 1-5 nm TEM : 0.1nm	AFM : 2-10 nm STM : 0.1nm
Vertical resolution	--	--	AFM : 0.1nm STM : 0.01nm
Magnification	~10^3	SEM: ~10^5 TEM: ~10^6	AFM: ~10^6 STM: ~10^9
Contrast	Absorption, Reflection	SEM: inclination effect, shadowing, edge contrast TEM: Scattering, Diffraction	STM: variations in charge/electron density AFM: variation in the interaction strength between surface and cantilever tip
Working mode		TEM: Bright field and dark field	AFM: Contact and Tapping STM: Constant current and constant height

5.4 CONFOCAL RAMAN IMAGING

Confocal microscopy is an optical imaging technique that employs a point light source (laser), which is directed onto the sample. The same objective usually collects the reflected (fluorescence) light, which is focused into a pinhole in front of the detector. This is made to make sure that only light from the image focal plane reaches the detector, which considerably increases image contrast; resolution can also be increased with appropriate selection of pinhole size. In Confocal Raman imaging the use of a spectroscopic electron multiplying charge couple devices (EMCCD) as the detector can considerably reduce the acquisition time down to a few milliseconds per spectrum, while enhancing sensitivity. Unique filters are utilized to restrain the reflected laser light while allowing the Raman scattered light to be detected with a combination of charge couple devices

(CCD) camera and spectrometer. In order to acquire an image, thousands of spectra are obtained in an extremely short time with less than 100ms integration time per spectrum. However, since the Raman scattering cross section is small and the excitation power is restricted to a few milliwatts, it becomes a challenging taste to enhance the overall sensitivity of a confocal Raman system. Choosing an appropriate detector with the highest sensitivity is very important.

An electron multiplying CCD (EMCCD) is a standard back-illuminated CCD with an extra readout register which is driven with a higher clock voltage than a normal CCD readout register. Owing to this high clock voltage, an electron multiplication via impact ionization is obtained by amplifying the signal up to 1000 times. With this arrangement, it is possible to increase the signal higher than the readout noise so that the signal-to-noise ratio is always restricted by the Poisson noise of the signal, even if an extremely fast readout amplifier is employed. For instance, a 1600 x 200 pixel EMCCD with a 2.5MHz readout amplifier can be read out in just 2.3 ms. For higher signals, wherein the intensity of signal is not readout limited, the excess noise factor of the EM process reduces the EMCCD's signal-to-noise ratio to less than that of a standard CCD. In such cases, the EM register can be switched off and the "normal" readout register is utilized.

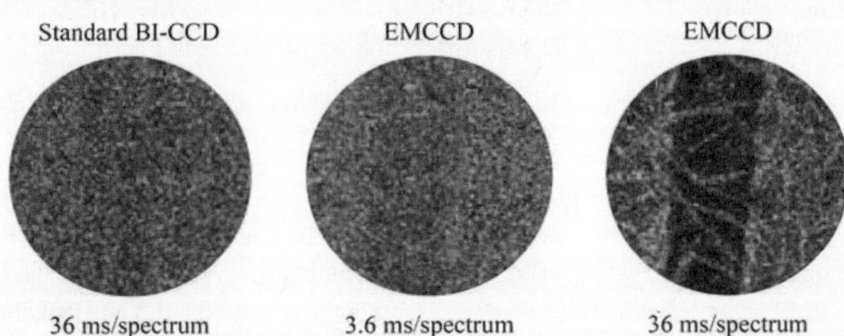

Fig. 5.20 a-c. Confocal Raman images of a 7.1 nm thin PMMA layer on glass obtained in the CH_2 stretching band around 3000 cm^{-1} (scale bar: 10 μm) in CCD and ECCD mode at different integration time

Figure 5.20 compares the Raman Imaging of a contaminated PMMA on a glass surface with standard CCD (5.20a) and with EMCCD (5.20b,c). The image 5 demonstrates almost the same signal-to-noise ratio, but with an EMCCD the integration time was only 3.6 ms, which is 10 times faster than Figure 5.19a. Image acquisition took 25 minutes with CCD while it took only 3.4 minutes for EMCCD. One can obviously see the scratch and contamination by way of a needle-like structure across the glass surface and PMMA. Thus, with the help of an EMCCD camera, speed and detection efficiency can be greatly increased, particularly for the short integration times required with a confocal Raman microscope. The distribution of a 7.1 nm PMMA can be easily identified with an integration time of 7ms per spectrum, which considerably reduced the overall acquisition time to 5.4 minutes for a 200 x 200 spectra confocal Raman image. For larger signals, the electron multiplying circuit can be switched off and all properties of a normal back-illuminated CCD are maintained, while for extremely small signals the EMCCD can greatly enhance the signal-to-noise ratio by a factor of 5 to 10.

Suggested Further Readings

1. S.Thomos, R thomos et al. (Ed). Microscopy Methods in nanomaterial Characterization, Elsevier Inc. 2017.
2. M. Ruhle, F. Phillipp, A. Seeger and J. Heydenreich, Ed., "High-Voltage and High-Resolution Electron Microscopy", *Ultramicroscopy*, 56 (1994) 1–3.
3. S. Horiuchi, Fundamentals of High Resolution Transmission Electron Microscopy, North Holland, Amsterdam, 1994.
4. A. Stroscio and W. J. Keiser, *Scanning Tunneling Microscopy*, Academic Press, 1993.
5. F. J. Giessibl, Advances in Atomic Force Microscopy, *Rev. Mod. Phys.* 75 (2003) 949-983.
6. L. Ding, Y. Li, H. Chu, X. Li and J. Liu, Creating cadmium sulphide nanostructures using AFM dip pen nanolithography, *J. Phys. Chem. B*, **109** (2005) 22337-22340.
7. T. Pradeep, Nano: The Essentials, Tata McGraw Hill, New Delhi, 2007.
8. D. Fairhurst, "An Overview of the Zeta Potential Part 3: Uses and Applications", Am. Pharm. Rev. Featured articles, Posted: June 13, 2013.

REVIEW QUESTIONS

Q1. What do you understand by terms: i. Resolution, ii. Resolving power, iii. Resolution limit.

Q2. What is the basic difference between microscopy and spectroscopy?

Q3. How the electron microscopy does is different from optical microscopy?

Q4. Which all signals may generate upon interaction of electron beam with matter and how are they being utilized in characterization of nanomaterials?

Q5. Illustrate basic difference between scanning mode and transmission mode of electron microscopy with the help of schematic diagrams.

Q6. Illustrate the basic principle of the techniques discovered in late eighties, led to revolutionary development in nanoscience and nanotechnology.

Q7. What are the basic components of Scanning Electron Microscope and their respective functions?

Q8. Discuss basic principle of Transmission Electron Microscopy (TEM).

Q9. Describe mechanism of contrast in an image recorded by TEM.

Q10. What is tunneling current? Discuss the operating modes of STM.

Q11. What is the basic difference between STM and AFM? Which of the two can be used for manipulation at atomic level in the designed synthesis of nanomaterials?

Q12. Discuss the basic principle and unique capabilities of Atomic Force Microscopy. Illustrate the answer with examples.

Q13. Discuss different modes of recording AFM images.

Q14. What do you understand by 'Confocal Microscopy'? How this concept has been used in Raman imaging technique?

Q15. How electron multiplying charge couple device (EMCCD) is superior to standard charge couple device (CCD) is Raman imaging?

■■■■

Unique Properties of Nanomaterials

6

6.1 ORIGIN OF UNIQUE PROPERTIES OF NANOMATERIALS

The great significance gained by nanotechnology in the last couple of decades is mainly due to the unique chemical and physical properties demonstrated by materials in the size regime of 1-100 nm in at least one dimension. Basically, nanomaterials represent a state of matter in transition between bulk and atomic or molecular level. Classical Newtonian physics is replaced by quantum mechanics to explain the unusual behavior and properties of materials at the nanoscale. At nanoscale. Size effects constitute a peculiar and fascinating aspects pertaining to the evolution of unique structural, thermodynamic, electronic, spectroscopic, electromagnetic, kinetic and chemical features.

To cite few examples, gold is known as a shiny, yellow noble metal that does not tarnish is non-magnetic and melts at 1336 K. However, when in nano size regime the same gold is quite different; 10 nm particles absorb green light and thus appear red. The melting temperature decreases dramatically as the size goes down. At 2–3 nm size, gold ceases to be noble, turns to be an excellent catalyst and also exhibits considerable magnetism. Rutile titanium oxide changes its phase to anatase at critical gain size of 50 nm; antiferromagnetic chromium (III) oxide turns super magnetic in nano size regime.

The present chapter intends to explain the origin of such special behavior of nanomaterial while addressing the obvious questions such as: Why does size influence the material's properties? How does size influence the material's performance? Why are properties of nanoscale objects different than those of the same materials in the bulk?

Generally speaking, the unique behaviour of nanomaterials is attributed to three basic phenomena:

1. **Quantum size effect:** At the nanometer scale the confinement of electrons, phonons, and photons leads to the appearance of new chemical, physical, and biological properties that makes them new materials in their own right. Quantum effect including quantum confinement, tunneling and surface plasmon resonance dominates the optical, electrical and magnetic properties of materials. Thus at nanoscale Newtonian physics is taken over by quantum physics to explain many of the physical properties.
2. **Large surface area**: Nanomaterials have a relatively high surface area to volume ratio resulting in the presence of a large number of atoms/ molecules at the surface as compared to the bulk material. As a consequence, surface atoms have higher energy than atoms in the bulk or in interior of the material thus making nanomaterials chemically more reactive. In some cases, inert materials in their bulk form may turn reactive when produced in their nanoscale form.

3. **Self-assembly**: Nanomaterials show ordered structure or patterns through self-assembly of atoms/molecules usually via non-covalent interactions viz. ionic, hydrophobic, van der Waals, hydrogen and coordination bonds along with kinetically labile covalent bonds. The intrinsic mobility of constituting atoms/molecules leads to ordered nanostructures upon equilibration between aggregated and non-aggregated states, thus providing a number of interesting designed properties such as error correction, self-healing, molecular recognition and high sensitivity to external stimuli.

6.1.1 Quantum Size Effects

According to **Gleiter's definition,** the size effects in microstructures arise when its size, *d,* is reduced to a critical value $d \sim d^*$, of the scale length of physical phenomenon, viz. free path length of electrons, phonons, coherent length, screening length etc. **(Table 6.1)**. Fascinating quantum size effects in nanomaterials leads to make their properties tunable by changing the size arising due to various factors such as tunnelling effect; quantum confinement in metals and semiconductors; superparamagnetism in magnetic materials (section 6.5). Quantum confinement is further understood in terms of surface plasmon resonance and density of states in metals and semiconductors as discussed below.

Table 6.1. Length scale of some critical physical phenomena

Phenomenon	Length scale
Electric Transport	Fermi wavelength, $\lambda F \sim 0.1$nm, Scattering length, $l = 10$-100 nm
Optical Interaction	$\frac{\lambda}{2n} = 100$-300 nm
Magnetic dipole Interaction	$l = 10$ nm
Range of exchange interaction	~ micron
Thermal	Average phonon free length is 100s of nm at 300 K

6. 1.1.1 Tunneling effect

Tunnelling is a quantum mechanical phenomenon. Classical physics states that a body can pass through a barrier (potential barrier) only if it has enough energy to “jump" over it. Therefore, if the object has lower energy than that needed to jump over the energy barrier (the “obstacle”); the probability of finding the object on the other side of the barrier is zero. In quantum physics, on the other hand, a particle with energy less than that required to jump the barrier has a finite probability of being found on the other side of the barrier mainly due to the tunneling effect (**Fig. 6.1a**). Tunneling is the penetration of an electron into an energy region that is classically forbidden. In terms of quantum tunneling, a barrier may be a form of energy state analogous to a "hill" or incline in classical mechanics, which classically suggests that passage through or over such a barrier would be impossible without sufficient energy. Electrons exhibit wave behavior and their position is presented by a wave (probability) function. The wave function represents a finite probability of finding an electron on the other side of the potential barrier. Since the electron does not possess enough kinetic energy to overcome the potential barrier, the only way the electron can appear on the other side is by tunneling through the barrier. In order to have tunnel effect the thickness of the barrier (*i.e.*, energy potential) must be comparable to the wavelength of the particle; in other words, electron (or quantum) tunneling is attained when a particle (an electron) with lower kinetic energy is able to exist on the other side of an energy barrier with higher potential energy, thus defying a fundamental law of classical physics. Therefore this effect is observed only at nanometer level.

In quantum mechanics, electrons are also assumed to behave as a wave thereby donot end abruptly at a wall or barrier, but taper off quickly. If the barrier is thin enough, the probability function may extend into the next region, through the barrier. Tunnelling is an effect of the wave- like nature of an electron.

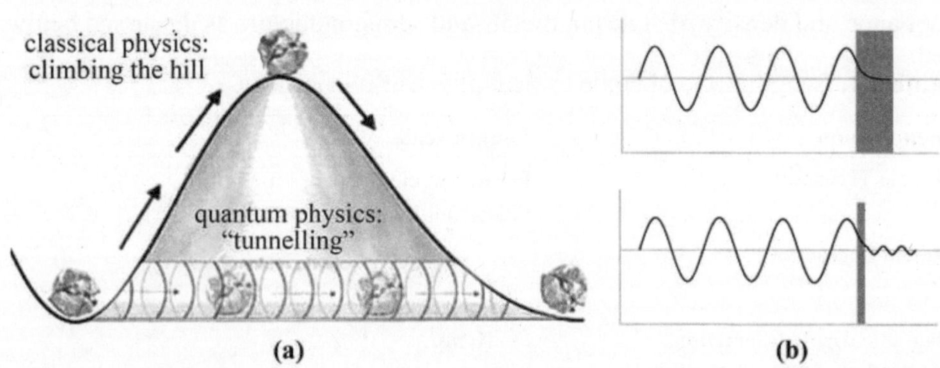

Fig. 6.1. Tunnelling effect

Figure 6.1(b), shows that when an electron (the wave) hits a thicker barrier, the wave doesn't abruptly end, but tapers off very quickly – exponentially-- without crossing the barrier, whereas if the barrier is quite thin (~ a nanometer), part of the wave does get through and therefore some electrons may appear on the other side of the barrier. Because of the sharp decay of the probability function through the barrier, the number of electrons that will actually tunnel is very much dependent upon the thickness of the barrier. The current through the barrier drops off exponentially with the barrier thickness. Scanning tunnelling microscope (STM) is an example of one of the nanotechnology developed based on the tunnelling electron and is used to characterize nanomaterials **(Fig. 6.2)**. A newly developed theory suggests quantum tunneling as an instantaneous process which could lead to faster and smaller electronic components, for which quantum tunneling is a significant factor. It will also lead to a better under-standing of diverse areas such as electron microscopy, nuclear fusion and DNA mutations.

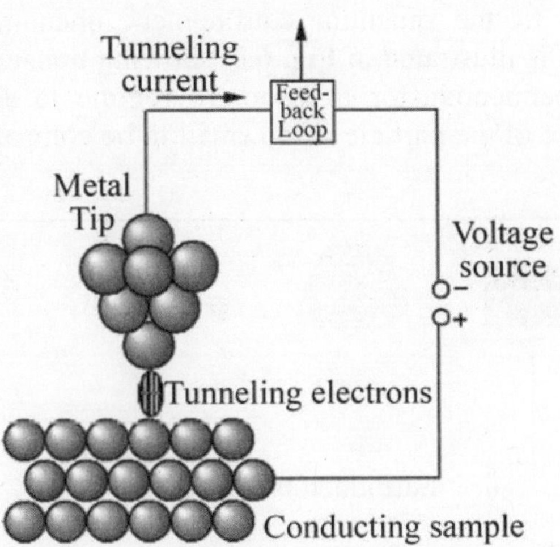

Figure 6.2 Tunnelling current in STM

6.1.1.2 Quantum confinement

Different types of materials viz. metals, semiconductors and insulators can be understood on the basis of the overlap or separation between valence and conduction bands **(Fig. 6.3).** Metals are those materials in which the electrons in the outermost atomic orbitals are bound loosely and behave very much like free electrons. Metal have their outermost electron filled band (valence band) overlapping with empty band (conduction band). On the other hand, if valence and conduction bands do not overlap and is characterized by a band gap energy $E_g>0$ then the material is either semiconducting or insulating. The band gap energy in semiconductors ranges between 1 and 3.5 eV while for insulators it is above 4eV. In semiconductors some electrons can still be promoted to conduction band, they show some conductivity which can be improved further on doping semiconductors by selected impurities (dopant). By introducing small amount of dopant to the host material, some localized energy states can be introduced in the energy gap. The electrical conductivity in many semiconductor devices is controlled by such dopants. Depending upon whether such levels are close to valence band or close to conduction band, the semiconductor becomes *p* or *n* type semiconductor.

As a material is reduced in size to that in nano size regime *i.e.* the size of a few tens of atoms we are pretty near to the size of a single atom. In such a scenario, band gap is formed in metal nanoparticles whereas band gap widens up in semiconducting nanoparticles together with the formation of discrete energy levels within both the conduction and valene bands of metal and semiconducting nanoparticles due to the quantum confinement phenomenon. The quantum confinement effect is illustrated in **Fig. 6.3**, showing transition of electronic state from bulk metal/semiconductor to nanosize regime to molecular/atomic size levels when the size of the particle is too small to be compared to the wavelength of the electron.

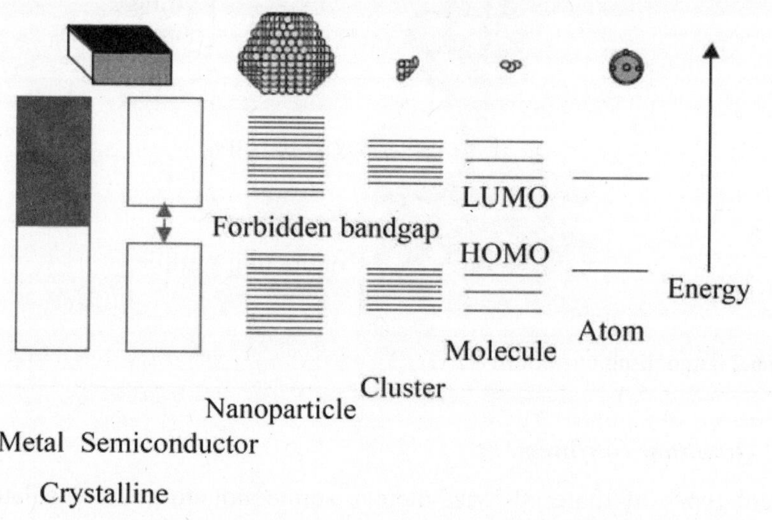

Fig. 6.3. Comparision of band energy diagram of metals and semiconductors in bulk, and size quantization effects in clusters, nanoparticles, molecules and atoms

The phenomena can be better understood on breaking the words like quantum and confinement; the word "confinement" means to confine the motion of randomly moving electron to restrict its motion in specific energy levels (discreteness) and quantum reflects the atomic realm of particles. So as the size of a particle decreases till we reach a nanoscale the decrease in confining dimension makes the energy levels discrete and this increases or widens up the band gap and ultimately the band gap energy also increases. An electron behaves as if it were free when the confining dimension (by the boundary of the particle) is large compared to the de Broglie wavelength of the electron, and its energy spectrum is (quasi-) continuous. However, as the confining dimension decreases and reaches a certain limit, typically in the nanoscale, the energy spectrum turns to discrete.

Since the electron levels in a quantum well are discrete (rather than continuous as in a bulk metal), this means the only allowed 'transitions of the outermost electrons are those between the discrete levels.

Of course this is also true for single atoms. Hence nanostructures are sometimes called artificial atoms. For the infinite quantum well the possible transition energies, Δ *E,* are of the type

$$\left(\Delta E = (m)^2 - n^2\right)\pi^2 \frac{h^2}{2ma^2}$$

where, *m* and *n* are integers with m > n > 0 and ђ is reduced Hinesburg constant. This means the lowest possible transition energy is

$$\pi^2 \frac{h^2}{2\,ma^2} \quad (m{=}2 \text{ and } n{=}1)$$

For the 3D infinite quantum well the possible transitions are more complicated, but the list of zeroes for the Bessel function given above shows that they are of the order of

$$\Delta E \sim \frac{h^2}{2\,ma^2}$$

Also, the lowest levels in finite quantum wells are actually of the same order as this. So let's use this simple formula to estimate the possible transition energies. Evaluating all the constants

$m = m_e$ gives

$$\Delta E \sim \frac{0.38\,\text{eV}}{(\text{a in mm})^2}$$

which means the transition would typically be 0.04 eV or less. Such low energy level transitions between discrete energy levels can be generated by photons or phonons (quanta of lattice vibrations), and have impact on optical or vibrational/ acoustic properties. Further, the $1/a^2$ dependence means that by changing the quantum well size (*i.e.* the size of a thin film or nanoparticle) it is possible to fine tune the energy at which the energy absorption or emission occurs.

Moreover, quantum confinement takes into account the aspect of the electronic structure of the nanoparticles that in turn critically depends on the size of the particles. Quantum confinement can be classified as strong, weak or no confinement depending on the particle size, a_j and its relation to the length of an exciton.

$a < \bar{l}$ strong confinement

$a > \bar{l}$ weak confinement

a >> $\bar{l}$ no confinement

A quantum confined structure is classified into three categories on the basis of four basic dimensions: quantum dot, quantum wire, quantum well or nano crystal **(Table 6.2)**.

Table 6.2. Classification of quantum confined nano structures

Structure	Quantum Confinement	Free dimensions
Quantum dots (0-D)	3	0
Quantum wires (1-D)	2	1
Quantum wells (2-D)	1	2
Nano crystal (3-D)	0	3

For 0-D nanomaterials the electrons are fully confined, whereas for 3-D nanomaterials the electrons are fully delocalized. In 1-D and 2-D nanomaterials, electron confinement and delocalization coexist. In the case of 2-D nanomaterials, the conduction electrons will be confined across the thickness but delocalized in the plane of the sheet. The effect of confinement on the resulting energy states can be calculated by quantum mechanics, as the "particle in the box" problem. An electron is considered to exist inside of an infinitely deep potential well (region of negative energies), from which it cannot escape and is confined by the dimensions of the nanostructure.

6.1.1.3 Density of states

In metals and semiconductors, the electronic wave functions of conduction electrons are delocalized over the entire particle. These electrons can, therefore, be described qualitatively as particles in a 'box'. Thus the densities of states (DOS) and the energies of the particles critically depend on the size of the 'box' which leads to some extent to a smooth size dependence. For example, ionization energies and electron affinities are tuned between the atomic values and the work function of the bulk material by variation of the cluster size. The density states, which essentially is the number of different states at a particular energy level (discrete energy level) that electrons are allowed to occupy, *i.e.* the number of electron states per unit volume per unit energy with the variation of nanoparticle size generates different electronic properties from bulk matter such as the conduction-insulator and nonmagnetic-magnetic transition of noble metals (*e.g.* Au at ~2-3 nm) at the nanoscale.

It may be noted that the formation of density of states in metal and semiconductor nanoparticles, is totally an electronic effect. In a metal, the quasi-continuous density of states in the valence and the conduction bands splits into discrete electronic energy levels; the spacing between these levels and the band gap increase with decrease particle size. Metal particles consisting of 50 to 100 atoms with a diameter between 1 and 2 nm start to lose their metallic character and tend to become semiconductors. Particles that show this size quantization effect are called Q-particles or quantum dots.

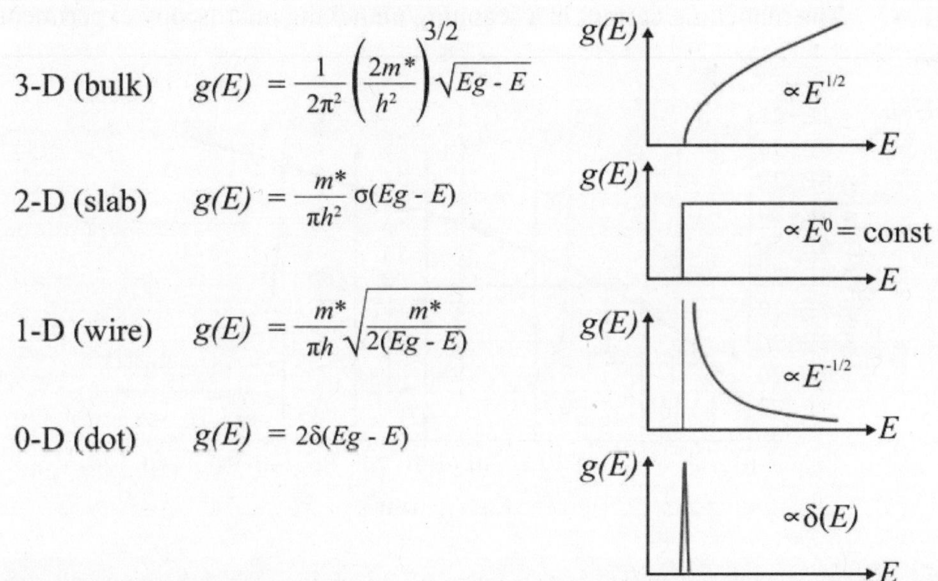

Fig. 6.4. Equations for the density of states in different dimensions and corresponding plots of DOS vs. energy

As illustrated in **Fig. 6.4** passing from three dimensions to two dimensions the density of states *g(E)* changes from a continuous dependence $g(E) \sim E^{1/2}$ to a step like dependence. The optical absorption edge for a quantum well is at a higher photon energy than for the bulk semiconductor and, above the absorption edge, the spectrum is stepped rather than smooth the steps corresponding to allowed transitions between valence-band states and conduction-band states, while, at each step, sharp peaks appear corresponding to electron-hole (exciton) pair states. In case of a zero dimensional systems, the density of states is illustrated as a delta function. **Figure 6.5** illustrates super-imposes plots of DOS vs. energy over a range of values for each dimension to further visualize different behavior between dimensions.

The density of states has a more peaked structure with decrease of dimensionality. This leads to a reduction of threshold current density and a reduction of the temperature dependence of the threshold current. A few examples of material properties attributed the DOS are:

- The specific heat of solids depends on both the phonon and electron density of state
- Photon absorption and emission in optical devices.
- Magnetic susceptibility of a magnetic material

- Concentration of electrons and holes in semiconductors
- Superconducting energy gap
- The tunnelling current in a scanning tunnelling microscopy experiment

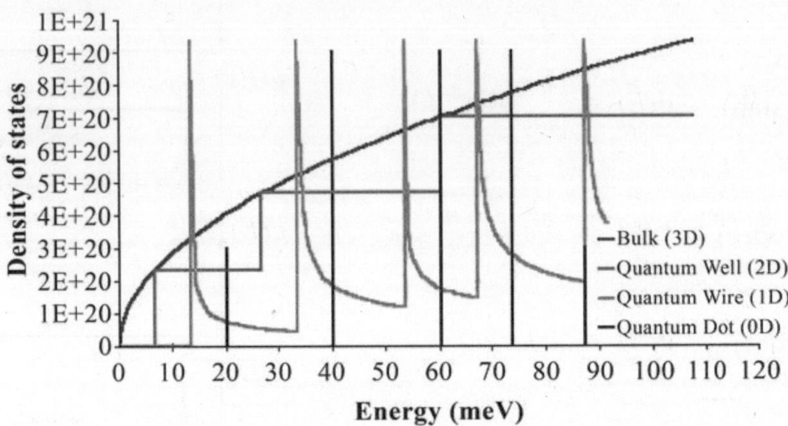

Fig. 6.5. Plots DOS vs. energy over a range of values for each dimension and superimposes the curves over each other to further visualize the different behavior between dimensions

It should be noted that a particle of few nanometers contains several hundred atoms so that discontinuities are no longer observed and the wavelength shift with size appears smooth. It is nevertheless more of a quantum effect since it is governed by the nature of standing waves of a confined system rather than by the fraction of atoms at the surface.

6.1.1.4 Surface plasmon resonance

When a small spherical metallic nanoparticle is irradiated by light, the oscillating electric field causes the conduction electrons to oscillate coherently. When the electron cloud is displaced relative to the nuclei, a restoring force arises from the Coulomb attraction between electrons and nuclei that results in oscillation of the electron cloud relative to the nuclear framework. The oscillation frequency is determined by four factors: the density of electrons, the effective electron mass, and the shape and size of the charge distribution. The collective oscillation of the electrons is called the dipole plasmon resonance of the particle (sometimes referred to as "dipole particle plasmon resonance" to distinguish from plasmon excitation that can occur in bulk metal or metal surfaces). Higher modes of plasmon excitation can occur, such as the quadrupole mode where half of the electron cloud moves parallel to the applied field and half moves antiparallel.

This is referred to as the surface plasmon resonance (SPR), since it is located at the surface.

Figure 6.6 shows schematic of surface plasmon absorption of spherical nanoparticles, illustrating the excitation of the dipole surface plasmon oscillation. The electric field of an incoming light wave induces a polarization of the (free) conduction electrons with respect to the much heavier ionic core of a spherical metal nanoparticle. A net charge difference is only felt at the nano particle surfaces, which in turn acts as a restoring force. In this way a dipolar oscillation of the electrons is created. Optical absorption spectra of spherical gold nanoparticles (20-100 nm) reveals broad absorption band in the range of 500-600 nm corresponding to the surface plasmon resonance.

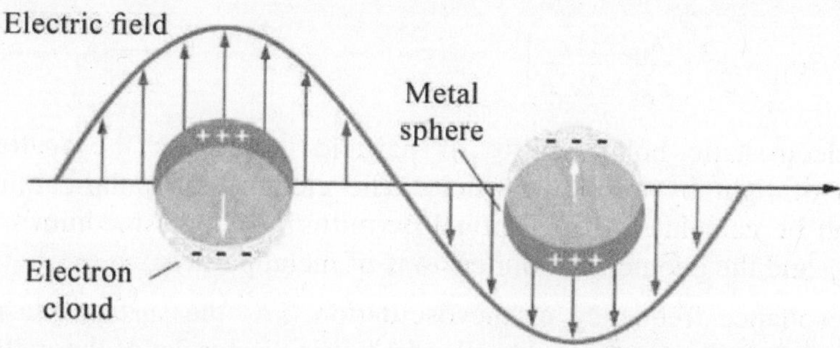

Fig. 6.6. Schematic of plasmon oscillation for a sphere, showing the displacement of the conduction electron charge cloud relative to the nuclei

The optical resonance related to these oscillations is called the localized SPR (LSPR). The origin of the term "surface" comes from the knowledge that the oscillations are caused by the polarization of the particle surface, and because the generated electric field is larger near the particle surface and decays farther away far from the surface, which makes that field similar to the evanescent field at a flat interface in the propagating SPR case.

Light in resonance with the surface plasmon oscillation causes the free electrons in the metal to oscillate. As the wave front of the light passes, the electron density in the particle is polarized to one surface and oscillates in resonance with the light's frequency causing a standing oscillation. The resonance condition is determined from absorption and scattering spectroscopy and is found to depend on the shape, size, and dielectric constants of both the metal and the surrounding material. The surface geometry changes with the change in shape or size of the nanoparticle causing a shift in the electric field density on the surface. This causes a change in the oscillation frequency of the electrons, generating different cross-sections for the optical properties including absorption and scattering.

Changing the dielectric constant of the surrounding material will have an effect on the oscillation frequency due to the varying ability of the surface to accommodate electron charge density from the nanoparticles. Changing the solvent will change the dielectric constant, but the capping material is most important in determining the shift of the plasmon resonance due to the local nature of its effect on the surface of the nanoparticle.

In general, the frequency of the collective oscillations does not coincide with the applied wave frequency and is determined by many factors such as electron concentration, electron effective mass, the particle shape and size, interaction between the neighbouring particles, and the influence of the environment. However, for an elementary description of plasmon resonance of nanoparticles it is sufficient to use the usual dipole approximation and the Drude model (Eq. 6.1).

$$\alpha_0 = \alpha^2 \left(\frac{\varepsilon_m - \varepsilon_d}{\varepsilon_m + 2\varepsilon_d} \right) \tag{6.1}$$

The electrostatic polarizability of particle α_0 assigns the scattering and absorption of light by a small particle. The electrostatic polarizability of the particle can be calculated using optical permittivity ε_m, the medium's dielectric constant ε_d, and the geometrical dimensions of metal particle.

The resonance frequency of the oscillation, *i.e.*, the surface plasmon (SP) energy, is essentially determined by the dielectric properties of the metal and the surrounding medium, and by the particle size and shape. The collective charge oscillation causes a large resonant enhancement of the local field inside and near the surface of nano particles. This field enhancement is used in surface-enhanced Raman scattering (SERS, see chapter 4) as one of the applications.

Calculation of the longitudinal plasmon resonance for gold nanorods generates an increase in the intensity and wavelength maximum as the aspect ratio (length divided by width) increases. Thus, the plasmon resonance can be tuned across the visible region by changing the aspect ratio. The increase in the intensity of the surface plasmon resonance absorption leads to an enhancement of the electric field, as exploited in many applications. For gold and silver nano particles, the resonance falls into the visible region. A striking consequence of this phenomenon is the bright colours exhibited by particles both in transmitted and reflected light, due to resonantly enhanced absorption and scattering. Surprisingly, this effect has found applications for many hundreds of years ago, though without any human knowledge at that time, in the staining of glass windows or ornamental cups.

6.1.2 High Surface Area

In general, surfaces and interfaces play an important role in the properties of materials as their size is reduced to nanometers. Owing to their small dimension, the surface-to-volume ratio increases which makes a large fraction of atoms of the materials at the surface (30% for a 1-nm crystal, 15% for a 10-nm crystal). **Figure 6.7** represents the dispersion of surface and core atoms in nanoparticles.

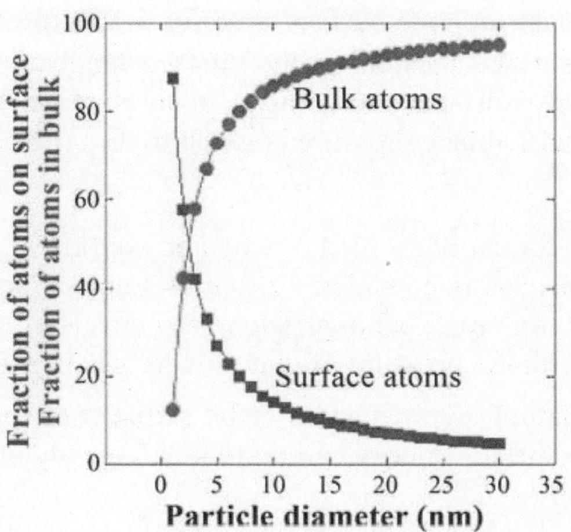

Fig. 6.7. Fractional composition of surface and core atoms vs. nanoparticle size

Surface atoms are characterized by more weak bonding and high surface energy and reactivity as compared to those atoms in the interior. The surface of a sphere scales with the square of its radius r^2, but its volume scales with cube of radius r^3. The total number of atoms N in this sphere increases linearly with volume. The fraction F of atom at the surface is called dispersion, and it increases with surface area divided by volume. Atoms or molecules on a solid surface possess fewer nearest neighbors or coordination numbers, and thus have dangling or unsatisfied bonds exposed to the surface. Because of the dangling bonds on the surface, surface atoms or molecules are under an inwardly directed force and the bond distance between the surface atoms or molecules and the subsurface atoms or molecules, is smaller than those between interior atoms or molecules. This situation arises irrespective of whether the bonding is covalent (as in a metal), ionic (in a salt), or non-covalent (in a liquid such as water). When solid particles are very small, such a decrease in bond length between the surface atoms and interior atoms becomes significant having consequential effect on a number of their physicochemical properties.

The presence of higher numbers of atoms/molcules at the surface of nanomaterials is important to understand surface effects, *i.e.* surface energy, dangling bonds, etc., and their implications in giving rise to the unique physical, chemical, and biological properties differing radically from their bulk counterparts.

The free surface energy (Gibbs energy, ΔG_S) in nanomaterial is given by Eq. (6.2):

$$\Delta G_s = \gamma . S_f \quad (6.2)$$

Where, γ is a specific surface energy (the energy per surface unit, 2 J/m^2), or a specific work to create the free surface area, or a specific tension. The surface energy causes the surface tension or the surface tangential force, $F=\gamma\ a$. The tension exists for any surface, including a plane surface. For nanoparticles of radii (r), an additional Laplace pressure is given by Eq. (6.3):

$$PL=2\ \gamma\ /r\ . \quad (6.3)$$

It is responsible for capillary phenomena and wetting of liquids. The Laplace tension for solid particles is considered often as Laplace pressure. For instance, the Laplace pressure for small nanoparticle (r=10 nm) is of the order of 200 MPa and is comparable with the pressure of gun powder gas in a gun barrel.

As per law of nature, every system to be stable tries to attain the minimum surface energy. The surface energy can be reduced by any one or combination of the following means:

- Surface relaxation: surface atoms or ions move inwardly.
- Surface restructuring: combining the dangling bonds into new chemical bond.
- Surface adsorption (physical or chemical).
- Segregation of impurity through solid state diffusion.

Any of the above means results into reduction of surface energy and give rise to novel physical, chemical and biological properties in nanomaterials. Especially, when the sizes of nanomaterials are comparable to Debye length (the distance over which a charge Q is shielded by the ions in a solution), the surface atoms are highly unsaturated, have higher surface energy, and their electronic contribution to the behaviour of the particles is totally different from that of the inner atoms. These effects may be even more pronounced when the surface atoms are ligated. This leads to different electronic transport and catalytic properties of the nanocrystalline particles. Thus nanostructured materials can be engineered with larger surface area and improved reactivity to create better catalysts; electrode material for batteries and fuel cells to produce cleaner, safer, and more affordable modes of producing and storing energy; to makes nanostructured membranes and materials ideal candidates for water treatment and desalination, among other uses. It also helps support “functionalization” of nanoscale material surfaces (adding particles for specific purposes), for applications ranging from drug delivery to thermal insulation of clothing.

6.2 ELECTRONIC PROPERTIES

The term electronic properties denote physical quantities which are directly related to the response of the charge carriers on electric, magnetic, and electromagnetic fields. To understand the electronic properties first we have to understand the electronic structure and band theory of the solid materials. The electronic properties of solids (energy levels, types of bonds between atoms, energy bands, energy gap, Fermi level, E_F) are of primary importance. The best description of the electronic properties is obtained using quantum mechanics, although classical descriptions can be used for macroscopic solids. The basic steps of quantum mechanics studies for solids include:

- The Schrödinger's equation
- The energy levels and atomic orbitals for the hydrogen atom
- Atoms with n electrons
- Bonds between atoms, hybridization of atomic orbitals
- Energy bands in solids

Electrons in solids are in localized or delocalized states, levels, or orbitals. The inner or core states are localized, and are very similar to the states in free atoms. The outer or valence electrons are delocalized in the sense that the can extend outside the region of just one atom. In metals, the delocalization of the valence electron is such that we can say that they are shared by all atoms in the solid. Two important properties of electronic states are their binding energy (U) and width ($\Gamma = \Delta U$). Binding energy, is the minimum energy needed to remove a particular electron from the atom. The width of the electronic state is related to t, the lifetime of a hole produced in the state when we remove the electron. This is the Heisenberg relationship: G = h/2pt. Since h/2p = 6.582 x 10^{-16} eV.sec, the widths of core levels are 0.001 - 0.1 eV, whereas the width of valence band states are 0.1 eV to several eV.

An operational definition of binding energy in electron spectroscopy is the energy that would be needed to put the electron at the Fermi level. The Fermi level of the material where the atom may not contain available delocalized empty levels (case of non-metals) but still the energy can be referred to that of a metal in electrical contact and in thermal equilibrium with the material, which shares the same Fermi level.

The electronic properties of a material are determined by its Fermi surface, a construction in momentum space. A 3D representation of Fermi surface is shown in **Fig. 6.8**; the grid pattern in the cutout denotes that there are discrete quantum states, defined by momentum (k_x, k_y, k_z) for electrons to occupy. Energy (E) of electrons is related to their momentum state (k) by Eq. (6.4)

$$E = h^2 k^2 / 2m \tag{6.4}$$

The observable electronic properties are manifestations of excitations from occupied electron states to unoccupied electron states, and the Fermi surface is the hard boundary between the two (at zero temperature) or the fuzzy boundary between the two (at finite temperature).

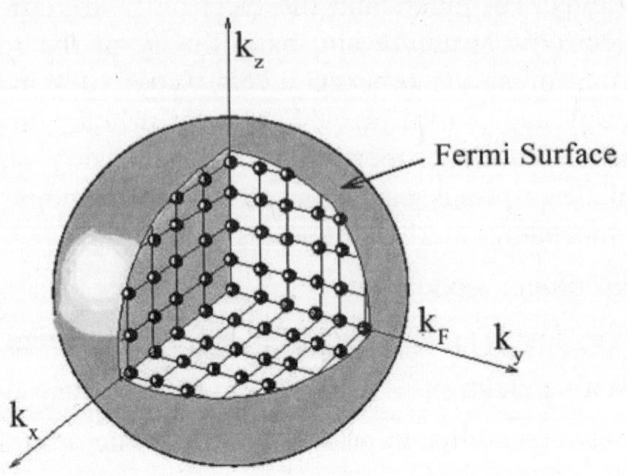

Fig. 6.8. Fermi surface in 3D

The band-gap in semiconductors and insulators results from the interplay between the electron wavelengths and the symmetric arrangement of atoms in the solid. At the surface, the symmetry is broken which relaxes the condition that inhibits electrons to propagate inside the lattice. The symmetry is also broken by defects and impurities. Broken symmetry results in electronic states which can exist in the band gap. The existence of states in the band have that pertain to the surface, or surface states, is the general rule.

With consideration of the small size of a semiconductor nanocrystal, its electronic properties are significantly affected by the transport of a single electron, giving the possibility of producing single electron devices. The complex interplay between charge transfer, hybridization, and charge transport is the key to their application potential and to design novel materials with interesting electronic, optical and magnetic properties such as one-dimensional metals or spin chains, two-dimensional devices based on the cuprate superconductors, and solar cells.

The electronic structure of nanomaterials certainly differ from atoms, molecules or solids in bulk form. Nanomaterials are in fact an intermediate form of solids. They are too small to be considered as bulk and too big to be considered as molecules. To start with an atom, bringing two atoms together to form a molecule changes the electronic structure. If three atoms come together there is further change. In fact, addition of each atom would change the electronic structure till the number of atoms becomes too large. Depending upon the type of atoms, at large numbers, the clusters of such atoms would reach bulk electronic structure. Electronic structures are useful to classify the materials as metals, semiconductors and insulators.

The electronic properties of nano-sized materials may differ strongly from their bulk counterparts owing to quantum size effect, quantum confinement leading to formation of density of states and electronic transition between the discrete energy levels and charge quantization. The significance of later effect is that the change in energy (E_c) is much greater than thermal energy (k_T). The energy interval between adjacent electronic levels changes a lot upon size reduction down to the nanometer. This feature gives a way to engineer the band gap in some semiconducting materials.

In the quantum world, electrons behave like waves, and the electrical transverse conductance of the two-dimensional material in the presence of a magnetic field is no longer like a stream. It rather changes in discrete steps. The phenomenon is called Quantum Hall Conductance. In metals, dimensionality affects electronic density of states and Fermi surface topology which has implications for conduction and tendencies towards interesting electronic phases. In semiconductors and insulators. Dimensionality can affect the size of band gaps and if they are direct or not.

The valence and conduction bands of all materials are drawn as continuous but are actually composed of closely spaced, discrete states **(Fig. 6.3)**. These states have a regular separation Δk inversely proportional to L, the macroscopic length of the sample, and for large L (compared to the atomic lattice spacing) the states are effectively continuous. At nanoscale, the discrete levels are pronounced and separation value Δk, may vary depending on the size and dimensionality of nanoparticles. Under these conditions, electrons no longer freely hop from one state to another: the change in either momentum or energy becomes too large for the available thermal fluctuations and the material is considered to be strongly quantized.

The overall changes in electronic properties at nanoscale are briefly summarized as under.

1. When the size of the system becomes comparable with the de Broglie wavelength of the electrons, the discrete nature of the energy states becomes apparent once again, although a fully discrete energy spectrum is only observed in systems that are confined in all three dimensions.
2. In certain cases, conducting materials become insulators below a critical length scale, as the energy bands cease to overlap. Owing to their intrinsic wave-like nature, electrons can tunnel quantum mechanically between two closely adjacent nanostructures, and if a voltage is applied between two nanostructures which aligns the discrete energy levels in the DOS, resonant tunneling occurs, which abruptly increases the tunneling current. In addition, some of 1D nanometals are transforms a insulators due to their susceptible to 'Peierls transition' (a density wave). Thus, 1D metals are not stable at zero temperature. Interactions can also render a 2D metal susceptible to charge density wave instabilities more easily than this can happen in 3D.

3. Conduction in highly confined structures, such as quantum dots, is very sensitive to the presence of other charge carriers and hence the charge state of the dot.
4. These Coulomb blockade effects result in conduction processes involving single electrons and as a result they require only a small amount of energy to operate a switch, transistor or memory element.

All these phenomena can be utilized to produce radically different types of components for electronic, optoelectronic and information processing applications, such as resonant tunneling transistors and single-electron transistors.

Materials of reduced dimensionality encompass a wide variety of interesting electronic phenomena. A quantum well is a semiconductor structure where electrons are restricted along one dimension such that only several quantized energy levels are available along the restricted dimension. These structures are extremely useful as diode lasers. A 2D electron gas (2DEG) is like a quantum well except it is explicitly metallic (with very high mobility at low temperature) and it is almost perfectly two dimensional. These structures are important for studying quantum phenomena at low temperature, and have yielded important discoveries such as the integer quantum hall effect and the fractional quantum hall effect. Molybdenum disulphide is a quasi-2D indirect band-gap semiconductor made of layers weakly bound together. When a single layer of this material is achieved, it has a direct band gap which leads to a wide variety of applications including low power photodetectors, LEDs (including ones that emit circularly polarized light), and transistors.

Electrical conduction in most nanomaterials is an outcome due to the motion of complex electronic excitations called quasiparticle. In bulk materials, these quasiparticles act like electrons (or holes) which have an effective mass that, because of these interactions, can be larger or smaller than that of an electron, but otherwise behave much the same. Quasiparticles have parabolic energy **(refer chapter 2 fig. 2.11).** However, something different happens in case of graphene. In graphene transport of charge behave as if they had no mass; that is, their effective mass is zero. The mass less change carriers move very fast with 0.3 per cent of the speed of light.

Charge carriers in graphene are described by a Dirac-like equation, rather than the usual Schrödinger equation. As a consequence of its perfect crystal structure, low-energy quasiparticles in it obey a linear dispersion relation, similar to massless relativistic particles. This essential characteristic of a gapless semiconductor has led to many observations of peculiar electronic properties in graphene including ballistic transport, pseudo spin chirality based on the "Berry phase", a room-temperature half-integer "chiral" quantum Hall effect, and conductivity without charge carriers. All these attributes make graphene a promising choice for future electronic materials, both as a device and as an interconnect. For example, graphene has the fastest electron mobility of ~15,000

$cm^2 V^{-1} cm^{-1}$ or 10^6 Ω cm (lower than Ag), a super high mobility of temperature-independent charge carriers of 200,000 $cm^2 V^{-1} s^{-1}$ (200 times higher than Si), and an effective Fermi velocity of 10^6 m s^{-1} at room temperature, close to the speed of light.

6.3 ELECTRICAL PROPERTIES

There are three categories of materials based on their electrical properties viz. conductors, semiconductors, and insulators. The energy separation between the valence band and the conduction band is called E_g (band gap). The ability to fill the conduction band with electrons and the energy of the band gap determine whether a material is conductor, a semiconductor or an insulator **(Fig. 6.3)**. In conducting materials like metals, the valence band and the conducting band overlap, so that the value of E_g is small: thermal energy is enough to stimulate electrons to move to the conduction band. In semiconductors, the band gap is few electron volts. If an applied voltage exceeds the band gap energy, electrons jump from the valence band to the conduction band, thereby forming electron-hole pairs called excitons. Insulators have large band gaps (E_g > Uev) that require an enormous amount of voltage to overcome the threshold. This is why these materials do not conduct electricity. As the size of a particle is brought down in the nanometer regime, the quantum confinement effect causes the band gap to increase, thus a metal become semiconductors on decreasing its size. Electrical properties of some nanomaterials are related to their unique structures and exhibit electrical properties that are absolutely exceptional. For instance, carbon nanotubes can be conductors or semiconductors depending on their nanostructure.

Some of the electrical properties such as electrical conductivity, dielectric properties, electrical strength, voltage endurance of nanomaterials are discussed below.

6.3.1 Electrical conductivity

The effects of size on electrical conductivity of nanomaterials are complex owing to the operation of distinct mechanisms. These mechanisms can be generally grouped into different categories *i.e.* surface scattering including grain boundary scattering and widening and discreteness of band gap. Reduction in material's dimensions would have two different contrasting effects on electrical conductivity. Firstly, at nanoscale a matrial demonstrate the crystal perfection and as well as reduction in defects. As a result electron scattering phenomenon due to crystal defects are reduced and a reduction in resistivity is experienced. However, at room temperature the defect scattering incident contributes a minor influence on the total electrical resistibility of various metals. On the other hand, surface scattering phenomenon which is highly increased due to reduction of particle size is one of the prominent reasons for increase of the total resistivity. In addition, a reduction in particle size below a critical dimension, *i.e.* (electron de

Broglie wavelength) result in an increase in electrical resistivity. Furthermore, increased perfection, such as reduced impurity, structural defects and dislocations, would enhance the electrical conductivity of nanostructures and nano-materials. The general trend followed is reverse in that a metal tends to be semiconducting and a semiconducting material moves toward insulating on size reduction.

Nanocarbon in different allotropic forms such as 2D (graphene), 1D (carbon nanotubes), or 0D (Buckeyball) exhibit distinct electrical conductivity. Graphene, a zero-overlap semimetal (with both holes and electrons as charge carriers) is well known for having very high electrical conductivity. While semimetals generally have far smaller electrical and thermal conductivity than metals, this however is not the case with graphene. It is a good electrical and thermal conductor having an intrinsic resistivity about two-thirds of the resistivity of silver. When the dimensionality is increased by putting several graphene layers together (eventually making graphite), the resulting band structure moves to that of a more trivial metal. Diamond is an insulator.

The electrical properties of carbon nanotubes depend on how the hexagons are oriented along the axis of the tube. **Figure 6.9** shows the three orientations that are possible: armchair, zigzag, and chiral. Carbon nanotubes with the hexagons oriented in the arm chain configuration labeled armchair (hexagons are lined up parallel to the axis of the nanotube) have electrical properties similar to that of metals. When voltage is applied between two ends of an armchair nanotube, current will flow. An armchair carbon nanotube is, in fact, a better conductor than the copper normally used in electrical wire, or any other metal.

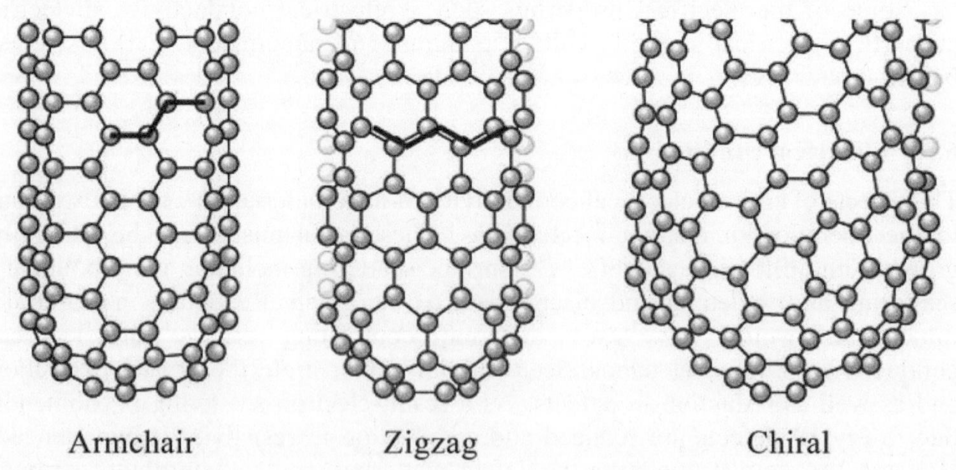

Fig. 6.9. Electrical properties depend on the orientation of the hexagons

This electrical conductivity of MWNTs can be comparable with that of independent graphene sheets. When the tube diameter (d_t) is smaller than the elastic mean free path (*le*), the one-dimensional ballistic transport predominates, while, if d_t is larger than *le*, the current flow could be described as diffusive/two-dimensional transport. Due to its unusual structure and extended π-electron system, CNTs exhibit a variety of outstanding electronic and electrical properties (1000–200,000 S/cm). The strong bonds between carbon atoms also allow carbon nano-tubes to withstand higher electric currents than copper. Electron transport occurs only along the axis of the tube. Single walled nanotubes can route electrical signals at speeds up to 10 GHz when used as interconnects on semi-conducting devices. Nanotubes also have a constant resistively. It has also been established that CNTs could exhibit amphoteric behavior by exchanging electrons with electron acceptors or donors to form the corresponding positively or negatively charged counter ions. One-dimensional nanomaterial (carbon nano-tubes, Titania nanotubes, silica nanowires, polymeric nanowires and nanofibers) are the most attractive materials due to having diameters sufficiently small to produce a quantum confinement of charge carriers. Such confinement permits free-electron behaviour in only one direction, along the length of the wire and therefore, improved electronic transport can be achieved compared to that of semiconductor quantum dots. Unique structure, optical and electrical properties of these one-dimensional (1D) nanostrucures make them the key building blocks for a new generation of nanoelectronics, sensors, and nanophotonic devices.

6.3.2 Dielectric Properties

A dielectric is an electrical insulator that can be made to exhibit an electric dipole structure (displace the negative and positive charge so that their center of gravity is different). When two parallel plates of area *A,* separated by a small distance *l*, are charged by $+Q$, $-Q$, an electric field (E) developed between the plates is given by equation 6.5.

$$E = D / \varepsilon \varepsilon_0 \tag{6.5}$$

where, $D = Q/A$. ε_0 is vacuum permittivity and ε the relative permittivity, or *dielectric constant* (ε = 1 for vacuum). In terms of the voltage between the plates, $V = E\,l$,

$$V = Dl/\ \varepsilon \varepsilon_0 = Q\ l/A\ \varepsilon \varepsilon_0 = Q / C \tag{6.6}$$

The constant $C = A\ \varepsilon \varepsilon_0 / l$ is called the capacitance of the plates.

The dielectric properties of materials are mainly due to contributions from electronic, ionic, dipolar and space charge polarizations. Among these, the most important present in the optical range of frequencies. The next contribution is from ionic polarization, which arises due to the relative displacement of positive and negative ions. Dipolar or orientation polarization arises from molecules having a permanent electric dipole moment that can change its orientation when an electric field is applied. Space charge polarization arises from molecules having a permanent electric dipole moment that can change its orientation when an electric field is applied. The main contribution for polycrystalline materials in bulk form is the electronic polarization, The dielectric constant, the response function of the measured external field (the displacement) to the local electric field, closely relates to the conductivity and optical properties of materials.

The real and imaginary parts of the dielectric constant ε' and ε'' of a material can be calculated from the measured values of capacitance and dielectric loss using the formulae (Eqs. 6.7 and 6.8).

$$\varepsilon' = C_p.d / \varepsilon_0 A \tag{6.7}$$

$$\varepsilon'' = \varepsilon^{T'} \tan\delta \tag{6.8}$$

where d is the thickness, A is the cross sectional area of the given sample, ε_0 is the permittivity of free space and tan δ is the dielectric loss factor.

A large volume of crystalline materials is consist of grain or interface boundaries. These boundaries contain defects such as dangling bonds, vacancies, vacancy clusters, etc. These defects can cause a positive or negative space charge distribution at interfaces. The space charge can move on applying an external field and when they are trapped by the defects lot of dipole moments are formed. This process is known as space charge polarization. Hence the space charge effect is a prominent factor, to decides the dielectric properties in materials with small particle sizes. In addition, ion jump polarization may also be greater in nanocrystalline materials since there will be a number of positions in the grain boundaries for the ions to occupy. The high values of the dielectric constant may be attributed to the increased ion jump orientation effect and the increased space charge effect exhibited by nanoparticles.

Most of the atoms in the nanocrystalline materials reside in grain boundaries, which become electrically active as a result of charge trapping. The dipole moment can easily follow the changes in electric field, especially at low frequencies. Hence, the contributions to dielectric constant increases through space charge polarization and rotation polarization, which occur mainly in interfaces. Therefore, dielectric constant of nanostructured materials should be larger than that of conventional materials. The negative dielectric constant of bulk metal decreases with the particle size and approaches positive in accord with the material that behaves as an insulator.

One of the reasons for the large dielectric constant of nanocrystalline materials at sufficiently high temperature is the increased space charge polarization due to the structure of their grain boundary interfaces. Also, at sufficiently high temperature the dielectric loss is dominated by the reason for the sharp increase of the dielectric constant at low frequencies and at lower temperatures.

As temperature increases, the space charge and ion jump polarization decreases, resulting a decrease in dielectric constant.

The dielectric loss of a material, very important property for MW absorption is given by Eq. 6.9 as

$$\tan\delta = (\in'')/(\in') = 4\pi\sigma / \in'\omega \tag{6.9}$$

where σ is the conductivity and ω is the frequency.

It has been observed that tan δ decreases with increase of frequency and at higher frequencies the loss angle has almost the same value at all temperatures. In dielectric materials, usually dielectric losses occur due to absorption current. The orientation of molecules along the direction of the applied electric field in polar dielectrics requires a part of electric energy to overcome the forces of internal friction. Another part of electric energy is utilized for rotations of dipolar molecules and other kinds of molecular transfer from one position to another, which also involve energy losses.

In nanomaterials, the grain boundaries have an amorphous or glassy structure. All the inhomogeneity, defects, space charge formation, etc. together produces an absorption current, which results in dielectric losses. Due to the presence of dangling bonds at the surface layers, nanoparticles is highly reactive and there is a chance of adsorption of gases like oxygen or nitrogen. These adsorbed gases can also cause an increase in dielectric loss in nanomaterials.

i. Electric Strength and Voltage Endurance

When a molecule is is subjected an electric field, it will respond in different ways depending on the type of material. Conducting materials have loosely bonded valence electrons, which when acted upon by an electric field, are free to move between the molecules within the material. In a capacitor material, the electron need to be strongly bonded not allowed to move between molecules. In an insulating or dielectric, the material has electron bonds that are stronger than that of conducting materials.

The magnitude of the electric field at which the dielectric material undergoes intrinsic breakdown is called the dielectric strength. There are many factors that contribute to the dielectric strength of a material such as density, impurities, geometry of the molecular structure, temperature, humidity, the frequency of the applied voltage, rise time of the voltage, and many other factors like whether the applied voltage is ac or dc.

Studies have shown that dielectric nano fillers can improve the dielectric breakdown strength, enhance the voltage endurance, depress the space charge formation and weaken the partial discharge degradation of the pristine polymers more significantly than the micro-sized fillers. These improvements are of significance to electrical insulation applications.

ii. Internal Charge Characteristics

A common feature of most of the nano dielectric systems is the behavior of the internal space charge build-up. The recent development of systems to measure the internal charge in dielectrics has allowed both the magnitude and dynamics of the internal charge to be assessed. It appears that there are substantial differences when nanoparticles are incorporated into a polymer matrix forming nanocomposites. Three features are usually predominant in most of the nano dielectric systems:

(a) The magnitude of the internal charge is much less,
(b) The dynamics of charge decay are much faster, and
(c) There is often a very different distribution of charge (and therefore of internal field).

6.4 OPTICAL PROPERTIES

6.4.1 Basic concept

Light has dual nature, consisting of particles called photons, which travels like a wave and its energy *E* is

$$E = h\nu = \frac{hc_o}{\lambda} \tag{6.10}$$

where, *h* is the Plank's constant, *v* is frequency of light waves, c_o is velocity and λ is the wavelength of light waves. Optical properties of material is governed by its interaction with light. If the energy of photons is transferred to the material, it is absorption; if photons of identical energy are immediately emitted by the material, it is reflection; when photons do not interact with the material, the process is transmission; or during transmission when photons change their velocity, the phenomena is known as refraction.

At any instance of interaction of light with a material, the total intensity of the incident light, I_0 striking a surface is equal to the sum of the absorbed I_A reflected I_R, and transmitted intensities I_T and expressed as

$$I_0 = I_A + I_R + I_T \tag{6.11}$$

Scattering: When the size of the particles of material are comparable to the wavelength of incident light scattering occurs. As the size of the matter is reduced, scattering (*S*) of light also contributes to its color or transparency. Both metals and non-metallic materials show optical properties. In metals, when photons are directed at metals, their energy is used to excite electrons into unoccupied states. Metals in general are opaque to the visible light; though transparent at high end frequencies *i.e.* x-rays and γ-rays. The absorbed radiation is emitted from the metallic surface in the form of visible light of the same wavelength as reflected light. Thus, all the above four optical phenomena are important.

Refraction: If the energy of the incoming light is the same as the vibration frequency of the electrons in the material, the light is able to go deep into the material, and cause small vibrations in the electrons. The vibrations are then passed on from atom to atom, each vibrating at the same frequency as the incoming light source. This makes light inside the material look bent. Example: a straw in a glass of water. The part under water will look bent. In other words when light photons are transmitted through a material, they cause polarization of the electrons and in turn the speed of light is reduced and as a consequence, light changes its direction in the medium. The refractive index η, is the ratio of light velocity in air c_o and in the medium c, as

$$n=\frac{c_o}{c} \tag{6.12}$$

Speed of light c can be related to its electrical and magnetic properties, as given in Eq. 6.13.

$$c=\frac{1}{\sqrt{\mu.\varepsilon}} \qquad n=\frac{c_0}{C}=\frac{\sqrt{\mu.\varepsilon}}{\mu_0.\varepsilon_0}=\sqrt{\mu_r.\varepsilon_r} \tag{6.13}$$

where μ is the magnetic permeability and ε is the dielectric permittivity. Since most materials are only slightly magnetic:

$$n \cong \sqrt{\varepsilon_r} \tag{6.14}$$

From, Snell's law of light refraction:

$$\eta=\frac{\sin\theta'}{\sin\theta}$$

Reflection (*R*): Most objects reflect light, but some are relatively more reflective than others, like metals. This is directly related to the number of free electrons that are able to pass from atom to atom with ease. Instead of absorbing energy from the light, the free electrons vibrate and the light energy is sent out of the material at the same frequency as that of the original light. Reflectivity is defined as fraction of light reflected at an interface (Eq. 6.15).

$$R=\frac{I_R}{I_0} \tag{6.15}$$

If the material of refractive index n under consideration is embeded in other material with refractive index η_i the reflection R of the composite is given by equation 6.16.

$$R = \left(\frac{n - n_i}{n + n_i}\right)^2 \tag{6.16}$$

Materials with a high refractive index have a higher reflectivity than materials with a low refractive index. Because the index of refraction varies with the wavelength of the photons, so does the reflectivity. In metals, the reflectivity is typically of the order of 0.90-0.95, whereas for glass it is close to 0.05.

Absorption (*A*): When there is no reflection (the object is opaque), then the incoming light source frequency is the same as, or very close to, the vibration frequency of the electrons in the given material. The electrons thus absorb most of the incoming energy, with little or no reflection. The energy levels of a substance determine the wavelengths of light that can be absorbed. It is a molecular phenomenon, dependent on the chemical identity and structure of the substance (not on the size of the molecules or clusters), and involves electronic transitions, vibrations and rotations. Chromophores and fluorophores are examples of organic materials that have specific transitions.

Transmission (*T*): If the incoming light energy is much lower or much higher than that required for the electrons comprising an object to vibrate, then the light source will pass through the material unchanged. This way matter will look *transparent* to the human eye, for instance glass. Fraction of light beam that is not reflected or absorbed is transmitted through the material.

Scattering (*S*): It is the phenomenon that occurs when radiation hits a surface composed of particles with dimensions comparable to the incident wavelength. In scattering no energy transformation occurs (as opposed to absorption); energy is simply redirected in many directions. The wavelength of the incoming light and that of the outgoing light are the same. Scattering simply contributes to the reflection (back scattering) and transmission (front scattering) parts. The light which has been absorbed cannot be scattered.

Optical properties of nanomaterials are significantly different from bulk materials, since absorption and/or emission wavelength is controlled by particle size and surface functionalization. At nano size, movement of electrons is restricted in a confined space, therefore, nano particles react differently with light. The optical emission and adsorption occurs when the transition of the electrons occur between highest occupied molecular orbital (HOMO) and lowest unoccupied molecular orbital (LUMO) states. The effect of reduced dimensionality on electronic structure has the most profound effect on the energies of HOMO which is the valence band and the LUMO, essentially the conduction band. As the system becomes more confined, energy level spacing is increased. Semiconductors and many metals show large changes in optical properties in terms of color and luminescence as a function of particle size. Surface plasmon resonance shown by specific metals also attributes to the unique optical properties of nanomaterials.

6.4.2 Color

Basically, color of any matter can be considered as an electromagnetic phenomenon, which depends on the way light is reflected by the objects. Every object absorbs a part of the incident light and reflects the rest towards our eyes: this reflected light is interpreted by our brain as a particular color. Apart from absorption, there can be a number of other distinct causes of color production such as reflection, scattering, interference, etc. **(Table 6.3).**

Table 6.3. Colour- causing phenomena in nanostructured objects

Nano objects	Colour causing phenomena
Butterfly wings (photonic crystal). Liquid crystals (*e.g.* soap)	Constructive interference of light wavelength with the nanomaterial
Colloids (*e.g.* milk).	Scattering of light of wavelength matching with particle size
Metal colloids (nano-gold and nano-silver).	Surface plasmons. This is a peculiar effect found in metal nanoparticles. Responsible for vivid colours of metal colloids
Semiconductor quantum dots (QDs)	Quantum fluorescence arises due to quantum confinement in nano-sized semiconductors which leads to discrete energy levels from which energy can be emitted (fluorescence) after it has been absorbed by the semiconductor.

Metal nanoparticles were supposed to be in use as decorative pigments since the time of the Romans. The most glaring example in glass of the famous Lycurgus cup (4th century AD). The cup kept still, at the British Museum possesses the unique feature of changing color depending upon the light in which it is viewed (**Fig. 1.3,** chapter 1). It appears green when viewed in reflected light, but looks red when a light is shone from inside and is transmitted through the glass. Analysis of the glass reveals that it contains a very small amount of nano (~70 nm) metal crystals containing Ag and Au in an approximate molar ratio of 14:1.

Metals with free electrons (essentially Au, Ag, Cu, and the alkali metals) possess plasmon resonances in the visible spectrum, which give rise to such intense colors. The color variations of nanoparticles arise from changes in the composition, size, and shape of nanoparticles, as well as from the proximity of other metal nanoparticles. Optical properties of metal nanoparticles depend strongly upon the particle size and shape. Bulk Au looks golden yellow in reflected light, but thin Au films look blue in transmission. This characteristic blue color steadily changes to orange, through several tones of purple and red, as the particle size is reduced down to ~3 nm (**Fig. 6.10**). These effects are the result of changes in the so-called surface plasmon resonance, the frequency at which conduction electrons oscillate in response to the alternating electric field of incident electromagnetic radiation.

Gold nanoparticles in different geometric shapes and their corresponding UV-Vis spectra is already illustrated in chapter 4 **(Fig. 4.5)**. A distinct red shift can be observed as the shape of particles changes from spherical to decahedral or rod shaped. The longitudinal oscillation is very sensitive to the aspect ratio of the particles, so that even a slight deviation from spherical geometry can lead to impressive colour changes. Apart from single particle properties, the environment in which the metal particles are dispersed is also of relevance to the optical properties of neighboring metal nanoparticles.

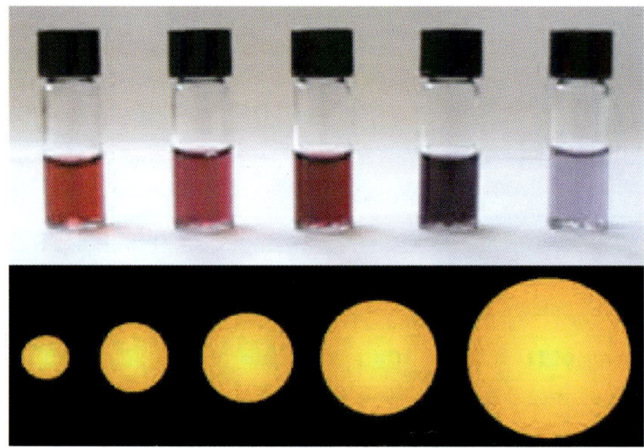

Fig. 6.10. Colour variation of gold nanoparticle of different sizes

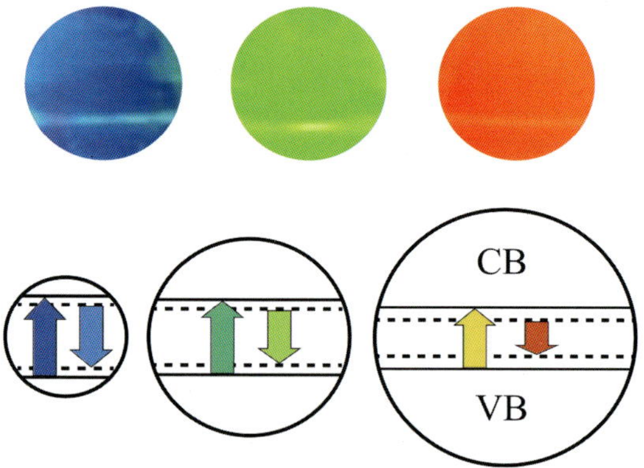

Fig. 6.11. Fluorescence of CdSe–CdS core-shell nanoparticles with a diameter of 1.7 nm (blue) up to 6 nm (red), giving evidence of the scaling of the semiconductor band gap with particle size

6.4.3 Luminescence

Luminescence is an emission phenomenon of light by a substance which may be caused by a number of factors such as chemical reactions, electrical energy, subatomic motions or stress on a crystal. Luminescence differs from incandescence. The later is related to the light emitted by a substance as a result of heating. Nanomaterials showing luminescence can generally be classified into four classes:

a. Metal nanoclusters

b. Semiconducting quantum dots (Q-dots) and carbon dots (C-dots)

c. Metal-ion doped nanomaterials

d. Hybrids/nano composites.

a. Metal Nanoclusters

Noble metals exhibit extremely weak photoluminescence (Φca.10^{-10}) in the bulk, assigned to recombination of the excited electrons with the holes within the D band. In metal nano particles the luminescence yield increases by as much as 6 orders of magnitude more because of the so-called lightning rod effect; the incoming (exciting) and outcoming (emitted) electric fields are amplified by the plasmon resonances around the particles. Hence, the luminescence energy and efficiency are affected to the size effect.

b. Semiconductors

Semiconductor nano particles show interesting linear absorption, photoluminescence emission, and nonlinear optical properties. The simplest example is the well-known blue-shift of absorption, *i.e.* the optical absorption peak of a semiconductor nanoparticle shifts to a short wavelength due to an increased band gap with the decrease in the particles size (**Fig. 6.11**).

Photoexcitation of a bulk semiconductor results in the transfer of an electron from valence band to conduction band creating an electron-hole pair called "exciton", bound by a weak columbic interaction. The minimum energy required to generate such an exciton is called the band -gap energy. For example, excitons can be treated as hydrogen-like system and the spatial separation between the charge carrier pair is termed as exciton Bohr radius (α_B) which can be deduced from the Bohr approximation. When the physical dimensions of matter become comparable or lower to the exciton Bohr diameter ($2a_B$), the functional properties of a semiconductor becomes sensitive to the size and shape due to confinement of excitons.

The size dependence of the HOMO–LUMO band gap is best seen in the luminescence properties of semiconductor nanoparticles. The fluorescence of colloidal CdSe–CdS core-shell nanoparticles, illustrated in **Fig. 6.11** shows blue fluorescence for particles with a diameter of 1.7 nm and particles of size up to 6 nm shows red fluorescence; giving an evidence of the scaling of the semiconductor band gap with particle size. Smaller particles have a wider band gap. By adjusting the particle size the fluorescence can be tuned between blue for particles of 1.7 nm diameter and red at 6 nm. A similar size effect but shifted to higher energy relative to fluorescence applies to the absorbed light. The difference in wavelengths between absorption and fluorescence occurs because after absorption the system relaxes to adapt to the new charge distribution, initiated by creation of the electron-hole pair that is also called exciton.

Photoluminescence that arises from recombination of the conduction band (CB) electron with the valence band (VB) hole is called Band edge luminescence. Such a luminescence is characterized by a very small Stokes shift relative to band edge absorption. Band edge emission is important in many applications. To achieve band edge emission, the surface of the nanoparticle can be passivated by chemically binding a capping agent. Another approach consists in over coating a nanocrystal of a given semiconductor with a thin layer of a semiconductor with higher band gap (core-shell particles). As an example a core-shell semiconductor consisting of a direct band gap CdSe core and indirect band gap Si as shell (CdSe@Si).

c. Metal doped-nanomaterials

Lanthanide ion doped nano phosphors for example Gd@ZnS are promising novel nanomaterials due to their excellent properties such as narrow emission lines (color sharpness), long radiative lifetimes, large Stokes shifts, high quantum yields of luminescence, non-photo bleaching and low toxicity.

d. Hybrids/Nanocomposites

For very small nanoparticles the main problem lie in that large number of dangling bonds of surface atoms induce additional states into the band gap. To remove this effect, often the nanoparticles are manufactured surrounded by a shell of another semiconductor, which passivates the dangling bonds. Such particles are called core-shell nanoparticles, for examples CdSe/CdS; CdSe/ZnS. The core usually consists of an inorganic phase. Another interesting examples of composites consist of an oxide core, a layer of organic fluorescence molecules and an outer protective layer **(Fig. 6.12).** These composite nano particles exhibit many advantages, which allows the various combination of different properties, formation of broad range of different colours in materials and dispersability of materials in many different liquids, independent of solubility.

6.4.4 Nonlinear Optical Properties

Nonlinear optical (NLO) properties arise as a consequence of interaction of intense electromagnetic field with a materials to produce modified fields that are different from the input field in phase, frequency or amplitude. Second harmonic generation (SHG) is a nonlinear optical process that results in the conversion of an input optical wave into an output wave of twice the input frequency. The light propagated through a crystalline solid, which lacks a center of symmetry, generates light at second and higher harmonics of the applied frequency. Such frequency- doubling processes are commonly used to produce green light (532 nm) from, for example, a Nd:YAG (yttrium-aluminium-garnet) laser operating at 1064 nm. This important nonlinear property of non-centrosymmetric crystals is called second harmonic generation. Lithium niobate ($LiNbO_3$), potassium dihydrogen phosphate (KDP) and some π-conjugated organic molecules are well documented second order NLO materials. Nonlinear optical properties of metal nanoparticles are also of great interest for applications in ultrafast optical switches. In particular, the third-order susceptibility of metal nanoparticles at wavelengths around the plasmon resonance achieves large values with very fast (<1 ps) response times. This is a consequence of the relaxation of the nonequilibrium electron distribution generated in the metal nanoparticles.

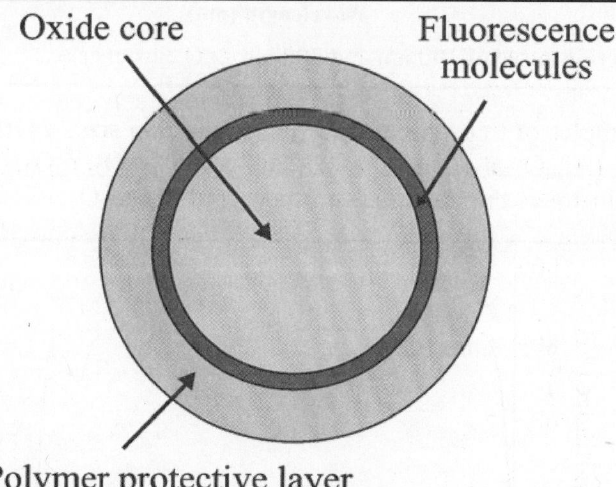

Fig. 6.12. Luminescent hybrids

6.4.5 Transparency in Nanoceramics

Ceramics are generally opaque as light hardly propagate through them. Scattering of visible light is responsible for the white appearance. A very common example is the white paste-like appearance of sunscreen cream containing ZnO and TiO_2 clusters of ~200nm. Visible light interacts with these clusters and all of its wavelength are scattered; combination of the visible spectrum is white. When dimension of cluster are reduced from 200 nm to 100 nm, maximum scattering occurs around 200 nm and the curve is shifted towards shorter wavelength (**Fig. 6.13**), below the visible region, rendering the cream as transparent.

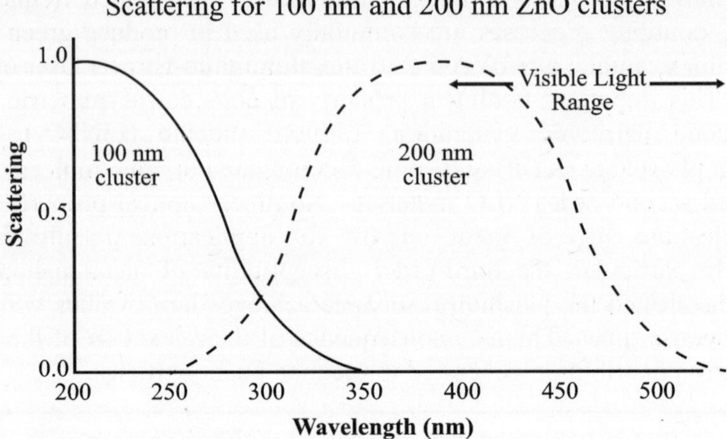

Fig. 6.13. Scattering curve of 100 nm and 200 nm ZnO clusters

Some examples of transparent ceramics with grain size, d<100 nm are Al_2O_3, YAG, ALON ($Al_{23}O_{27}N_5$), Mg-spinel ($MgAl_2O_4$), ZrO_2:Y_2O_3, Y_2O_3:ThO_2, etc. **Figure 6.14** illustrates the transmission spectra of $MgAl_2O_4$.

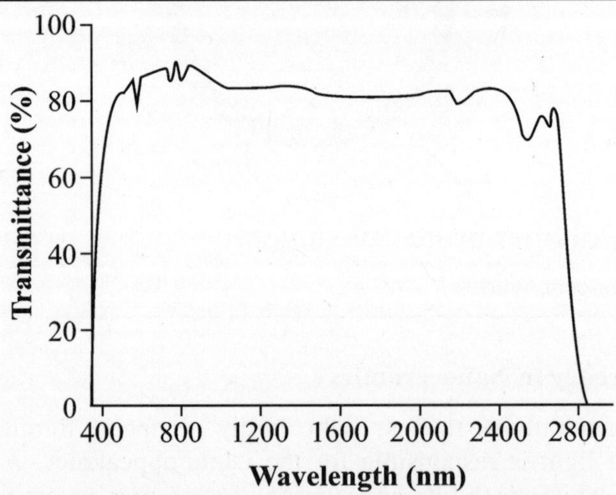

Fig. 6.14. Transmission spectra of glass ceramics-Mg-spinel ($MgAl_2O_4$)

Transparent ceramics having excellent mechanical and good optical properties are used as passive optical components such as waveguides, lenses, mirror supports, optical filters for high power optical beams, etc. Further, such ceramics can be potential replacement of glasses in armors, house appliances, lamps, LED's, etc. Increasing number of active optical devices incorporate ceramics as phosphors, scintillators, optical amplifiers, in high power lasers and multifunctional materials in general.

6.5 MAGNETIC PROPERTIES

In general, the magnetic behavior of a material depends on the structure of the material and on its temperature. In order to feel a magnetic field a material need to have non-zero net spin (*e.g.* transition metals). The typical size of classically expected domain is around 1 μm. When the size of a magnetic material is reduced, the number of surface atoms becomes an important fraction of the total number of atoms, surface effects become important, and quantum effects starts to prevail. When the size of these domains reaches the nanoscale, these materials show new properties. The large surface area to volume ratio results in a substantial proportion of atoms having different magnetic coupling with neighboring atoms which lead to different magnetic properties. Some illustrated examples are given below:

1. Bulk gold and platinum are nonmagnetic but in the nano size regime they act as magnetic particles. Au nanoparticles become ferromagnetic when they are capped with the appropriate molecules such as thiol.
2. Giant magnetoresistance (GMR) is a phenomenon observed in nanoscale multilayers consisting of strong ferromagnet (Fe, Co, Ni) and a weaker magnetic or nonmagnetic buffer (Cr, Cu). It is usually employed in data storage and sensing.
3. Magnetic nanoparticles showing superpara magnetism are used in a range of applications like imaging, bioprocessing, refrigeration as well as high storage density magnetic memory media.

Two key issues that dominate the magnetic properties of nanoparticles are finite-size effects and surface effects. Finite-size effects result, for example, from the quantum confinement of the electrons, whereas typical surface effects are related to the symmetry breaking of the crystal structure at the boundary of each particle. Because there is no general agreement on the size limits for nanoparticles, in the following we use this term for particles with diameters ranging from 1 to 100 nm.

6.5.1 Finite-size effects

The two most studied finite-size effects in nanoparticles are the single domain limit and the super paramagnetic limit models. In bulk magnetic material, it is well known that there is a multi-domain structure, where regions of uniform magnetization are separated by domain walls. The formation of the domain walls is a process driven by the balance between the magnetostatic energy (ΔE_{MS}), and the domain-wall energy (E_{dw}), which increases proportionally to the interfacial area between domains. If the size of particles of a material is reduced, there is a critical volume below which it costs more energy to create a domain wall than to support the external magne-tostatic energy (stray field) of the single-domain state. This critical diameter typically lies in the range of a few tens of nanometers and depends on the nature of the material. It is influenced by the contribution from various anisotropy energy terms. The critical diameter of a spherical particle (*Dc*), below which, it exists in a single-domain state, when $\Delta E_{MS}=E_{dw}$ is defined by equation 6.17

$$D_C \approx 18 \frac{\sqrt{AK_{eff}}}{\mu_o M^2} \tag{6.17}$$

where, A is the exchange constant, K_{eff} is anisotropy constant, μ_o is the vacuum permeability, and M is the saturation magnetization.

Figure 6.15 gives a simplistic view of some phenomena present in small magnetic particles. In reality, a competition between various effects establishes the overall magnetic behavior. Typical values of D_c for some important magnetic materials are listed in **Table 6.2.** A single-domain particle is uniformly magnetized with all the spins aligned in the same direction.

Table 6.2. Critical diameter of spherical magnetic nanoparticles#

Magnetic NPs	**D_c (nm)**
hcp Co	15
fcc Co	07
Fe	15
Ni	55
SmCo	750
Fe_2O_3	128

The magnetization will be reversed by spin rotation since there are no domain walls to move. This is the reason for the very **high coercivity** observed in small nanoparticles. Another source for the high coercivity in a system of small particles is the shape anisotropy. The departure from sphericity to shape with aspect ratio >1, for single-domain particles is significant and has an influence on the coercivity as is shown, in **Table 6.3** for Fe NPs.

Table 6.3. Influence of Shape of Fe NPs on Coercivity

Aspect ratio	**H_c (Oe)**
1.1	820
1.5	3300
2.0	5200
5.0	9000
10.0	10100

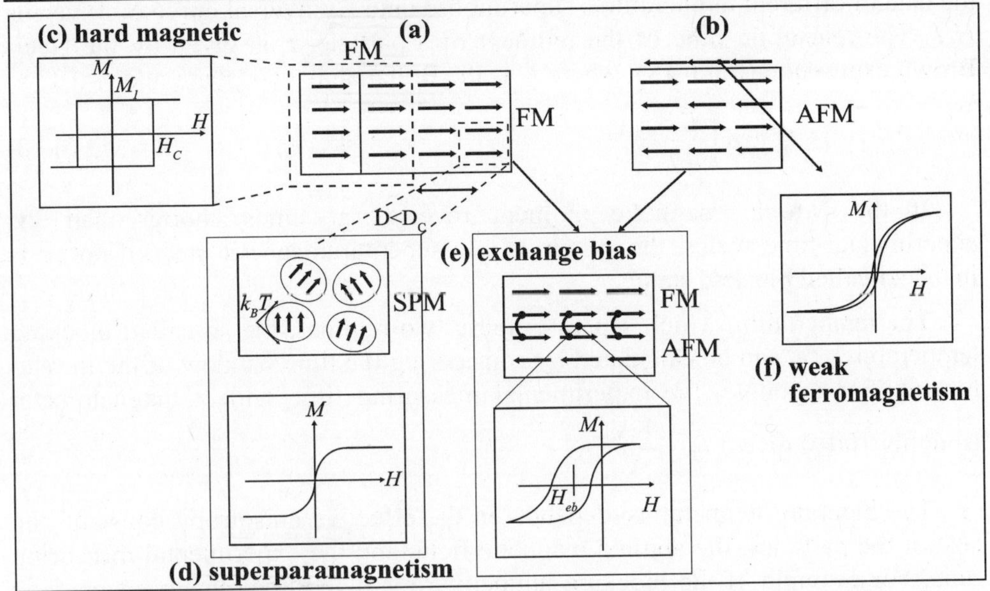

Fig. 6.15 The different magnetic effects occurring in magnetic nanoparticles. The spin arrangement in (a) ferromagnet (FM); (b) antiferromagnet (AFM); D=diameter, Dc=critical diameter; (c) combination of two different ferromagnetic phases; (d) illustration of the magnetic moments in a super paramagnet (SPM); (e) The interaction (exchange coupling); (f) pure antiferromagnetic

It must be noted that the estimation of the critical diameter holds only for spherical and non-interacting particles. Particles with large shape anisotropy lead to larger critical diameters.

The second important phenomenon which takes place in nanoscale magnetic particles is the superparamagnetism. A super magnet is defined as an assembly of giant magnetic moments which are not interacting and which can fluctuate when the thermal energy, k_BT, is larger than the anisotropy energy. The superparamagnetism may better be understood by considering the behavior of a well-isolated single-domain particle. The magnetic anisotropy energy per particle which is responsible for holding the magnetic moments along a certain direction can be expressed as follows: $E(\theta)=K_{eff}\,V\sin^2\theta$, where V is the particle volume is anisotropy constant, and θ is the angle between the magnetization and the easy axis K_{eff}.

The energy barrier $K_{eff}V$ separates the two energetically equivalent easy directions of magnetization. With decreasing particle size, the thermal energy kT, exceeds the energy, barrier, $K_{eff}V$ and the magnetization is easily flipped. For $k_BT > K_{eff}V$, the system behaves like a paramagnet, instead of atomic magnetic moments, there is now a giant (super) moment inside each particle (**Fig. 6.15d**). This system is named a superparamagnet. Such a system has no hysteresis and the data of different temperatures superimpose onto a universal curve of *M* versus *H/T*. The relaxation time of the moment of a particle, *t*, is given by the Neel-Brown expression (Eq. 6.18); where k_B is the Boltzmann's constant, and $t_0 \approx 10^{-9}$ s.

$$t = t_0 \, exp\left(\frac{KV}{kT}\right) \tag{6.18}$$

If the particle magnetic moment reverses at times shorter than the experimental time scales, the system is in a superparamagnetic state, if not, it is in the so-called blocked state.

The temperature, which separates these two regimes, the so-called blocking temperature, T_B, can be calculated by considering the time window of the measurement. For example, the experimental measuring time with a magnetometer (roughly 100 s) gives: $T_B = \frac{KV}{k}$.

The blocking temperature depends on the effective anisotropy constant, the size of the particles, the applied magnetic field, and the experimental measuring time. For example, if the blocking temperature is determined using a technique with a shorter time window, such as ferromagnetic resonance which has a $t_0 \approx 10^{-9}$s, a larger value of T_B is obtained than the value obtained from dc magnetization measurements. Moreover, a factor of two in particle diameter can change the reversal time from 100 years to 100 nanoseconds. While in the first case the magnetism of the particles is stable, in the latter case the assembly of the particles has no remanence and is superparamagnetic.

The most common techniques used to measure the magnetic properties of an assembly of nanoparticles include: SQID magnetometery and vibrating sample magnetometery. Ferromagnetic resonance (FMR) probes magnetic properties in the ground state and provides information about magnetic anisotropy, magnetic moment, relaxation mechanism of magnetization, and *g*-factor. X-ray absorption magnetic circular dichroism (XMCD) is the method of choice to determine the optical and spin magnetic moments.

6.5.2 Surface effects

As the particle size decreases, a large percentage of atoms in a nano-particle reside at the surface. It implies that surface and interface effects become predominant. For example, for face-centered cubic (fcc) cobalt with a diameter of around 1.6 nm, about 60% of the total number of spins are surface spins. Owing to this large surface atoms/bulk atoms ratio, the surface spins make an important contribution to the magnetization. This local breaking of the symmetry might lead to changes in the band structure, lattice constant or/and atom coordination. Under these conditions, some surface and/ or interface related effects occur, such as surface anisotropy and, under certain conditions, core–surface exchange anisotropy.

a. Non-magnetic coatings on magnetic nanoparticles

Surface effects can lead to a decrease of magnetization of small particles, for instance oxide nanoparticles, with respect to the bulk value. On the other hand, metallic nanoparticles, for example cobalt, shows an enhancement of the magnetic moment with decreasing size. This result is associated with a high surface-to-volume ratio.

Another surface-driven effect is the enhancement of the magnetic anisotropy, K_{eff}, with decreasing particle size. K_{eff} changes when the surfaces are modified or adsorb different molecules, which explains very well the contribution of the surface anisotropy to K_{eff}. For uncoated antiferromagnetic nanoparticles, weak ferromagnetism can occur at low temperatures **(Fig. 6.15f)**, which has been attributed to the existence of uncompensated surface spins of the antiferromagnet. A non-magnetic layer of precious-metal around the magnetic nanoparticles will have an influence on the magnetic properties. For example, it is shown that gold-coated cobalt nanoparticles have a lower magnetic anisotropy than uncoated particles, whereas gold coating of iron particles enhances the anisotropy, an effect which was attributed to alloy formation with the gold. Organic ligands, used to stabilize the magnetic nanoparticles, also have an influence on their magnetic properties, that is, ligands can modify the anisotropy and magnetic moment of the magnetic metal atoms located at the surface of the particles.

b. Magnetic coatings on magnetic nanoparticles

A magnetic coating on a magnetic nanoparticle usually has a dramatic influence on the magnetic properties. The combination of two different magnetic phases lead to new magnetic nanocomposites. The most striking feature which takes place when two magnetic phases are in close contact is the exchange bias effect. The exchange coupling across the interface between a ferromagnetic core and an antiferromagnetic shell or vice versa causes this effect. Exchange bias is the shift of the hysteresis loop along the field axis in systems with ferromagnetic (FM)–antiferromagnetic (AFM) interfaces. This shift is induced by a unidirectional exchange anisotropy created when the system is cooled below the Neel temperature of the antiferromagnet. This exchange coupling can provide an extra source of anisotropy leading to magnetization stabilization.

Exchange biased nanostructures have found applications in many fields such as permanent magnets, recording media and spintronics. A new approach to produce high-performance permanent magnets is the combination of a soft magnetic phase (easily magnetized), such as FePt, and a hard magnetic phase (difficult to magnetize and thus having high coercivity), such as Fe_3O_4 which interact through exchange bias effect, and the so-called training effect. The right choice of ferromagnetic and antiferromagnetic components can provide a structure suitable for use as a recording medium.

Above all, magnetic behavior of an assembly of nanoparticles is a result of both the intrinsic properties of the particles and the interactions among them. The distribution of the sizes, shapes, surface defects, and phase purity are only a few of the parameters influencing the magnetic properties.

6.6 MECHANICAL PROPERTIES

Almost all the nanomaterials possess high mechanical strength as compared to their bulk counterparts. The mechanical strength of nanomaterials may be one or two times higher in magnitude than that of single crystals in the bulk form. Defects in the form of atomic vacancies can lower the tensile strength of the materials by up to 85%. Conversion of materials into nanoscale increases crystal perfection or reduction of defects, resulting the enhancement in mechanical strength. Cutting tools which should be harder than the material in bulk state are made of nano materials; examplex includes tungsten carbide, tantalum carbide, and titanium carbide. These cutting tools are much more erosion-resistant and durable than their conventional large grained bulk materials. In general, hardness of metals (Cu) increases linearly with increase of grain size. However, in case of nanomaterial, shown in **Fig. 6.16** the hardness increases linearly with decrease in particle size **(Fig. 6.16).** At the nanoscale, surface and interface forces such as adhesion forces, capillary forces and strain forces become dominant. These forces can exceed those which are normally dominant at macroscopic length scales.

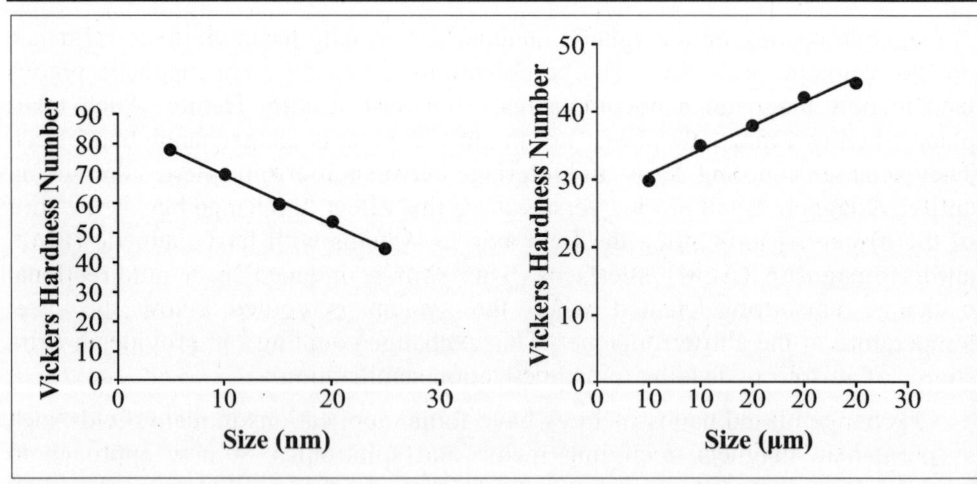

Fig. 6.16. Hardness variation for metal in nanometer and micrometre scale

Mechanical properties of nanoparticles consisting of metallic and ceramic materials is greatly influenced by porosity, grain size and composition and show exciting mechanical properties such as plasticity, enhancement of hardness, strength, fracture ductility and wear resistance.

Polymers reinforced with nanoparticles, nanorods and nanotubes, leads to significant improvements in their mechanical properties. Particulate filled polymer-based nanocomposites exhibit a broad range of failure strengths and strains. Although, the best composites are those filled with nanofibers or nanotubes; however, sometimes such composites have the least ductility. On the other hand, by using carbon nanotubes it is possible to produce composite fibers with extremely high strength and strain at rupture. Among the most exciting nanocomposites are the polymer-ceramic nanocomposites, where the ceramic phase is platelet-shaped. This type of composite is preferred in nature, and is found in the structure of bones and abalone where it consists of crystallized mineral platelets of a few nanometers thickness that are bound together with collagen as the matrix. Composites consisting of a polymer matrix and defoliated phyllosilicates exhibit excellent mechanical and thermal properties.

In nanocrystals, the elastic strain energy is reduced. The forces on dislocations due to externally applied stresses are reduced by a factor of about three and the inter-active forces between dislocations are reduced by a factor of about 10. Hence, recovery rates and the annealing out of dislocations to free surfaces are expected to be reduced, as well. Dislocations are positioned closer together and their movement in the net is hindered by interaction between them. Together with the reduced elastic strain energy, this fact results in dislocations that are relatively immobile and the imposed stress necessary to deform a material which increases with decrease in grain size. Moreover, nanostructures allow alloying of components that are commonly immiscible in the solid and/or molten state.

6.7 THERMAL PROPERTIES

Thermal properties represent the behavior of a material when it is subjected to heat fluctuation. The major thermal properties of a material are heat capacity, specific heat, thermal expansion, thermal conductivity and phase changes in the material. Both the aspects of reduction in size of a material to nanoscale *i.e.* high surface to volume ratio and quantum effects influence its thermal properties. Most of the unique thermal properties of nanomaterials such as the abnormal heat conduction, the size dependence of thermal conductivity, phonon boundary/edge scatterings arise due to change in atomic structures. It is to be noted that electrons are the primary thermal energy carriers in bulk metals, and their distribution can be altered by nano structuring. For example, in metallic-type carbon nanotubes, the free electron density is low because of spatial confinement, and phonons dominate their thermal transport behavior. Phonons transport super-diffusively in low dimensional structures, in other words, Fourier's law is not applicable. Quantum confinement changes the distribution of carriers by altering the availability of energy states. As a consequence, the carriers tend to be found in narrower energy bands which can be used to enhance or diminish magnitude of thermal properties such as melting point, thermal conductivity, etc. Further, melting point of a nanomaterial is also greatly affected due to presence of high fraction of its constituent atoms or molecules at the surface with higher free energy. Some of the thermal properties of nanomaterials are discussed here.

6.7.1 Melting point

Melting point of the nanoparticle decrease with reduction in size. Its physical origin is attributed to: (i) the increase of surface energy, (ii) the increase of amplitude of atomic vibrations, and (iii) the additional surface growth of thermal vibration energy. Theoretically, size-dependent melting point of a nanomaterial can be calculated through classical thermodynamic analysis. The result is the Gibbs–Thomson equation, Eq. (6.19)

$$T_M(d) = \frac{4T_{MB}}{H_f d}\left(\sigma_{sv} - \sigma_{lv}\left(\frac{\rho s}{\rho l}\right)^{2/3}\right) \tag{6.19}$$

Where, T_M is the melting point of the nanoparticle; d = particle diameter; T_{MB} is bulk melting temperature, σ_s = solid-liquid interface energy, H_f = Bulk heat of fusion, ρ_s = density of solid

A faceted solid nanocrystal melts into a spherical droplet in order to minimize its surface energy resulting in a dramatic depression of its melting point. For example, a 2 nm diameter silver nanocrystal melts at 127°C which is about 800° degrees below that of the bulk and is in close agreement with the theoretically derived melting point as illustrated below.

A silver nanocrystal of 2 nm in diameter, has about half of its atoms on the surface. A faceted crystal of nano silver has about 1.24 times higher surface area as compared to the spherical liquid droplet of the same volume. If the decrease in surface area is about 20% upon melting, and the surface energy is about 1/4 of the bulk bonding energy of the atoms, then for a 2 nm diameter nanocrystal we would estimate that:

$$\begin{aligned}\Delta H^\circ_{\text{fusion}} &\approx \Delta H^\circ_{\text{fusion, bulk}} - (1/2)(0.25)(1/4)\Delta H^\circ_{\text{vap}} \\ &= \Delta H^\circ_{\text{fusion, bulk}} - 0.025\,\Delta H^\circ_{\text{vap, bulk}}\end{aligned} \tag{6.20}$$

where $\Delta H^\circ_{\text{vap, bulk}}$, the heat of vaporization, is the total bonding energy of atoms in a bulk crystal. For silver, $\Delta H^\circ_{\text{fusion, bulk}}$ = 11.3 kJ/mol and $\Delta H^\circ_{\text{vap, bulk}}$ = 250 kJ/mol. From this we can calculate the heat of fusion of a 2 nm Ag nanocrystal:

$$\Delta H^\circ_{\text{fusion}} \approx 11.3 - (0.025)(250) = 5.1 \text{ kJ/mol} \tag{6.21}$$

The melting point of bulk silver is 962°C (~1235 K). Assuming that the entropy of fusion is the same in the bulk and in the nanocrystal, the melting point of the nanocrystal should be 1235 K x (5.1/11.3) = 557 K = 284°C, a drop of almost 700 degrees from the bulk value. Similarly, the melting point of 2.5 nm gold nanoparticles is reported to be about 657°C (930 K), much lower than its bulk value of 1063°C (1336 K) and the melting point of 10nm CdSe 427°C (700 K) compared to the bulk CdSe at 1407°C (1680 K). This is a drastic lowering in the melting point on going to nano regime. The same effect *i.e.* energetic destabilization of the surface atoms relative to bulk atoms results in the lower boiling point, higher vapour pressure and higher solubility of nanomaterials as compared to their bulk counterparts.

6.7.2 Thermal conductivity

In solid materials, heat is transported by the momentum transfer of interatomic vibrations (phonons) to nearby atoms (lattice transport) and/or by mobile electric charges (electrons or holes). Therefore, the thermal conductivity (*K*) is represented by a simple linear contribution of the lattice thermal conductivity (*KL*) and the electronic thermal conductivity (*Ke*); Mathematically, $K = KL + Ke$. The lattice heat transport is greatly simplified when it is represented by a hypothetical particle called phonon. *KL* is function of the acoustic phonon scattering processes such as phonon-phonon scattering, phonon-electric carrier scattering, and phonon-impurity scattering. All the phonon scattering mechanisms dictate the phonon mean free path.

The thermal conductivity of nanostructured materials is generally smaller than their bulk counterpart due to the effective scattering of acoustic phonons with mean free paths longer than the characteristic dimensions of the nanostructured material. **Figure 6.17** illustrates the decrease in thermal conductivity specifically of copper nanoparticles with decrease in size.

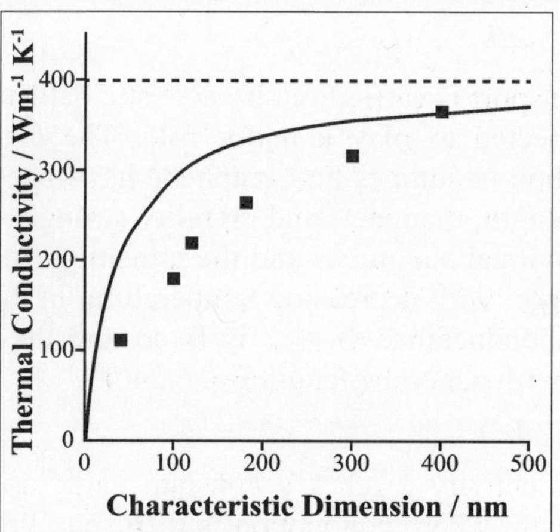

Fig. 6.17. Size- dependent thermal conductivity of copper

The solid line represents the thermal conductivity of copper nanoparticles and the dashed line represents the thermal conductivity of bulk copper at 298 K.

The scattering phenomenon of phonon, known as phonon-boundary scattering play a dominant role in thermal conductivity of NPs. While phonons can be scattered in nanostructured materials, electronic charge carriers transport could remain unaffected because their mean free path is several orders of magnitude smaller than phonons. Assuming the Wiedemann-Franz Law (Eq. 6.22):

$$K_e = L\,\sigma\,T, \tag{6.22}$$

where, L is the Lorenz number; σ is the electrical conductivity; T is the absolute temperature and K_e is the electronic thermal conductivity. K_e will be affected by nanostructuring if σ changes due to carrier scattering. For example, the thermal conductivity of a single-crystal of p-type silicon-germanium alloy nanowires of about 250 nm in diameter is 3 times smaller than its bulk counterpart with similar electrical conductivity.

Amorphous solid materials are known to have the smallest thermal conductivities because the phonons are scattered by the short order arrangement of atoms in the materials. Amorphous materials are also known as "phonon glasses". Theoretical studies based on phonon transport in amorphous materials provide a lower limit for the thermal conductivity. However, nanostructuring could reduce the thermal conductivity of crystalline materials to values comparable to amorphous non-electrically conductive materials such as SiO_2. This property is of great significance in realizing thermoelectric applications of conducting/semiconducting nanomaterials showing high figure of merit (Z), defined in equation 6.23.

$$ZT = \frac{\sigma S^2}{k} \tag{6.23}$$

$Z \geq 1$, when $\sigma S^2 \geq k$

If the thermal transport is carried out by acoustic ballistic phonons, boundary scattering is not expected to play a major role. The extremely high thermal conductivities of carbon nanotubes and graphene have opened up a new area of research. Nanotube length, diameter, and chirality strongly influence the thermal conductivities of individual nanotubes and the transition from primarily diffusive to ballistic heat transport with decreasing temperature. In the ballistic conduction regime, the thermal conductance $G_{\text{ballistic}}$ is fixed and the thermal conductivity apparently increases with nanotube length Eq. 6.24.#

$$k = G_{\text{ballistic}}\,L/A \tag{6.24}$$

The thermal conductivity reaches a constant value at lengths much longer than the mean free path, where conduction is diffusive. Presence of defects and isotopic impurities reduces the thermal conductivity of CNTs.

The thermal conductivity of a single atomic layer of graphene is very high, between 4000 and 5000 W/mK at room temperature. It is a factor of 5 larger than the thermal conductivity of graphite. Even though graphene has a large σ (large

Ke), acoustic ballistics phonons dominate the heat transport. Due to the low degree of scattering of low-energy phonons in 2-dimensional graphene, *K* actually increases compared to bulk graphite. The thermal conductivity of graphene is several orders of magnitude larger than most metals (*e.g.* copper, stainless steel, etc.), so it could replace metals in some heat exchange applications for corrosive fluids. Figure 1B provides a comparison among the thermal conductivities of some key materials and nanostructured materials.

6.7.3 Thermodynamics Laws

Thermodynamics can be defined as the study of thermal energy, energy transformations and its relation to matter. Whenever there is an interaction between energy and matter, thermodynamics is involved.

First law of thermodynamics: One of the most fundamental laws of nature is the conservation of energy. It simply states that during an interaction, energy can change from one form to another but the total amount of energy remains constant.

Second law of thermodynamics: Energy has quality as well as quantity, and actual processes occur in the direction of decreasing quality of energy.

Third Law of thermodynamics: All processes cease as temperature approaches absolute zero. This is the temperature at which molecules cease movement, cease producing kinetic energy. In other words, there is no energy involved.

From the second law we can write that the change in the internal energy, u, or a system is equal to heat supplied to the system, Q, minus any work done by the system W (Eq. 6.25).

$$du = \delta Q - \delta W \tag{6.25}$$

From the definition of entropy above we can replace δQ, and we can also make the replacement as $\delta .W = P\ dv$, given by Eq. 6.26

$$du = TdS - P\,dv \tag{6.26}$$

Now if we have a system of particles that are different then we may get chemical reactions occurring, so we need to add one more term to take this into account (Eq. 6.27)

$$du = T\,dS - P\,dv + \sum_i \mu d\,Ni \tag{6.27}$$

Thermodynamic laws have been the key for the design of useful everyday devices from car engines and refrigerators to power plants and solar cells. Conventional thermodynamics for bulk substances encounters challenges when one considers materials on the nanometer scale. Quantities such as entropy, enthalpy, free energy, melting temperature, ordering temperature, Debye temperature, and specific heat no longer remain constant but change with the crystal dimension, size, and morphology. Often, one phenomenon is associated with a variety of theories from different perspectives. The most viable model that can reconcile the size and shape dependence of the thermal properties of nanoscale substances can be achieved by using Hill's nano thermodynamics law (Eq. 6.28):

$$\frac{N}{V} \xrightarrow[\substack{N\to\infty \\ V\to\infty}]{} n \tag{6.28}$$

where, N is the number of particles of the system, V is the volume of the system and n is its number density, considered fixed and constant for any defined system. Hill developed a theory to introduce the finite-size effects in the macroscopic thermodynamics. He obtained this result by introducing a new thermodynamic potential, named subdivision potential U_S, defined as (Eqs. 6.29, 6.30):

$$U_S = \left(\frac{\partial U}{\partial N_s}\right)_{S,V,N} \tag{6.29}$$

$$dU = T_{dS} - p\, dV + \mu\, dN \tag{6.30}$$

where, T is the temperature, p is the pressure and μ is the chemical potential.

Nanomaterials are also characterized by lower phase transition temperatures as compared to their bulk counterparts. Phase transitions are collective phenomena. With a lower number of atoms in a cluster, a phase transition is less well defined; it is therefore no longer sharp. Small nanosized clusters behave more like molecules than as bulk matter. It may be more prudent to think of different structural isomers to coexist rather than the phases over a temperatures range. For example, part of a cluster may look frozen since its atoms do not exchange, while another part exhibits liquid-like dynamic behavior. At a given temperature, the solid-like regime may fluctuate in position and in size.

When phases are no longer well defined the Gibbs phase rule loses its meaning. In case of nano material having narrow pores the conventional fluid-liquid critical point may shift to very low temperature, or it may even vanish. However, new phases may appear which are characterized by discrete chemical potentials and separated by so-called layering transitions.

6.8 CHEMICAL PROPERTIES

Chemical properties of materials change at nanoscale. A nanomaterial is either a cluster of atoms or a cluster of molecules. Thus all types of bonding that are important in bulk chemistry are also important in nanoscience. However, intermolecular bonding (*i.e.* physical interaction) predominates the intramolecular bonding (*i.e.* chemical bonding). They include ion-ion and ion-dipole interactions, Van der Waals interactions, hydrogen bonds, hydrophobic interactions, repulsive forces such as steric repulsions. The interactions between nanoparticles depend on the chemical nature of the surface. Due to large surface area, large amount of charge species, defects and impurities may be easily attracted to surfaces and interfaces of nano particles and thus chemical nature of the surfaces alter abruptly as compared to their bulk counterpart.

The presence of a large fraction of atoms at the surface of a nanomaterial increases its reactivity and catalytic activity. Apart from large surface area to volume ratio, the variations in geometry and the electronic structure of nano particles have a strong effect on catalytic and chemical sensing properties. Molecular monolayer may also change or control surface properties and to mediate the interaction between nanoparticles.

6.8.1 Chemical Reactivity

One of the most common characteristics of nanomaterials is that the atoms in nano-materials have much lower coordination numbers than those of the bulk materials. Surface atoms with lower coordination numbers are responsible for enhanced chemical reactivities. As an example nanometal oxide surfaces, coordinatively unsaturated metal atoms allow catalytic reactions, whereas coordinatively saturated sites are inert. Further, oxygen vacancies on oxide surfaces are responsible for the catalytic reactions, which do not occur on the stoichiometric oxide surfaces. On metal single crystal surfaces, atoms at steps, kinks and edges often act as active centres to catalyze chemical reactions, whereas terrace atoms do not catalyze the same reaction. It further demonstrates that atoms with lower coordination numbers are chemically more reactive. Some of the reactions employing nnanoparticle catalysts are outlined here.

i. Dehalogenation and hydrogenation reactions

Nanoparticle catalysts can be used in the hydrogenolysis of C-Cl bonds such as polychlorinated biphenyls. Rhodium nanoparticles catalyzes the dehalogenation of aromatic compounds as well as the hydrogenation of benzene to cyclohexane. Polymer-stabilized nanoparticles can also be used for the hydrogenation of cinnamaldehyde and citronellal. The ruthenium nanocatalysts are more selective in the hydrogenation of citronellal compared to the traditional catalysts used

ii. Hydrosilylation reactions

R + R'_3SiH ⟶ SiR'_3

The nanoparticles of gold, cobalt, nickel, palladium, or platinum catalyze the hydrosilylation reaction, which is important for the synthesis of optically active alcohols. Palladium functionalized with 2,2-bis (di phenyl phosphino)-1,1'-birapthye (BINAP) and gold nanoparticles have been used for the hydrosily-laytion of styrene under mild conditions; they were found to be more catalytically active and more stable than bulk Pd-BINAP complexes. The reaction may also be catalyzed by a nanoparticle that consists of two metals.

iii. Organic redox reactions

Cobalt nanoparticles catalyse oxidation of cyclohexane to synthesize adiapic acid.

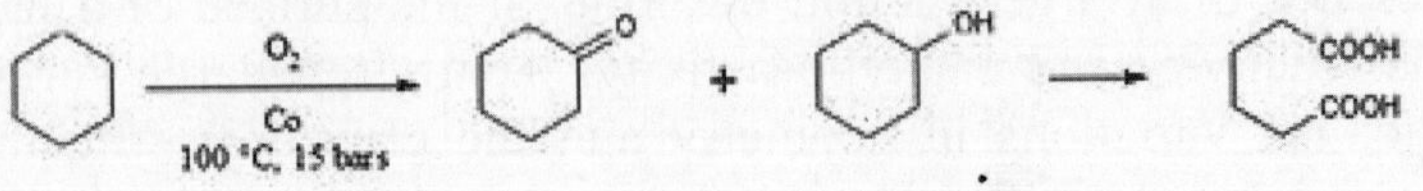

This is used in an industrial scale to produce the nylon 6,6 polymer. Other examples of oxidation reactions that are catalyzed by metallic nanoparticles include the oxidation of cyclooctane, ethene and glucose.

iv. C-C coupling reactions

Palladium nanoparticles are found to efficiently catalyze Heck coupling reactions. It is observed that increased electronegativity of the ligands on the palladium nanoparticles increased their catalytic activity.

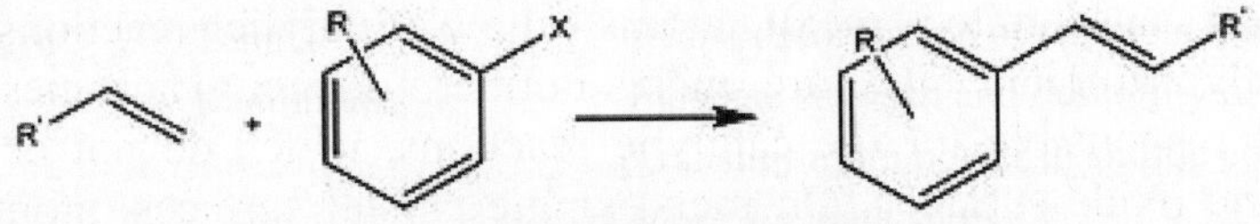

Heck coupling reaction

v. Photocatalysis

Photocatalysis is a reaction which uses light to activate a substance to modifies the rate of a chemical reaction without being involved itself. Chlorophyll of plants is the most significant example of natural photocatalyst. The most important difference between chlorophyll and man-made nano TiO_2 photocatalyst (here below mentioned as photocatalyst) is that usually chlorophyll captures sunlight to turn water and carbon dioxide into oxygen and glucose, but on the contrary, nano TiO_2 photocatalyst act as strong oxidation agent and a p-type semiconductor to break down the organic matter to carbon dioxide and water in the presence of light and water. The Au/Pt co-decoration on the TiO_2 nanoparticles shows much higher visible and ultraviolet (UV) light absorption response, which leads to remarkably enhanced photocatalytic activities on both the dye degradation and photo electrochemical (PEC) performance.

Nano titanium dioxide (TiO_2, anatase) absorbs ultraviolet (UV)* radiation from sunlight or an illuminated light source (fluorescent lamps) to produce pairs of electrons and holes (**Fig. 6.18**). The electron of the valence band of titanium dioxide becomes excited on absorption of UV radiation from sun light or a light source. The excess energy of this excited electron allow it is jump to the conduction band thereby creating the negative-electron (e-) and positive-hole (h+) pair. This stage is referred to as the semiconductor's "photo-excitation state". The energy difference between the valence band and the conduction band is known as the 'Band Gap'. Wavelength of the light necessary for photo-excitation is: 1240 (Planck's constant, *h*) / 3.2 ev (band gap energy) = 388 nm.

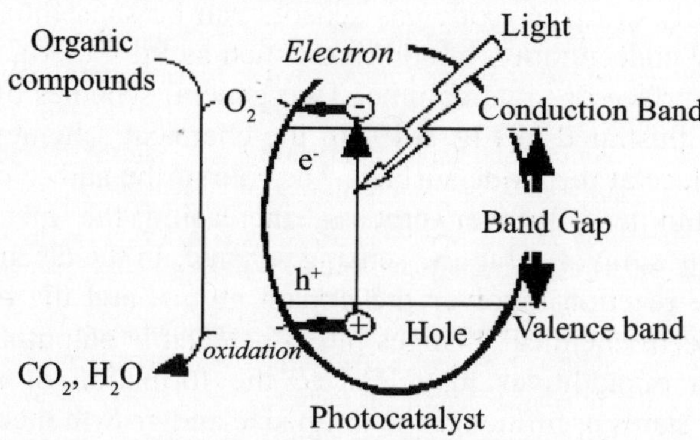

Figure 6.18. Mechanism of Photocatalysis

The positive-hole of titanium dioxide breaks apart the water molecule to form hydrogen gas and hydroxyl radical. The negative-electron reacts with oxygen molecule to form super oxide anion. This cycle continues in presence of light.

vi. Chemical sensing

A number of nano metal oxide have been derived to improve the gas sensing properties. When the particle size of metal oxide is close to or less than double thickness of the space-charge layer, the sensitivity of the sensor increase remarkably and the effect called "small size effect".

The change of electrical properties (conductivity) of the semiconducting metal-oxide due to adsorption of gas molecules is primarily connected with the chemisorption of oxygen. Molecular oxygen adsorbs on the surface by trapping an electron from the conduction band of the semiconductor. At temperatures between 100 and 500°C the ionized molecular and atomic (O^-, O^{2-}) forms of oxygen can be present at the surface. The molecular form of oxygen dominates below 150°C, whereas above this temperature, ionic species prevail. The presence of charged oxygen species on the surface of a semiconductor induces band bending and formation of a depletion layer. Depending on the type of semiconductor; the concentration of charge carriers on the surface layer can be either increased or decreased. Depending on the content of the atmosphere, the concentration of the surface oxygen ions and therefore the occupation of the surface states can be changed, leading to the change in conductivity. As a measure of gas sensitivity one can use either the conductivity change of the sample, exposed to the analytes containing atmosphere in relation to its conductivity in the reference gas, or the slope of the dependence of conductivity on analytes concentration. The gas sensitivity of metal-oxide sensors are enhanced on utilizing metal doped nano metal oxides.

The sensitivity of metal-oxide gas sensors can be substantially improved by dispersing a low concentration of additives, such as Pd, Pt, Au, Ag, Cu, Co, and F on the oxide surface or in its volume. Two general schemes of the gas sensing mechanism are illustrated in **Fig. 6.19**. In the chemical scheme **(Fig. 6.19a)** the reaction takes place at the oxide surface. The role of the additive nanoparticles is considered within a spill over process, increasing the metal-oxide surface coverage of the gas, involved in the sensing scheme. In the electronic mechanism **(Fig. 6.19b)** the reaction involves the dopant atoms, and the oxide material to transduce the electrochemical changes into a detectable output signal. Moreover, the introduction of additives may lead to the formation of a new donor or acceptor energy states or influence the grain size and growth mechanism.

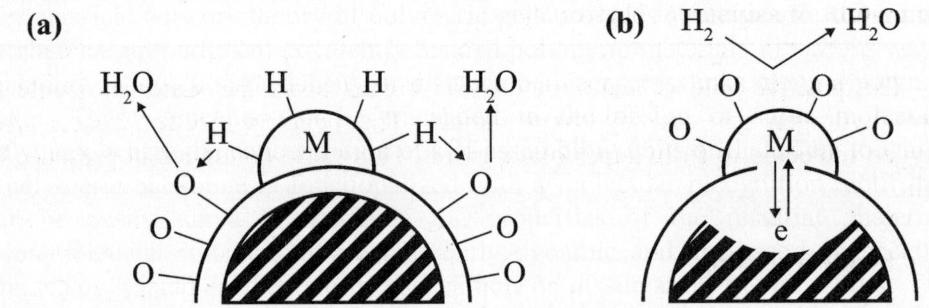

Fig. 6.19. (a) Chemical and (b) electronic sensitization schemes in metal-doped SnO, gas sensor

The chemical scheme is usually considered in the case of catalytic additives, for example Pt- or Pd-doped SnO_2 material exposed to hydrogen, hydrocarbons, or carbon monoxide. It is supposed that the reduction of gas molecules is first activated by the metal surface, forming the active surface species that then react via a spill over process with the charged oxygen molecules, adsorbed on tin oxide. This reaction leads to the re-injection of the localized electrons back to the bulk, thus increasing the conductivity of the material.

For instance, the sensing mechanism proposed for the Pt/SnO_2 + CO system involves two main processes. First, at elevated temperature, Pt is oxidized by the chemisorbed oxygen:

$$Pt + 2O_{ads} \text{----------------}\rightarrow PtO_2$$

Second, exposure to CO leads to a reduction of platinum oxide:

$$PtO_2 + 2CO\text{---------------------}\rightarrow Pt + 2CO_2$$

6.9 SELF-ASSEMBLED SUPRAMOLECULAR NANO STRUCTURES AND THEIR PROPERTIES

Nanomaterials often arise from a number of molecules held together or from large molecules that assume specific three-dimensional structures through intermolecular physical bonding. Therefore, nanoscience also deals with supramolecular chemistry.

Supramolecular self-assembly can be defined as the spontaneous formation of organized structures from building blocks with the aid of multiplc intermolecular non-covalent forces including hydrophobic or solvophobic effect, electrostatic forces, hydrogen bonding, π-π stacking, coordination interactions, and ion-dipole interactions. Nature's power to efficiently create diverse complex biological functions is based on the non-covalent interactions of covalently prefabricated building blocks including bi-layered lipid liposomes, protein tertiary structures, DNA double helixes, as well as the complex biological process. For chemists, supramolecular self-assembly represents a powerful synthetic methodology in the creation of large, discrete, ordered structures from relatively simple units. Inspired by nature, chemists have endeavored to explore molecular or supramolecular self-assembly in order to:

- Reveal the mechanism of biological self-assembly and understand the way in which biological objects are organized;
- Mimic the complicated but efficient self-assembly process in nature to design hierarchical structures possessing bioactivities;
- Create novel structures or materials capable of novel biological, physical and chemical properties.

Self-assembled monolayers are probably the most versatile and useful example of supramolecular assemblies on surfaces. Its applications range from micro patterning and cell patterning, electrochemistry, corrosion protection, the study of molecular junctions, controlling the growth and surface properties of various types of nanoparticles, to aligning and nucleating crystals, etc. Depending on the specific attachment chemistry, SAMs can form on various types of substrates (*e.g.* thiols, disulphides, or dithiolanes on Au, Ag, Cu, Pt, Pd, or Ni; silanes on silica or titania; and phosphoric or carboxylic acids on iron oxides, etc.). Head groups SAMs can impart surface properties ranging from hydrophobicity/hydro-philicity, to redox activity and light switchability, to biospecificity or bioresistance.

Because of the specific, tunable and reversible character, supra-molecular interactions confer a number of unique properties on materials across various length scales, from nano to macroscopic. The properties such as modularity, mechanical tunability, responsiveness and biomimicry highlight some of the specific utility and benefits of supramolecular biomaterials over traditional biomaterials.

a. Modularity

Supramolecular interactions are highly specific and allow systems to be created by conjugating a diverse array of functionalities without disrupting the self-assembly mechanism, kinetics or thermodynamics. This feature allows for the development of modular and generalizable material platforms which are conducive to simple 'mix-and-match' assembly without significant alteration of the basic material properties. The modularity of supramolecular materials is a feature based on both stacking motifs and polymeric crosslinking motifs and allows for precise control over material composition, functionality and bioactivity, while also enabling signals or targeting groups to be 'multiplexed'. This method to multiplex activity in a biomaterial has been leveraged for targeted nano-particles, two-dimensional substrates and hydrogels.

b. Mechanical properties

The classical network theory of polymeric gels provides strategies for tuning the mechanical properties of covalently bended polymeric materials in various ways, including modulation of density or distance between crosslinks along a polymer chain. These concepts can be translated to the design of supra molecular polymers, where the concentration of self-assembling units or the extent to which a polymeric backbone is functionalized with a molecular recognition motif that can be easily tuned to modulate the properties of the resulting material. Supramolecular interactions are inherently dynamic and are based on specific timescales related to the rate of association or dissociation of the motifs. The interchange between active and inactive interactions, as well as the lifetime of a bound complex, can have a large impact on the overall properties of the resulting material. Many factors such as the molecular weight of the polymer (or effective length of a one-dimensional assembly), valence and spacing of interacting units, binding mode for the interaction, and the thermodynamics and kinetics of binding interactions; all of them contribute to the mechanical properties of supramolecular materials. Because these factors can be controlled, in part, through the rational design and selection of recognition motifs, supra-molecular materials have exceptional potential for tunability. The properties that arise from the dynamic and reversible nature of supramolecular interactions could prove particularly useful for a number of biomedical applications, such as the generation of shear-thinning and self-healing gels.

c. Responsiveness and dynamic reciprocity

Because of the dynamic nature of non-covalent interactions, supramolecular materials can rapidly respond to a large number of external stimuli, thereby recreating aspects of the dynamics present in living systems. Stimuli that have been used to control assembly include physical parameters such as temperature, light, voltage or magnetic field, chemical parameters such as pH, ionic strength, redox agent or competitive host–guest interactions and biological cues (from enzymes or proteins, for example). Such responsiveness could be used to engineer supra-molecular smart materials that allow to tune properties in real time in response to externally applied stimuli, or those that are capable of autonomously sensing and responding to environmental stimuli.

d. Biomimicry

Biological systems are exceptionally complex scientific entity always inspiring to develop synthetic materials capable of recapitulating the structural and functional complexity of biological materials. Biomimetic supramolecular materials can replicate aspects of structural and/or functional features of biological materials, such as signal transduction. This is especially true for materials prepared from peptide or protein building blocks, whether assembled through molecular stacking or from polymeric precursors. Biominietic materials that can replace or recapitulate deficient native materials or signalling pathways could be especially useful for applications in regenerative medicine or tissue engineering.

6.10 NANO CLUSTERS AND THEIR PROPERTIES

Atomic clusters are aggregates of atoms containing a few to several tens of atoms. The properties of atomic clusters, in general, are different from those of the corresponding material in the macroscopic bulk phase. Molecules have fixed composition, structure, bond lengths, bond angles and, hence, fixed properties. The clusters, on the other hand, are the aggregates of atoms and may or may not be stable. The structure, bonds, etc. vary with the number of atoms and so do the properties. To make a comparison, a large size metal atom like uranium (atomic no. 92), in which the positive charge is concentrated and the nucleus occupy very small volume and 92 electrons have some specific electron configuration. On the other hand, in a cluster of a light metal atom, *e.g.* 90-atom cluster of sodium (Na_{90}), each atom's positive charge is spatially separated from the other atoms and each one would have its own ionic shell configuration; only 90 electrons would be free to form an electron gas of the cluster. In terms of chemistry, small clusters are treated as molecules; bands as delocalized molecular orbitals; and the Kubo gap becomes the HOMO–LUMO gap. In such a scenario it is obvious that high cluster symmetry leads to correspondingly degenerate molecular orbitals. Incompletely filled degenerate HOMO leads to magnetism, but often also to symmetry lowering by Jahn–Teller distortion.

Considerable effort has been devoted to understand evolution of the structural properties of nanoclusters. For example, gold can adopt a wide variety of structures; clusters with less than about twelve atoms prefer to form flat flakes, whereas spherical structures are stable when the total number of valence electrons in the cluster (each gold atom contributes about one) is equal to or near the value of 2, 8, 18, 32, 50, 72,..., $2\ (L+1)^2$ electrons, where L is an integer. These 'magic numbers' are defined as the number of electrons necessary to fill, the s, p, d, f and g orbitals of atoms respectively. Just as in real atoms; the s, p and d orbitals of clusters do not have to fill sequentially and to the contrary more complex filling arrangements can occur. For instance, the 20 electrons in the compound Si@Au_{16} fill two s orbitals, one p and one d orbital. Also, the same electron count can be satisfied by two different filling arrangements, giving rise to so-called 'false magic numbers'. In addition, certain highly symmetric geometric structures are associated with particular numbers of atoms. For example, Au_{20} is a tetrahedron and Au_{32} is predicted to be an empty icosahedron.

Ultra-small noble metal nanoclusters (NCs), typically composed of several to a hundred metal atoms, present unique physical and chemical properties between single atoms and large nanocrystals (>2 nm). Owing to the strong quantum size confinement in this size range, NCs possess discrete and size-tunable electronic transitions and display unique molecule-like properties, such as quantized charging and luminescence.

Unique Properties of Clusters

In general, the physical properties of materials are dependent on the dimensions of the material in question; their properties (*e.g.* conductivity, elasticity, etc.) are scalable with respect to the amount of atoms in the material. In scalable size-regime, the properties of nanoclusters vary greatly with every addition or subtraction of an atom; whereas, in non-scalable size-regime every atom counts for its properties.

i. Reactivity

In nanoclusters a large fraction of atoms is on the surface. For example in a cluster of 55 atoms like that of argon or sodium 32 atoms are on the surface and only 23 are inside. This has important implication for the design of catalytic agents and epitaxial materials. Reactivity is highly dependent on the electronic structure, leading to large variations in the reactivity of clusters even with sizes differing by only a single atom. In general, the reactivity of all nanoclusters is non as compared to that of bulk. The underlying reason is attributed to their higher surface to volume ratio, *e.g.* gold nanoclusters are highly reactive if compared to the fairly inert bulk material. Metal and metal oxides, metal sulphide nano clusters are reported to exhibit versatile acid-base and redox properties; catalyze many reactions (dehydration, partial/selective reduction or oxidation, photo- or electro-chemical reactions, etc.

ii. Melting point

One important effect caused by a high surface to volume ratio in clusters, is a substantially lower melting temperature as compared to bulk counterpart. Unusual phenomena such as negative heat capacities have also been observed in clusters. At the melting point, an increase of the internal energy of Na^{+}_{147} by 1 eV leads to a decrease in of melting point by about 10 K. In large enough systems, the added energy is converted completely into potential energy, reducing continuo-usly the fraction of its solid phase. The kinetic energy and thus the temperature remain constant. A small system, on the other hand, tries to avoid partly molten states and prefers to convert some of its kinetic energy into potential energy instead. Therefore, the cluster can become colder, while its total energy increases.

iii. Optical properties

In semiconductor clusters the band gap increases as cluster size decreases, causing a blue shift in the absorption spectrum. Energy levels and energy level separations are also dependent on the size of the clusters, which in turn affect the energies needed for the transitions of electrons to excited states. Clusters of different sizes will therefore have different absorption spectra and hence different colors.

Nobel metal nanoclusters possess discrete and size-tunable electronic transit-ions and display unique molecule-like properties, such as quantized charging and luminescence. Strong luminescence is one of the most attractive features of these NCs owing to their ultrafine size, good photostability and low toxicity. Some of these properties are realized superior over other luminophores, such as organic dyes.

iv. Magnetic properties

Interesting magnetic properties have also been observed in small clusters. In a cluster the magnetic moment of each atom will interact with the moments of the other atoms, and can force all the moments to align in one direction with respect to the symmetry axis of the cluster just to show magnetism in a non-magnetic material. As cluster size decreases it becomes easier to exhibit ferromagnetic behavior. For instance, rhenium clusters show a pronounced increase in their magnetic moment when they contain less than 20 atoms. For clusters with less than 15 atoms these moments are fairly large. Enhanced magnetism in clusters of elements that are ferromagnetic as bulk solids such as iron, cobalt or nickel is well known.

Theoretical studies predicted high spin ground states for clusters of up to 13 atoms even in non-magnetic elements such as Pd and Pt nanoparticles comprising several hundred atoms of Au, Pd and Pt embedded in a polymer reveals magnetic moments corresponding to several unpaired electron spins per entire particle. To understand the basic mechanism bulk Pt metal has an incompletely filled d-band corresponding to a $5d^9\ 6s^1$ electronic configuration and a Fermi level of 6.10 eV. In the Pt_{13} cluster the Fermi level rises considerably because of the shorter interatomic distances and concomitant higher orbital overlap. Density functional calculations of the discrete energy level diagram predict near degenerate frontier orbitals with 8 unpaired electrons. Such studies provide insight into the electronic and magnetic structure of clusters and are fundamental for an understanding of how magnetism develops in non-magnetic elements.

Pseudo-atoms

Metal nanocrystallites are often called pseudo-atoms or super atoms, because of the existence of quantized states of electrons and holes. Experimental and theoretical studies support the pseudo-atom concept of small metal clusters. Properties such as ionization potentials and electron affinities reveal shell structures akin to those of conventional elements. An example of a super atom is $AlPb_{12}^{+}$ where 12 lead atoms are placed in a perfectly icosahedral shell around a core Al atom. The 50 valence electrons occupy molecular orbitals which are delocalized over the cluster and remind very much of conventional hydrogen-like atomic orbitals. It may be noted that there is a HOMO–LUMO gap of ca. 3 eV, indicating the large band-gap semiconductor nature of this species. The orbitals of this super-atom resemble the s-, p-, d- and higher hydrogen-like orbitals and show basically the energetic sequence which is also obtained from the three-dimensional "particle-in-a-box" model.

An instructive example of a composite super atom is the magic aluminum cluster anion Al_{13}^{-}, which has 40 valence electrons and forms a closed 2p shell. The neutral Al_{13} open shell cluster has distorted icosahedral structure. Al_{13} is a halogen-like super atom and can form a molecular ion with an iodine atom and which has an Al_{13}–I bond dissociation energy of 2.46 eV finding perspective pseudo-atoms that are best regarded as new elements which can be combined with others, leading to a completely new chemistry. Applications of pseudo-atoms range from light emitting diodes (LEDs) and novel solid-state lasers to potential future devices for quantum electronics and quantum computing.

Suggested further readings

1. E. Roduner, Nanoscopic materials: Size dependent phenomena, Cambridge, UK; RSC Publishing, 2006.
2. M. Quinten, Optical properties of nanoparticle systems: Mie and beyond, Publ. Wiley- VCH Verlag GmbH, 2011
3. K. L. Kelly, E. Coronado, L. L. Zhao, and G. C. Schatz, The optical properties of metal NPs: The influence of size, shape and dielectric environment, *J. Phys. Chem. B*, **43:** #(2003) 668- 677.
4. C. Z. Li, K. B. Male, S. Hrapovic and J. H. T. Luong, Fluorescence properties of gold nanorods and their application for DNA biosensing, *Chem. Commun*, (2005) 3924-3926.
5. B. Bhushan, D. Luo, S. R. Schricker, W. Sigmund, S. Zauscher (Ed.), *Handbook of Nanomaterials Properties*, Springer, 2014.

6. D. Vollath, Nanomaterials: An Introduction to Synthesis, Properties and Applications, Second edition, Wiley, 2013.

REVIEW QUESTIONS

Q1. Give the factors responsible for the novel properties of nanomaterials.

Q2. What do you understand by quantum confinement?

Q3. What is Tunneling effect? What are the implications of tunneling effect in nanotechnology?

Q4. What is band gap? How does it differ in conductors, semiconductors and insulators?

Q5. How does change in band gap with the size of NPs affect the electrical behaviour of a material?

Q6. How does size of nanoparticles affect the surface properties of nanomaterials in terms of 1. Fraction of atoms at the surface 2. Smooth scaling properties?

Q7. How are thermal, optical and electrical properties of materials get affected when material transforms to nano size ?

Q8. Write a short note on Surface plasmon resonance in metal nanoparticles and quantum fluorescence.

Q9. Give a plausible cause for photoluminescence in semiconducting material. Explain with the help of band gap energy model.

Q10. What is dielectric constant? How is dielectric constant measured? How does it affect the conductivity and optical properties of nano materiasl?

Q11. Discuss important dielectric properties of nanomaterials.

Q12. What new properties arise when the size of a magnetic material reaches in nanosize regime? Explain giving examples.

Q13. Give the characteristic features of superparamagnetic materials.

Q14. How is superparamagnetism related to thermal energy (k_BT) and anisotropy constant (K_{eff})?

Q15. What will be the impact of coating of a magnetic over magnetic nanoparticles?

Q16. Explain why mechanical properties of nanocomposites are better than their constituent materials?

Q17. Does bonding change when bulk materials transform into nanomaterial? Give the significance of physical bonding in nanomaterials?

Q18. Discuss the various types of bonding leading to the formation of supramolecular structures. Give some examples of biological supramolecules.

Q19. What features of NPs make them attractive for catalysis and sensor applications? Explain with some examples.

Q20. What do you understand by pseudo atom, give some of its potentials applications?

■■■■

Impact of Nanoscience on Technology 7

7.1 INTRODUCTION

Nanoscience is distinguished by its interdisciplinary nature and regarded as a science with huge potential and great expectations. Since the 1980s the basic research in nanoscience followed by more than two decades of focused technological applications are ever increasing.

Benefits of nanotechnology depend on the fact that it is possible to tailor the essential structures of materials at the nanoscale to achieve specific properties, thus greatly extending the well-used protocols of materials science. Products where the addition of a relatively small amount of functionalized nanoparticles leads to a major change in the properties are proving revolutionary in many commercial technologies. Using nanotechnology, materials can effectively be made to be stronger, lighter, more durable, more reactive, more sieve-like, or better electrical conductors, among many other traits. It is well established now that size, shape and morphological characteristics such as surface charge, area, porosity together with composition and structure of the nanomaterials are solely responsible for their unique properties and associated applications.

Nanotechnology has revolutionized a number of technological and industrial sectors such as, information technology, energy, environment, healthcare, security, food safety, and transportation, besides many others. The following list includes some specific examples of the nanomaterials that may exist naturally or may be engineered to realize various technological application.

- Carbon nanomaterials, such as fullerenes, carbon nanotubes and graphenes existing as hollow spheres (buckyballs), tubes and films, respectively, are entirely composed of carbon. They are stable structures, strong antioxidants, have limited reactivity and are excellent thermal and electrical conductors. These materials find several biomedical, super-capacitor, sensor, and photovoltaic applications.
- Nanosized metal oxides such as titanium dioxide (TiO_2), zinc oxide (ZnO), cerium oxide (CeO_2), and iron oxide (Fe_3O_4) can be used as photocatalysts, in pigments, controled release of drugs, medical diagnostics, and in sunscreens.
- Zero-valent metals, such as nanoscale zero-valent iron (nZVI), have high surface reactivity and are used in the remediation of contaminated water, sediments, and soils.
- Quantum dots range in size from 10 - 50 nanometers. Find applications in medical imaging, photovoltaics, telecommunications, and sensors.

- Dendrimers are highly branched polymers that can be designed and manufactured to incorporate a variety of functional groups. Common shapes include cones, spheres, and disc-like structures. Dendrimers are used in drug delivery, chemical sensors, modified electrodes, and as DNA transferring agents.
- Composite nanoscale materials may be integrated with biological and synthetic molecules, which provide novel electrical, catalytic, magnetic, mechanical, thermal, or imaging capabilities. Potential applications include drug delivery and cancer detection. They are also used in auto parts and packaging materials to enhance the mechanical and flame-retardant properties.

7.2 CONSUMER GOODS

A technology's popularity is known by its applicability in areas concerned with the daily use of a common man. Nanotechnology has proven its worth by its increasing use even in many of the consumer products beside many high end applications to health care, energy, secunty etc. Firstly, we discuss their use in consumer goods.

More and more consumer products are branded with the buzzword "nano" or nanotechnology. In general, the products claimed to contain nanotechnology exploit nanoscale effects, primarily interface effects and quantum effects.

7.2.1. Sports goods

The degree of competitiveness in sports has been remarkably impacted by nanotechnology same as any other innovative idea in materials science. Within the niche of sport equipment, nanotechnology offers a number of advantages and immense potential to improve sports equipment making athletes safer, comfortable and more agiler. Baseball bats, tennis and badminton racquets, hockey sticks, racing bicycles, golf balls/clubs, skis, fly-fishing rods, archery arrows, etc. are some of the sporting equipment whose performance and durability are being improved with the help of nanotechnology. Nanomaterials such as carbon nanotubes (CNTs), silica nanoparticles (SNPs), nanoclays fullerenes, etc. are being incorporated into various sports equipment to improve the performance of athletes as well as the equipment (**Table 7.1**). Each of these nanomaterials is responsible for an added advantage such as high strength, stiffness, durability, reduced weight and abrasion resistance in sports equipment.

Footballs/tennis balls with nano-clay lining as a barrier material reassure a longer game-play. Incorporation of nanomaterials (*e.g.* fullerene) lightens golf clubs by lowcring thc ccntrc of gravity, which increascs the power and accuracy during the game-play. Graphene oxide and bucky-paper (sheet of CNTs) are incorporated into race-boat hulls/masts to increase glide while making them stronger and much lighter. In ice-skating, ultra-hard nanoceramics make the edges of skates super-sharp, whereas self-assembling bionic nanoparticles are used in speed cars to improve the fuel efficiency and speed even with 50% less nitro methane fuel mixture. Among various nanomaterials, CNTs are the most prominent nanomaterial being used in nano-enhanced sporting equipment such as golf shafts, rackets, skis, snowboards, etc.

Table 7.1. Attributes of nanomaterials in sports equipment and associated benefits

Nanomaterials	Sports	Attribute
Silica nanoparticles	Tennis/ Badminton	Increase stability, power, and durability of racquets
	Skiing	Decrease torsion index and facilitate transition in skis
	Fly-fishing	Enhance hoops and flex strength of rods
Nano- titanium	Tennis/ Badminton	Resist deformation of racquets frames Increase strength and durability of racquets Transmit more power to shuttle cock/ball More accurate shots
Nano-nickel	Golf	Increase moment of inertia, stability of clubs
Carbon nanotubes	Tennis/ Badminton	Improve stiffness, consistency, durability, resiliencey, impact, repulsion power and vibration control in racquets
	Golf	Reduced weight, Lower torque/spin of clubs
	Kayaking	Enhanced abrasion/crack resistance Easy paddling
	Archery	Better vibration control in arrows
Fullerenes	Tennis/ Badminton	Reduce weight and resist twisting of racquet frames
	Golf	Facilitate flexible club whipping
	Bowling	Reduce chipping and cracking of balls
Carbon nanofibers	Cycling	Reduce weight Increase stiffness
Carbon nanoparticles	Motor racing	Decrease rolling resistance of tires Increase grip and mileage of tires
Nanoclays	Tennis/ Golf	Increase resilience and bounce of balls
	Watercraft	Reduce weight Enhance speed of water- boats

7.2.2. Cosmetics

Nano cosmetics aims at products intended for application to the facial and body skin, with anti-aging action and photo protection, capable of penetrating into the deep layers of the skin, showing superior properties regarding its performance as compared to the conventional products. In the cosmetic industry, nanoparticles are present in shampoos, conditioners, toothpastes, anti-wrinkle creams, anti-cellulite creams, skin whitening creams, moisturizers, face powders, aftershave lotions, deodorants, soaps, sunscreens, make-up materials in general, perfumes and nail polishes. Nano emulsions, in turn, constitute a class of emulsions with uniform droplets of small size in the range between 20 and 500 nm, have become increasingly popular as the materials for controlled release and optimized dispersion of active ingredients. Nanoparticles in cosmetic preparations are found to improve the stability of various cosmetic ingredients such as unsaturated fatty acids, vitamins and antioxidants by encapsulating them; increase the efficacy and tolerance of UV filters on the skin surface; make the product aesthetically more pleasing; and enhance the penetration of certain active ingredients to the epidermis.

Nano titanium dioxide (TiO_2) and nano zinc oxide (ZnO) are efficient photo stable UV filters. They absorb UVB and UVA radiations and re-emit less damaging UVA through visible fluorescence or heat. Formulations that utilize TiO_2 or ZnO as the only active sunscreen agents provide photo protective properties and reduced risk of irritation compared to other sunscreen ingredients. Uncoated TiO_2 absorbs photons of light and emits an excited electron, which can be transferred to free radicals and absorbed into the dermal layers, resulting in oxidative damage. To check the formation of reactive oxygen species (ROS) and prevent the agglomeration of particles, TiO_2 and ZnO are generally coated with aluminium oxide, silicon dioxide or silicon oils. Organic alternatives to these have also been developed.

Another important use of nanotechnology in cosmetics is for delivery vehicles. Liposomes and isomers are used in the cosmetic industry as delivery vehicles. Newer structures such as solid lipid nanoparticles (SLN) and nanostructured lipid carriers (NLC) have been found to be better performers than liposomes. In particular, NLCs have been identified as a potential next generation cosmetic delivery agent that can provide enhanced skin hydration, bioavailability, stability of the agent and controlled occlusion. Nano crystals and nanoemulsions are also being investigated for cosmetic applications.

7.2.3. Textiles

Addition of nanomaterials to textiles bring new functionalities. Some of the new functionalities aesthetics (*e.g.*, luminescence), shrink resistance, wrinkle resistance, soil resistance, antimicrobial, oil and stain resistance, electrostatic protection, fire resistance, UV protection, fragrance release, high strength, water repellent (hydrophobic), self-cleaning and improvement of dye ability. Textiles with such functionalities significantly reduce or even eliminate traditional textile-care operations viz. washing, drying and ironing. Nanoparticles and nano whiskers that cause water to bead up and dirt to rinse off, provide a "lotus effect" making the fabric water and stain resistant. Silver nanoparticles in the fabric kill bacteria, making clothing odor-resistant. Nanopores provides superior insulation for shoe inserts in cold weather. Antimicrobial textiles may impart anti-odour, antifungal and antibacterial properties. In such products fibers possess intrinsic antimicrobial properties or will acquire them through subsequent treatments through chemical finishing, addition of bactericidal nano metals, etc. Products enriched with nano materials have potential applications in apparel, intimate apparel and in a number of non-apparel applications like furniture, curtain carpets, etc. **Table 7.2** includes a representative list of nanomaterials used for textiles, their functionality and potential areas of applications. Many products use, or will use, chemical, physical or electronic stimulai to respond passively or actively to thermal, chemical, biological, electromagnetic and mechanical stress. These products include warming and cooling textiles, conductive textiles, communicating textiles, textile sensors and actuators, digital fashion, smart textiles, etc. Integration of nanotechnology into textile products is being realized in coatings, treatments, fibre material composites and nanoscale fibres. Four categories, viz. nanofinished textiles, nanocomposite textiles, nanofibrous textiles and nano-enabled nonwoven nanotextiles are described below.

i. Nanofinished textiles

Nanofinished textiles are those that add a nanoscale property after the base textile has been fabricated. This includes post-manufacture treatments and coatings to apply nanomaterials or create nanostructured surfaces on fibre media. Nanomaterials used as additives in textiles include metal nanoparticles (such as silver for antimicrobial functionality) or clay nanoparticles (for fire resistance). The Lotus effect has been an inspiration for several innovative materials, such as coatings and textiles. The realization that certain surface properties can induce water repellence is important in numerous applications. Nanomaterials are being engineered to render superhydrophobic surfaces for self-cleaning property **(Fig. 7.1)**.

ii. Nanocomposite textiles

Nanocomposite textiles have composite fibre materials containing one or more nanostructured or nanoscale components. This type of nanotextile are based upon pre-manufacture integration of nanoscale properties into fibrous components. Important source materials added to produce the nanocomposite textile are carbon nanotubes (to enhance fibre strength) and rare earth metal doped nanoparticles (for luminescence). Polymer matrices are the most prevalent in nanocomposite fibres; however, other matrices may also be useful.

Fig. 7.1. Nanomaterial coated non-staining textile

Similar to nanofinished textiles, nanocomposite fibres may not require significant changes to the manufacturing process. If the matrix material is the same shape and size as the process has been designed for, only small alterations may be necessary to integrate the composite component. If the fibre material is entirely changed by choosing a different matrix material or significantly altered by the composite properties, significant reconfiguration may be needed. These nanotextiles have promise, but currently represent a relative minority in commercialized form.

iii. Nanofibrous textiles

Nanofibrous textiles have fibres with nanoscale dimensions. These true nanofibres have a nanoscale cross-sectional area and may or may not have a nanoscale length. Fibre material may be either a single material or a composite (which, based on the fibre dimensions, may also be a nanocomposite). Nanofibres may also be nanofinished. These nanotextiles focus on fabricating fibres to exploit nanoscale properties. These properties may emerge either from the nanomaterial composition (such as fibres made of carbon nanotubes, giving them very high strength) or from the scale of the individual fibres (such as filter media, giving them increased fibre surface area and nanoscale porosity). Fibre fabrication for nanofibrous textiles typically would be new. Initial fabrication of nanofibres require such as electro-spinning or force spinning, which otherwise are not conventional drawing methods processes, but all of the subsequent steps in the manufacturing process must accommodate these smaller fibres. For woven textiles, the technology does not exist on an industrial scale to weave fibres of this size. Beyond that, dyestuffs and laminations may have to be reformulated for nanofibres. Some means of incorporating nanofibres into textiles may be accomplished through entangling or encapsulating the fibres in a larger fibre. While research into nanofibrous textiles is widespread, little has reached significant commercialization. Its development, however, may afford exciting applications and opportunities.

iv. Nano-enabled nonwovens textiles

Nanotechnologies in nonwovens may make use of improvements in properties to benefit textile processing; for example, adhesive properties may be enhanced or conventional methods of securing layers replaced with adhering nanostructured surfaces. Other nanoscale functionalities may utilize nanofilms or coatings in layers or barriers; for example, for antibacterial properties, energy production (for renewable solar power) and luminescence (for colour control).

Table 7.2. Nanomaterials for textiles, their functionality and areas of application

Nanomaterial	Property/ functionality	Functionality	Application areas
Silver nanoparticles	Antibacterial, odour control	Silver nanoparticles inhibit bacterial metabolism, which causes infection and odour, causing the bacteria to die	Textile such as socks, gloves, inner wears, hospital linens, apron
Drug carrying polymeric nanocapsules: PLA poly (lactide); PCL, poly (e-caprolactone); PVP: polyvinyl pyrrolidone	Targeted bioactivity	Fabric embedded with drug filled nanocapsules.	Medical textiles such as wound dressings; used for sustained release of drug
Organic conductive polymers and metal nanoparticles	Electrical conductivity	Conductive nano coatings on textile fibres and transistors	Smart textiles for sensing hazardous environment for military and mining personnels
Montmorillonite nanoplatelets	Flame resistance	Intumescent nanomaterials are coated onto fibres and expand to insulate fibres when heat from flame reaches the textile.	Fire- proof upholstery, clothing for fire fighters
Mesoporous carbon (biochar), Polysaccharide confined silver nanoparticles such as nanofibrous media (with added functionality)	High filter efficiency	High surface-to-weight ratios from nanoscale fibre thickness increase efficiency with possible nanoparticle and nanocomposite functionality for antibacterial and chemisorption properties.	Water treatment, environmental management
Carbon nanotube polymer composite	High strength	Carbon nanotubes provide very high strength properties due to their graphitic structure. Enables reinforcing a polymer matrix for a fibre.	Textile for aviation and naval applications such as sail, parachutes, etc.

Nanomaterial	Property/ functionality	Functionality	Application areas
Rare earth nanocomposites	Luminescence	Electroluminescent tagging systems.	Fluoresce to detect infrared radiation
Hydrophilic/ hydrophobic titanium dioxide nanocoating	Moisture transport	transforming hydrophobic to hydrophilic coating on exposure to light	Moisture shifting textiles
Keratin nanocoating	Shrink resistance	Resists shrinking due to fibre rearrangement and compaction during machine washing.	Textiles with increased structural strength
Polymer nanofilaments such as nano titania (anantase) coated PVC, HDPE	"Lotus effect" due to superhydroph-obic surface	Polymer nanofilaments grown on textile fibres create a rough geometry that repels water due to low surface contact. The high contact angle causes water to roll off the surface, picking up dirt particles, thereby cleaning the surface and less likely to stain.	Stain resistant, self-cleaning, water repellent textiles
Zinc oxide nanoparticles	Static protection	Conductive zinc oxide nanoparticles disperse any static charge developed on textile fibres.	Anti-static textiles for ordinance factories, ammunition depots, etc.
Silica aerogels	Thermal insulation	Silica aerogel shows extremely high thermal insulating properties.	Super-lightweight thermally insulating textiles for cold weather
Titanium dioxide nanoparticles	UV protection	Semiconductor nanoparticles blocks UV rays by photocatalysis.	Sunscreen cream against UV radiation; textiles especially in snow bound areas against degradation .

7.3 HEALTHCARE

The application of nanotechnology in the healthcare sector is referred to as nanomedicine. Specifically, this area of application uses nanometer-scale materials and nano-enabled techniques for early diagnosis, point-of-care monitoring, treatment and prevention of a plethora of ailments such as cardiovascular diseases, cancer, musculoskeletal and inflammatory conditions, neurodegenerative and psychiatric diseases, diabetes, and infections. The potential contribution of nanotechnology in the medical sector is extremely broad and includes new diagnostic tools; imaging agents and methods; drug delivery systems and pharmaceuticals; therapies; implants; tissue-engineered constructs; etc. Nanomaterials by definition are the materials at a nanoscale level, which, in nanomedicine, often goes beyond 100 nm and up to about 500 nm. This is the size range of biomolecules (*e.g.*, proteins, enzymes, DNA) and molecular complexes such as the ion pump. These natural nanomaterials are the constituents of larger hierarchical structures that regulate the multiple function of the cell. Bacteria are larger (up to a few micrometres), but their functions including toxicity to healthy cells derive from the interaction between the biomolecules and the surrounding cells. Basically, nanotechnology makes it possible to create engineering materials such as drug delivery systems, diagnostic probes, or even tissue-engineered constructs of dimensions of biomolecules which, in turn, is the scale that regulates the functions of cells. In a way, nanotechnology has the great potential to improve the entire healthcare process beginning from diagnosis to drug delivery to therapy and tissue engineering. Some of the impacts of nanotechnology in healthcare are briefly discussed.

7.3.1 Diagnosis

Diagnosis of a suspected disease is one of the most critical step in healthcare and medicine. Diagnoses must be quick, reliable, specific, accurate together with the minimum risk of 'false positives'. Nanotechnology based diagnostic techniques provide two major advantages, *i.e.* early detection and rapid testing, which facilitates start of early treatment with minimum damage to the patient. The accurate targeting and quantification of molecules indicative of cellular disorders at the single-molecule level is a challenging task for current high-throughput analysis systems. The combination of nanoparticles with other nanotechnology-based materials has the potential to address this emerging challenge and provide technologies that enable diagnoses at the level of a single cells and single molecule. The possibilities for nanoparticle applications in diagnostics are almost unlimited because nanoparticles enable the selective tagging of a wide range of medically important targets, including bacteria, biomarkers, and individual molecules such as proteins and DNA.

In recent years, there has been a trend to make diagnostic devices smaller by using nanoparticles. Nanomaterials can be engineered to be very specific, so their use will make the device even more accurate and capable of carrying out even more tests simultaneously. For example, carbon nanotubes and gold nanoparticles are being used in a sensors, specific for detection of proteins indicative of cancer and infections. Nanofibers coated with antibodies are also being used for specific binding and trapping the cancer cell for analysis. Silver nanorods are being used in a diagnostic system to separate viruses, bacteria and other microscopic components of blood samples, allowing clear Raman spectroscopy signals for quick identification of viruses and bacteria. Nanomaterials have the characteristic of exhibiting some peculiar quantum effects that can be used to amplify the signal arising from the detection. Thus, the use of nanomaterials in miniaturized *in vitro* diagnostic devices makes it possible to improve the specificity of the analysis, its throughput (the number of tests that can be done simultaneously), and its read-out. These types of devices are likely to make it possible to perform 'point-of-care diagnostics' *i.e.* it will be possible to make a diagnostic test anywhere, and not just in the hospital.

The nature of the sample to be tested may also change and become saliva rather than blood, which is much more convenient and safer to handle. This will allow a large number of patients to be tested. For example, in the event of an epidemic, or a large number of diseases, or the many parameters needed for one specific disease to be considered for the diagnosis of complex medical conditions. Miniaturized diagnostic devices include biosensors, microarrays, and "lab-on-a-chip" (LOC) devices, also called miniaturized total analysis systems (μTAS). The first two are based on a parallel processing technique, whereas LOC devices are based on a serial processing technique. In addition, nanoparticles are also playing an important role in diagnostic imaging technologies to enhance contrast. LOCs devices are 'miniaturized integrated laboratories' that allow the separation and analysis of biological samples (*e.g.* blood) in a single device. They are made of micro fluidic systems, including micro pumps and micro-valves, integrated with microelectronic components. The devices can also integrate one or more sensors. As with microarray technology, the impact of nanotechnologies in this area is in further miniaturization of these devices, although the handling of ultra-small volumes of samples could pose a problem. Such devices are conceptually shown in **Figure 7.2,** which shows integrated dielectrophoresis and antibody-based capture; growth of bacterial cells, cell lysing and PCR in a silicon-based platform. Such devices can be of great importance to public health applications.

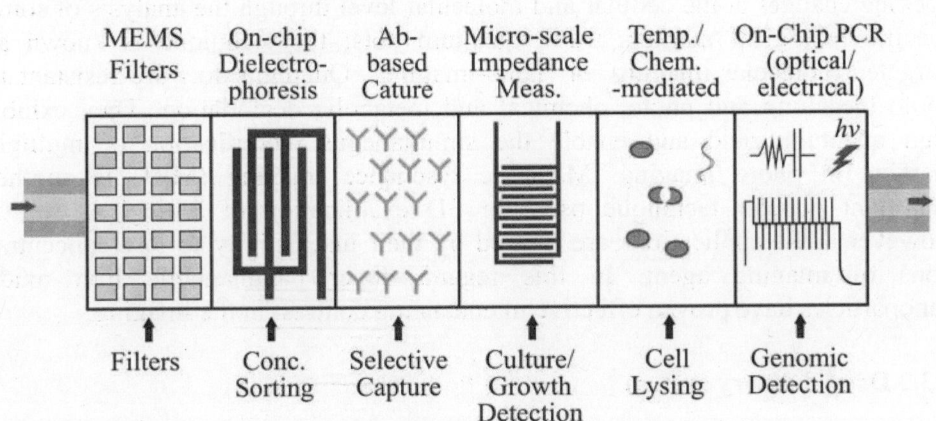

Fig. 7.2. Integrated lab on chip for detection of cells and microorganisms

A **biomarker** is an indicator of a biological process or state, such as a disease, or the response to a therapeutic intervention. This can be an altered gene, or a change in protein production, or even a physical feature of a cell. The aim is to detect biomarkers of disease and diagnose illnesses before, or at the onset of, the first symptoms, in this way making *in vivo* imaging a tool for the early detection of a disease. Effective early detection is crucial for planning a therapy with less severe and costly therapeutic demands, especially in diseases such as cancers, where timing is vital for the success of the treatment. **Figure 7.3** illustrates a nano gold-based biosensor for dengue detection.

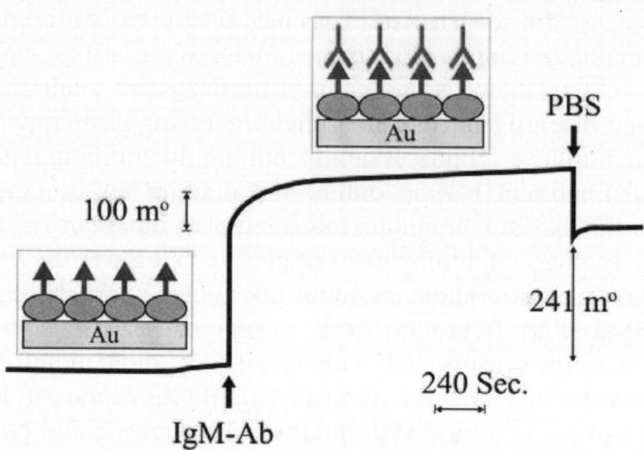

Fig. 7.3. Nanogold surface immobilized with dengue antigen-BSA conjugate for detection of dengue infection using SPR technique

The latest improvements in the area of imaging deal with the capability of tracking changes at the cellular and molecular level through the analysis of some specific biological markers using quantum dots; the technique is known as 'targeted molecular imaging' or 'nano-imaging'. Quantum dots are resistant to photo bleaching and photo, chemical and metabolic degradation. They exhibit high quantum yield and enable the simultaneous identification of multiple markers in photo imaging. Magnetic resonance imaging (MRI) is another important popular technique used for 3D examination of biological events. However, MRI applications are limited by their insensitivity to low concentrations of imaging agent. In this regard, super paramagnetic iron oxide nanoparticles have proven effective in enhancing contrast in the imaging.

7.3.2 Drug delivery system

Advances in the field of pharmacological aspect of nanomedicine stem from two main concepts: the development of new biologically active drugs and the development of new drug delivery systems able to reach the specific site of the disease. Drug delivery systems are in the same form as the "pills" that are frequently taken and that release their active component gradually (slowly release drugs) or that dissolve based on some physiological conditions (*e.g.*, acidity of the environment). Drug delivery systems also exist in the form of implants, inserts, or other drug-releasing systems.

a. Advanced Drug Delivery System

Nanoparticle-based drug delivery provides many advantages, such as enhancing drug-therapeutic efficiency and pharmacological characteristics. For example, nanoparticles improve the solubility of poorly water-soluble drugs, modify pharmacokinetics, increase drug half-life by reducing immunogenicity, increase specificity toward the target cell or tissue (thus reducing side effects), improve bioavailability, diminish drug metabolism, and enable a more controllable release of therapeutic compounds and the delivery of two or more drugs simultaneously for combination therapy. Advanced drug delivery systems aim to improve bioavailability and pharmacokinetics of pharmaceuticals and to replace invasive intervention by non-invasive routes of administration. Examples for advanced drug delivery systems are controlled-release formulations or pulmonary dosage forms of proteins such as insulin. Nano-drug delivery systems (NDDS) are a subclass of advanced drug delivery systems that consists of drug carriers with a size of less than 1 μm and mostly less than 200 nm. Examples for NDDS are liposomes, nano suspensions, polymeric nanoparticles, dendrimers, fullerenes, carbon nanotubes, and inorganic nanoparticles.

b. Targeted Drug Delivery

In future, this could lead to targeted therapies and personalized medicine. The aim is to design and deliver drugs in such a way that they can recognize the "bad" cells at a molecular level, penetrate the cell membrane, and act inside the infected cell. This is often crucial for the efficacy of a drug, since most virus replication and other disease conditions take place across the cell membrane and inside the cell. This way, the treatment will be delivered where it is needed and will be specific, eliminating the problem of the drug killing healthy cells. An example of this approach is siRNA drug delivery. Targeted drugs and targeted drug delivery system (DDS) could allow the creation of drug formulations with optimal loading, which deliver to the body only the necessary amount of the drugs and reduce side effects for the patients. In addition to the possibility of having nano-DDS that are biodegradable inside the body, this will help to reduce drug toxicity. Drug safety can be further enhanced by the possibility of introducing inside the drug formulation a label that changes color when the drug reaches its expiry date or is no longer functioning. This will allow for the improvement of drug shelf-life and better monitoring of drug safety.

c. Remotely Triggered Delivery Systems

Systems that release drugs in response to a remote trigger, that is, systems that are actuated by an external stimulus, could offer distinct clinical advantages over those that release drugs passively or are triggered internally, for example, by a chemical stimulus. Drug-release profiles could be tailored to the specific therapy. For example, insulin is most effective when delivered to a diabetic in short bursts, whereas an anaesthetic should be delivered in a steady, continuous manner. Most importantly, it is expected that remotely triggered systems would allow the patient or doctor to control the dose, timing, and duration of drug release. Numerous stimuli such as visible or near-infrared (NIR) light, electric or magnetic fields, and ultra sound (US) could be used for drug triggering. Some of these techniques, such as US, MRI, and NIR fluorescence microscopy, are already used in the clinic or lab for imaging. NIR (~650–900 nm) light is a promising tool for drug delivery because it can penetrate the skin and tissue, enabling triggering deep within the body. Recently, nanofunctional materials have been developed that are sensitive to these stimuli and that have been used to design drug delivery systems. A variety of nanostructures such as gold nanorods, nanoshells, and nanocages absorb strongly in this range and then convert light energy into heat. Hence, gold nanoparticles can behave as localized heat sources, and they have been used to trigger drug release from temperature-sensitive materials. One temperature-sensitive material that has been used extensively is poly(*N*-isopropyl acrylamide) (pNIPAm), a polymer that abruptly and reversibly collapses when heated beyond a critical temperature.

7.3.3 Therapy

A therapy usually involves a pharmaceutical route *i.e.* drugs to treat the disease from inside of the body or, when a pharmaceutical therapy is not possible or not effective, other routes to fight the disease from the "outside" of the body are radiative therapies. In some circumstances, surgery is required, and an organ substitute is inserted into the body in the form of an implant or a donated organ. In all these approaches, which are often used in combination, the aim is always the same: to selectively eliminate the source of the disease as a long-lasting solution. Nanotechnologies are making a tremendous impact in this field, with new drugs and new types of treatments under development, some of which have already proven clinically effective and have entered the market. Following are the examples of some of the newly emerging nanotechnology-based therapeutic treatments.

a. Theranostics

One of the most exciting opportunities that nanotechnology offers to the therapeutic field is the possibility of integrating the diagnosis and therapy, and follow-up of a disease. This is referred to as theranostics and could be enabled by nanoparticles (*e.g.*, quantum dots) incorporated inside a drug that can change some property such as color, once the drug has reached the target. Drugs could therefore, have a feedback action. In addition to a slow, targeted release system, the nanoparticles could gradually change color during the drug action, thereby informing the progress of a therapy to the doctors. This approach is called "find, fight and follow" and could become a reality as a result of the parallel progress in the field of molecular imaging. It is believed that in times to come the entire process of diagnosis, pre-imaging, therapy, and post-imaging of a specific disease will be integrated. Examples of theranostics for cancer treatment are: Photodynamic therapy using NaYF:Yb^{3+}, $Er3^{+}$-RB/FA nanoshells (**Fig. 7.4a**) and hyperthermia using superparamagnetic nano particles (Fe@Co NPs) (**Fig.7.4b**) for imaging and treating cancer cells at the same time.

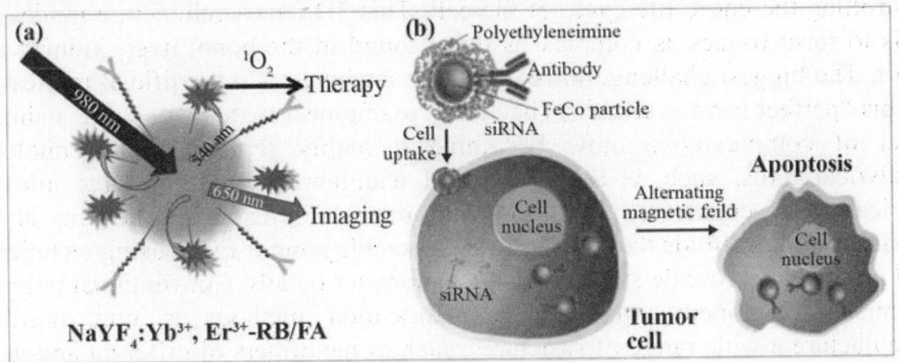

Fig. 7.4. a. Simultaneous luminescence (UCL) imaging and photodynamic therapy; b. Magnetic Hyperthermia for treatment of cancer

b. Regenerative Medicine

At times, the only way to treat a disease is the removal of the infected organ or tissue. Such loss can also derive from an injury or a congenital condition (*e.g.*, vision or hearing impairment). To compensate for the lost or impaired body function, an artificial construct is implanted in the body. Depending on the type, site, and extent of the loss, this construct can be in the form of an engineered tissue or an implant. For tissue engineering and regenerative medicine, self-assembled peptide nanofibers are the most suitable materials. They are obtained by the self-assembly of relatively simple peptide building blocks, readily obtained by fast and reliable synthesis, which in turn facilitates the introduction of peptidic or non-peptidic epitope for specific applications. In addition, their peptidic constitution confers them unique biocompatibility and degradability as they can, in principle, be used as nutrients by cells.

7.3.4 Tissue and Biomaterial Engineering

Tissue engineering deals with the fabrication of artificial scaffolds to support the growth of donor cells. It differentiate and grow into a tissue that mimics the lost natural one. This tissue-engineered construct is then implanted in the patient and, in time, resorbed by the body and fully integrated by the host tissue. Current applications of tissue-engineered constructs include engineering of skin, cartilage, and bone for autologous implantation (*i.e.*, the implantation of a tissue regenerated by seeding cells of the patient). In the body, cells are supported in their growth and function by a natural scaffold, called the extracellular matrix (ECM). This is a very complex and intricate "web" of nanofibers that provide the mechanical architecture for cellular growth. Moreover, the ECM is filled with small molecules (*e.g.*, growth factors) and proteins that direct many cell processes, such as adhesion, migration, growth, differentiation, secretion, and gene expression. The 3D spatial organization of these "cues" is critical in controlling the entire life cycle of the cell. This 3D nano-architecture guides the cells to form tissues as complex as those found in the bone, liver, kidney, and heart. The biggest challenge in regenerative medicine is the artificial replication of this "perfect nano-scaffold." The ability to engineer materials having a similar level of complexity is now becoming a reality through nanotechnology. Analytical tools, such as the AFM, and nanofabrication tools have allowed fabrication of engineered scaffolds with nanoscale features. The aim is the fabrication of scaffolds that most closely resemble natural ECM using techniques that produce macroscale structures with nanometer details. Conventional polymer chemistry combined with new nanofabrication methods is now used to manufacture a wide range of structures, such as nanofibers of different and well-defined diameters and surface properties; nanofibrous and porous scaffolds; nanowires, nanotubes, nanospheres, and nanocomposites.

One of the distinguishing features of nanotechnology is its ability to create new functional and bio-compatible materials. Which can be exploited in the fabrication of new biomaterials that have better mechanical properties. For example, to increase the implant stability and reduce fatigue failure in orthopaedic implants, and materials that have enhanced electrical properties, for example, needed in neural prostheses. Nanocoatings are also used for better integration of synthetic implants with the biological tissue, in order to prevent tissue inflammation and the onset of rejection.

7.4 INFORMATION AND COMPUTER TECHNOLOGIES

Nanotechnology in many respects, has proven its key role in Information and Communication Technologies (ICT). Computer microprocessors and memory storage devices have followed a path of miniaturization over the last 20 years that has 'naturally' brought transistors to have dimensions lower than 100 nm. There are challenges now to further lower this size. Nanotechnology is already in use in many computing, communications, and other electronics applications to provide faster, smaller, and more portable systems that can manage and store larger and larger amounts of information. These continuously evolving applications in ICT sector include:

- Nanoscale transistors that are faster, more powerful, and increasingly energy-efficient; soon your computer's entire memory may be stored on a single tiny chip.

- Magnetic random access memory (MRAM) enabled by nanometer-scale magnetic tunnel junctions that can quickly and effectively save even encrypted data during a system shutdown or crash, enable resume-play features, and gather vehicle accident data.
- Displays for many new TVs, laptop computers, cell phones, digital cameras, and other devices incorporating nanostructured polymer films known as organic light-emitting diodes (OLEDs). OLED screens offer brighter images in a flat format, as well as wider viewing angles, lighter weight, better picture density, lower power consumption, and longer life.
- Other computing and electronic products include Flash memory chips for iPod nanos; ultra-responsive hearing aids; antimicrobial/ antibacterial coatings on mouse/keyboard/cell phone casings; conductive inks for printed electronics for RFID/smart cards/smart packaging; more life-like video games; and flexible displays for e-book readers.

7.4.1 Integrated circuits

As semiconductors, metals and insulators are reduced to nano-size ($\leq$100 nm), quantum effects start to predominate and determine their properties. Nanotechnology offers the opportunity to exploit, the quantum effects in the development of the next generation of integrated circuits (ICs). In the late 1960s, Gordon E. Moore, co-founder of Intel Corporation, made a memorable observation, later known as Moore's Law. It states, 'The number of transistors on a chip roughly doubled every 18 months'. The progression in transistor density on a chip over the last 40 years has, indeed, followed Moore's law. In order to keep up with this law, transistors have become smaller and smaller. The first transistor was about 1cm high and made of two gold wires 50μm apart on a germanium crystal. The latest transistors from Intel® have 45 nm feature sizes and a quad core configuration have over 731 million transistors. These numbers give an idea about the enormous miniaturization efforts that have been achieved by the semiconductor industry. The scaling-down of transistors has been the goal for a number of reasons. Transistors are the basic building blocks of the elements in an integrated circuit (microprocessor, mass memory, logic gates, etc.). As the size of the transistor is reduced, its density on a chip increases, which increases the speed, the amount of memory stored per unit area, and the number of functions that can be integrated on a single device.

7.4.2 Data storage

Data storage is a key component of many devices of everyday use such as computers which contain both a hard drive and RAM, mobile devices and digital cameras with memory card, portable media players *e.g.* iPods, pen drives and mobile phones that use Flash memory. Data storage technologies include two main groups, hard disk drives (having a mechanical component) and solid-state data storage devices, which can be further divided in volatile and non-volatile types. 'Volatile' means that memory is lost once the power is turned off; 'non-volatile' means that memory is retained when the power of the device is switched off. Volatile memory storage devices today include mainly static random access memory (SRAM) and dynamic random access memory (DRAM). Non-volatile memory storage devices, such as Flash memory, are used in mobile devices and portable media players. Each type of memory has its advantages and disadvantages. Microprocessors work by processing information through the passage of an electrical signal. Other information processing techniques use variables such as spin (spintronics), molecular states (molecular electronics), photons (photonics), mechanical state, resistance, quantum state (including phase) and magnetic flux. These new state variables require new materials such as nanomaterials (*e.g.* relying on spin implies using magnetic materials rather than a semiconductor) and new functional organization of the device (nano architecture).

Spintronics: This is a new type of technology that exploits the spin of the electron (rather than only its charge) to store and process information. It requires thin layers of magnetic materials. The giant magneto résistance effect (GMR) represents the first type of this new technology, and is used in the new-generation memory devices such as the Magnetic Random Access Memory or magneto resistive random access memory (MRAM) Apple iPod is an example of making use of spintronics. Magnetic materials are used in MRAM, in which each memory cell consists of two magnetic thin-film materials separated by an isolating layer. Magnetic multilayer nanocomposites (very thin films made of magnetic materials) display magneto resistance properties. Magnetoresistance is a phenomenon whereby the application of a DC magnetic field changes the resistance of a material. In metals, this effect occurs only at very high magnetic fields and low temperatures

Photonics: Another possibility is the use of photons in the visible or IR range to transmit and process data (optical communication). Photonic materials (photonic crystals) engineered at the nanoscale are the materials of choice. A photonic crystal consists of a periodic structure made of dielectric materials that affects the propagation of light. Essentially, photonic crystals contain regularly repeating internal regions of high and low dielectric constant. Photons (behaving as waves) whether propagate through this structure or not depend on their wavelength. The periodicity of the photonic crystal structure has to be of the same length scale as half the wavelength of the electromagnetic waves (*i.e.* approximately 200 nm (blue) to 350 nm (red) for photonic crystals operating in the visible part of the spectrum. Such crystals have to be artificially fabricated by methods such as electron-beam lithography and X-ray lithography.

Quantum electronics: This term means the explicit use of the tunnelling effect in nanomaterial to transport electrons from the source to the drain of the transistor. One such transistor exists and is called the single-electron transistor (SET). In this type of transistor, there is a quantum dot between the source and the drain: electrons must tunnel through this dot in order to get from the source to the drain. In simple terms, this is like having two electrodes (source and drain) separated by a thin insulating barrier with a third electrode (the quantum dot, or quantum island) in the middle of this insulating gap. The transport of charge from the source to the quantum island and then from the quantum island to the drain occurs via the quantum mechanical process of tunnelling (see section 6.1.1.1. in chapter 6 for more details).

Carbon nanotube transistors: In many ways, carbon nanotubes are considered a 'dream material' when it comes to electronics. They can be conductive or insulating depending on their chirality (and when they are conductive, they are extremely good electric conductors, with little resistance and, consequently, little heat dissipation); they are strong (mechanically) and chemically inert; they are resistant to high temperatures; and they can be functionalized with specific molecules to act as anchoring points. Carbon nanotubes are 1–2 nm in width and as narrow as the double stranded DNA molecule (the molecule that carries our genetic information). So arranging nanotubes into electronic circuitry could allow miniaturization by a factor of about 100 over the current limit.

Molecular electronics: Molecular electronics is a branch of nanoscience that aims to make explicit use of molecular assemblies for the transmission and storage of data. The field includes molecular wires, molecular switchers, molecular sensors and other 'hardware' components of electronics.

7.4.3 Displays

A display is an electronic device for the visual perception of data or images. Display technology has progressed enormously in the last decade. Until a few years ago, there were bulky televisions with cathode ray tube (CTR) technology, and mobile phones with black and white displays that could only show text. Nowadays, coloured liquid Crystal displays (LCD) in televisions and mobile phones, many even with touch displays, can show photos and movies are the norm. Further progress has been driven by constant need for more functional devices combining image and video quality, low power consumption and low cost. Lower thickness and flexibility are also becoming an important requirement. Emerging display technologies which make use of nanomaterials include, organic LEDs (OLEDs), quantum dot LEDs, field emission displays and electronic paper.

a. OLEDs and OLETs

OLEDs are full of promise for a range of practical applications. OLED technology is based on the phenomenon that certain organic materials emit light when fed by an electric current and it is already used in small electronic device displays in mobile phones, MP3 players, digital cameras, and also some TV screens. Areas where nanomaterials and nanofabrication techniques are used in OLEDs are transparent electrodes (where for instance carbon nanotubes thin-films are gaining popularity) and nanoparticles-based coatings for packing the OLEDs to protect them from environmental damages (*e.g.* water). Nanoparticle-based deposition methods may also overcome OLED fabrication issues such as material damage, yield, and thickness uniformity. Recently, new concept of OLEDs with a few nanometer of graphene as transparent conductor has been developed. This paved the way for inexpensive mass production of OLEDs on large-area low-cost flexible plastic substrate, which could be rolled up like wallpaper and virtually applied to anywhere you want.

Organic light-emitting transistors OLETs, are alternative, planar light sources combining in the same architecture, the switching mechanism of a thin-film transistor and an electroluminescent device. OLET is a new light-emission concept, providing planar light sources that can be easily integrated in substrates of different nature such as silicon, glass, plastic and even paper, using standard microelectronic techniques. Thus, OLETs could open a new era in organic optoelectronics and serve as test beds to address general fundamental optoelectronic and photonic issues.

b. QD-LEDs

Quantum dots (QD) the nanoscale semiconductor particles characterized by emitting a specific color based on the size of the nanoparticle are a new class of nanomaterials under investigation in the manufacture of more efficient displays and light sources (QD-LEDs). A minute change in particle size results in a totally different color being emitted; for instance a 6 nm-diameter particle would glow red, while another of the same material but only 2 nm wide would glow blue. Light emission from a QD is monochromatic, therefore it is very pure. As a consequence, their use in displays would lead to images of exceptional quality. The most exciting property of QD-LEDs, however, is that they use much less power than the currently used LCDs where light is filtered by numerous polarizers. Like OLEDS, QD-LEDs emit light, rather than filtering it, so, for this reason, QD-LEDs are expected to be more energy efficient.

c. Field Emission Displays

Carbon nanotubes have shown promise to create a new class of large area, high resolution, low cost flat panel displays, known as field emission display (FED) technology. The later technology is considered the biggest competitor to LCD's dominance in the panel display arena. FED is the technology of choice for ultra-high definition, wide-screen televisions. FEDs, in a sense, are a hybrid of CRT and LCD technologies. Field emission display technology makes possible the thin panel of today's liquid crystal displays (LCD) and offers a wider field-of-view, provides the high image quality of today's cathode ray tube (CRT) displays, and requires less power than today's CRT displays.

d. Electronic Paper display

Electronic paper (e-paper) is a kind of display technology that aims to mimic the appearance of ordinary ink on paper. There are various technologies to create e-paper; among them electrophoretic paper is the most established. Unlike conventional flat panel displays, electrophoretic displays rely on reflected ambient light, rather than the emission of light from the display, and can retain text or images without constant refreshing, thereby requiring much less power. This also means that they can be used under sunlight without the image fading. e-paper display, is also very light and flexible. Some new materials such as transparent carbon nanotubes, photonic crystals, liquid crystals and other nanomaterials are emerging as elements of e-paper. It has been shown that organic ink nanoparticles could provide an improved electronic ink fabrication technology resulting in e-paper with high brightness, good contrast ratio, and lower manufacturing cost

7.5. NANOELECTROMECHANICAL SYSTEMS (NEMS)

The past decade has brought tremendous growth in the field of micro-electromechanical systems (MEMS) devices such as sensors, computers, electronics, and machines at the micro scale. The design of these devices utilizes a combination of traditional semiconductor processing and advanced mechanical engineering. Typically, the output of an electromechanical device is the movement of the mechanical component; a transducer is used to convert the mechanical energy into electrical/optical signals, or vice versa. Some outstanding MEMS applications include:

(i) Automotive: air flow and tire pressure sensors, "smart" suspension, head light levelling navigation, vehicle security, automatic seatbelt restraint, etc.

(ii) Micro nozzles for inkjet printers

(iii) Micro tweezers

(iv) Aerospace navigational gyroscopes

(v) Disposable blood pressure transducers

(vi) Portable skin analysis sensors for cosmetics applications

The extension of MEMS to the nano regime is referred to as nanoelectromechanical systems (NEMS), representing the ultimate in future devices, with benefits such as lower power dissipation and ultra-sensitive and localized responses. Further, due to the size of NEMS, it should be possible to directly incorporate a number of auxiliary functionalities alongside transistors within a single chip. Indeed, the applications for NEMS will span the fields of sensors, electronics, biotechnology, affecting virtually every aspect of our lives. Though the technology is in place to fabricate NEMS, there are some following enlisted issues need to be addressed primary prior to the realization of widespread commercial applications:

- To communicate signals from the nanoscale to the macroscopic world?
- Understanding and controlling mechanical responses at the nanoscale.
- Developing methods for low-cost, high-volume, and reproducible nanofabrication.

One recent NEMS device is a mass sensor with a resolution at the zeptogram level (1 zg = 10^{-21} g). Thus far, the lowest detection limit for this device is 7 zg, or ca. 30 Xe atoms. Incredibly, it is suggested that this technology may be combined with nano fluidics for the genetic analysis of the DNA present within a single cell. In comparison, current methods utilize PCR amplification, whereby small samples of DNA are repeatedly replicated in order to facilitate detection. Such analyses will be of use for the detection of genetic markers for cancer or other diseases such as HIV. The ability to identify proteins will be essential for future diagnostic and forensics applications.

Another area of NEMS that is receiving tremendous attention is the mimicry of biological systems, aptly referred to as biomimetic. For instance, in the development of linear molecular muscles that undergoes contraction and extension movements.

Initial work in this field utilized transition metal complexes containing **rotaxanes and catenanes**, due to the non-destructive redox processes occurring on the metal centres. Though these complexes were actuated by a chemical reaction, the movement was in a noncoherent manner. In order to better mimic skeletal muscle movement, one has to look at the mode of motion within the most efficient molecular machines in our human bodies.

7.6 ENERGY

Today, the global energy demands are satisfied mainly via the combustion of fossil fuels. Of the 210 million barrels of oil equivalent per day used worldwide, about 85 million barrels come from oil; the rest comes from coal (23%), gas (17%), biomass (17%), some fission (5%), a small amount from hydroelectric power (6%) and almost none from renewable resources. It is estimated that by 2050, twice the amount of energy that is burned or consumed today will be needed.

Because of the limited availability of fossil fuel the world is presently facing an 'energy challenge' and there is an urgent need for alternative energy resources to fossil fuels. Every day, the Earth is hit by 165000 TW of solar power; in the words of Nobel Prize winner Richard E. Smalley, every day 'we are bathed in energy'. Solar energy collection, conversion, storage and distribution are major challenges. Currently, photovoltaic solar panels have about 15–20% energy conversion efficiency and they are very expensive. Another alternative energy carrier is hydrogen, but hydrogen fuel cell technology still faces a number of issues (*e.g.* hydrogen extraction, hydrogen storage, fuel cell lifetime and cost) before a hydrogen economy can become a reality.

It is in this context that nanoscience and nanotechnology are poised to play a transformative role in providing clean and sustainable energy. With control of phenomenon arising out of nanoscale, it is expected to put its foot hold in the energy sector through harnessing solar energy (Photovoltac and thermal devices) conversion of chemical to electrical (batteries and fuel cells), thermal to electrical (Thermo electrical devices), and storage (hydrogen). Some of these technologies are briefly discussed in following sub sections.

7.6.1 Photovoltaic Technologies for Solar-Energy Harvesting

Solar energy is by far the most abundant exploitable renewable energy resource. More energy is provided to the earth by sunlight irradiation **(Fig. 7.5)** within one hour than is consumed by human society globally in one year. Semiconductor materials that exhibit a photovoltaic (PV) effect can be used to convert solar radiation into electricity through a photovoltaic process. PV technology has been growing and expanding rapidly. The total global energy production by PV processes has reached close to 100 GW (1 GW=10^9 W). Despite this considerable capacity, however, PV technology only accounts for 0.1% of electricity generation globally, largely as a result of the inability of existing PV technologies to produce electricity with an efficiency that fulfills the grid parity set by conventional power-generation routes.

Practically all PV devices, or solar cells, incorporate a p-n junction. Such junctions occur in various possible configurations. Solar cells containing multiple p–n junctions have recently been investigated intensively for the more efficient absorption of light with different wavelengths the main aim behind is in reducing the inherent sources of energy loss in conventional single-junction cells. A conversion efficiency of 42.3% was achieved by combining multijunction cells and concentrator technology. However, the high efficiency of multijunction cells is offset by their increased complexity and manufacturing cost, which limit their application mainly to aerospace exploration, for which a high power-to-weight ratio is desirable. Some latest developments in photovoltaic devices using semiconducting nano materials are enumerated below:

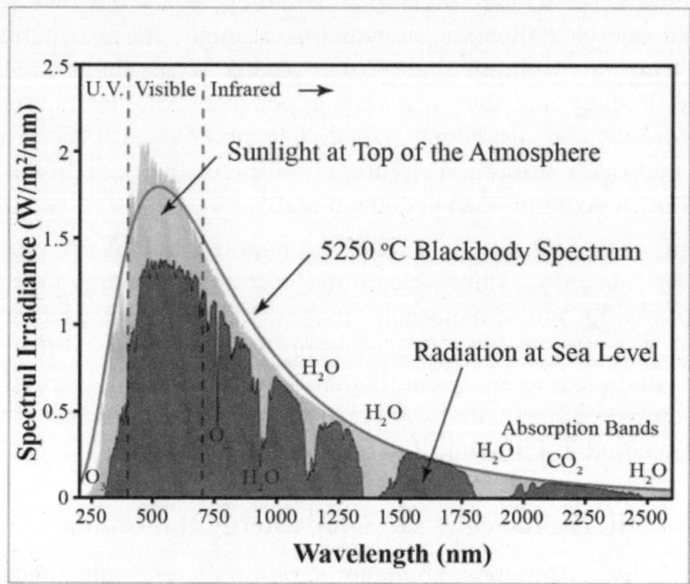

Fig. 7.5 Solar radiations spectrum reaching the Earth's surface

a. Dye-Sensitized Solar Cells (DSSC)

Dye-sensitized solar cell is a thin film solar cell, representing a photoelectrochemical cell. Modern version of dye-sensitized solar cell (DSSC) was invented by Michael Grtzel and his co-workers in 1991. During the past two decades, the efficiency of DSSC could be improved by optimizing the porosity of the electrode prepared from fine oxide powder However the instability needs to be addressed.

Dye-sensitized solar cells comprising two electrodes, the photo anode and the counter-electrode, in a sandwich arrangement are shown in **(Fig. 7.6).** The photo anode is a conducting glass covered by a mesoporous and nanocrystalline TiO_2 film, sensitized by the dye-sensitizers which is one of the most important components of this device since it is responsible for the sunlight harvesting and electron injection, the first steps of energy conversion. 2,2-bipyridine ruthenium salts are used as dye sensitizers. The counter electrode is a conducting glass covered by a thin film of catalyst, such as platinum or graphite. In between these electrodes is placed a mediator layer, usually a solution of $I^{3-}I^{-}$ in acetonitrile.

A three-dimensional network of randomly dispersed TiO_2 nanoparticles has typically been adopted as the photo anode in DSSCs since their first introduction. The disorderly network of TiO_2 nanoparticles presents numerous grain boundaries and hence multiple trapping sites. These trapping sites not only promote the increased recombination of electrons but also hinder the collection of the excited electrons and thus compromise the overall conversion efficiency.

To overcome these problems structural configurations with higher degrees of order than the random assemblies of nanoparticles are expected to facilitate the transport/collection of electrons and hence improve the efficiency of solar cells.

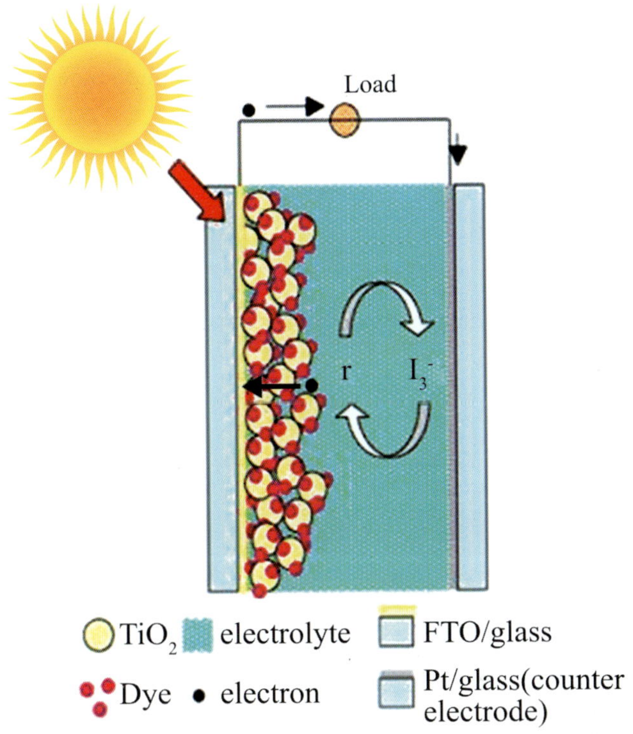

Fig. 7.6. A typical sandwich-type dye-sensitized TiO_2 nanocrystal solar cell

As an alternative to disorderly network of TiO_2 nanoparticles, ZnO nanomaterials are considered as transparent photo anodes in DSSCs. The combination of piezoelectric, semiconductor, and photo excitation properties of ZnO together with the straightforward methods available for the synthesis of various ZnO nanostructures at low temperatures could potentially facilitate the wide application of ZnO in DSSCs.

One-dimensional nanostructures have been investigated intensively and utilized in all types of PV devices mentioned above. For example, energy-conversion efficiencies of about 10 and 3% have been reported for DSSCs based onTiO_2 NWs and ZnO NWs, respectively. A solar cell based on an array of Si NWs with a projected efficiency of 17.4% and another based on an array of InP nanopillars with an efficiency of 8.1% have also been demonstrated. Moreover, the performance of solar cells based on NW arrays has been improved strategically by enhancing dye adsorption and reducing surface recombination.

Novel strategies, such as the integration of ZnO or TiO_2 nanowires with 3D optical fibers or planar waveguides, have been shown to address issues such as insufficient internal surface area and deficient dye loading, which are shortcomings of DSSCs based on 1D nanostructures. Such strategies offer the potential for the manufacture of DSSCs with high flexibility, adaptability, and enhanced efficiency for concealed applications. DSSCs could be suitable for powering distributed MNSs for self-sufficient, maintenance-free applications, since ideally DSSCs convert sunlight into electricity in a self-sustainable manner, without suffering any permanent chemical transformation or consumption of the materials in the system.

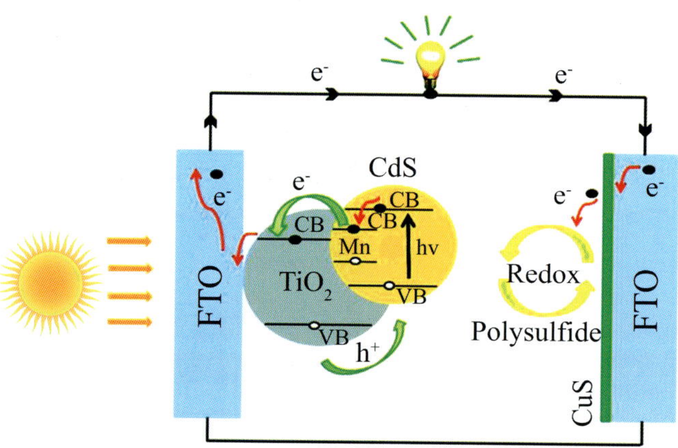

Fig. 7.7. Schematic illustration of the structure of a quantum dot sensitized solar cell

b. Semiconductor quantum dot-sensitized solar cells (QDSCs)

Traditional photovoltaic devices (*i.e.* the p-n junction silicon crystalline solar cells) are costly. As an alternative to silicon-based photo-voltaic, semiconductor quantum dot -sensitized solar cells (QDSCs) have attracted considerable attention as the next generation solar cells. Narrow-band-gap semiconductor QDs, such as CdS, CdSe, PbS, and InAs, have been used as the photosensitizer, instead of organic dyes, due to their versatile optical and electrical properties such as, tunable band gap depending on the QD size; high extinction coefficient; higher stability toward water and oxygen; and multiple exciton generation (MEG) with single-photon absorption. The theoretical photovoltaic conversion efficiency of QDSCs can reach up to 42% in view of the MEG effect of QDs. **Figure 7.7** shows the cell structure of a QDSC, which consists of a wide-band-gap mesoporous oxide film (a photo electrode, such as the commonly used TiO_2 or ZnO), QDs (the sensitizer), an electrolyte, and a counter-electrode. During operation, photons are captured by QDs, yielding electron-hole pairs that are rapidly separated into electrons and holes at the interface between the nanocrystalline oxide and QDs. The electrons jump into the oxide film, and the holes are released by redox couples in the electrolyte.

c. Nanostructured Organic–Inorganic Hybrid Solar Cells: Heterojunction Solar Cells

Although considerable progress has been made in improving the Photoconducting Electrodes (PCEs) of fully solid-state semiconductor-sensitized solar cells, they still suffer from low efficiency mainly because of rapid carrier recombination at the device interfaces.

A number of inorganic–organic heterojunction solar cells have been fabricated to overcome this drawback and to achieve stable, higher efficiency and cost-effective solid-state solar cells. Notable example among them is the use of inorganic–organic perovskite compounds ($CH_3NH_3PbI_3$) as light harvesters on mesoporous-TiO_2, and poly-3-hexylthiophene (P3HT) or poly-triarylamine (PTAA) as organic hole-transport materials (HTMs) or electron-blocking layers. Because the interfacial charge separation is the critical step in the photovoltaic process, the larger the interfaces, the more the opportunity for the excitons to reach the interfaces, and probably the higher the conversion efficiency. Therefore, most of the hybrid photovoltaic materials that have been studied are nano-structured composites. There are many types of nanostructures, including particles, rods, tubes, tetrapods, sheets, needles, and porous network, etc., and the way to mix the components together can be either disordered or ordered.

7.6.2 Artificial Photosynthesis: Production of Solar Fuel

The conversion of light into stored chemical energy by artificial photosynthesis is a chemical process which replicates natural photosynthesis and has been the focus of much attention recently. Photoelectrolysis, or water splitting, is a form of artificial photosynthesis that converts water into hydrogen and oxygen through the use of sunlight. An artificial photosynthetic system for producing solar fuel typically requires antenna/reaction center complexes for the generation of electrochemical potential from sunlight and appropriate catalysts for the oxidization of water and reduction of precursors to hydrogen.

There has been growing interest in the use of nanostructured materials as the photo anodes in photosynthesis applications as a result of the benefits offered by the large surface area, short lateral diffusion length, and low reflectivity. The use of structured photo electrodes based on semiconductor NWs with a preferred band gap or an intermediate energy level introduced within the band gap has also been investigated in efforts to improve the water-splitting efficiency through enhanced light absorption in the visible region of interest and enhanced electron-transfer efficiency due to the single-crystalline nature of low-dimensional semiconductor nanostructures. The creation of high-aspect-ratio photo electrode surfaces by engineering the morphology of nanostructure also enables the use of less expensive catalysts with lower catalytic activities. Furthermore, the ability to modulate the composition of nanomaterials at the atomic scale and to synthesize nanostructures of various morphologies, both in a controlled manner is likely to improve conversion efficiency in artificial photosynthesis applications through more effective management of the light absorption.

7.6.3 Thermoelectric Energy

The direct conversion of heat into electricity, or electricity into heat, is known as thermoelectricity. This conversion process is based on the Peltier-Seebeck effect, named after two physicists, Jean Peltier and Thomas Seebeck. Thermo-electric materials generate a current when placed in a temperature gradient as a consequence of charge carrier diffusion from the hot to the cold side, thus allowing the regeneration of waste heat into electricity. Thermoelectricity is an extremely interesting source of electric power because of its ability to convert heat flow directly into electricity. Thermoelectric devices **(Fig. 7.8)** are energy converters that are easily scalable and have no moving parts or liquid fuels, making them applicable in almost any situation where large quantities of heat tend to go to waste. Whilst thermoelectric materials have been known and understood for quite some time they have so far not been efficient enough to use commercially.

Nanostructures materials are expected to provide the basic requirements of good electrical conductivity and reduce thermal conductivity leading to improvements in figure of merit (ZT). The usage of nanostructured materials with size less than the wavelength of light enhances the scattering of photons and thus decreases thermal conductivity. Decrease in thermal conductivity is the most vital benefit of nanostructuring for thermoelectric materials. Most state-of-the-art thermoelectric materials are characterized with various nanostructures from fine grains, dispersed particles, nano-inclusions to atomic defects, and therefore are referenced as nanocomposite. Remarkable advances have been achieved in various material systems through the introduction of different nanostructures to tune the transport of phonons and electrons.

Thermoelectric (TE) materials based on bismuth telluride are best suited for room temperature applications with ZT (figure of merit) of ~1. Carbon nanotubes, graphene, semiconductor nanowires (NWs), PbTe, $CoSb_3$, Bi_2Te_3, etc. are other nanomaterials showing improved thermoelectric properties. Among these low-dimensional TE materials, silicon NWs with small diameters exhibit interesting and promising TE properties, with reported a ZT value of 0.6 at room temperature and a ZT value approaching those of commercial devices at 200 K.

7.6.4 Piezoelectric nanomaterials

A nanogenerator is a device, which harvests external mechanical energy and converts it into electricity as a result of bending or stretching nanostructured piezoelectric materials such as zinc oxide nanowires. Piezoelectric is a term that comes from pressure and electricity. When walking, mechanical force is generated this is "captured or harvested" to generate the energy that is "stored" in the pill embedded in a shoe sole for further use. The mechanical energy may be provided in a countless number of ways, *e.g.* virtually any body movement, a rolling tire, vibrations, or airflow. Nanogenerators capable of powering commercial liquid-crystal displays or light-emitting diodes have been demonstrated.

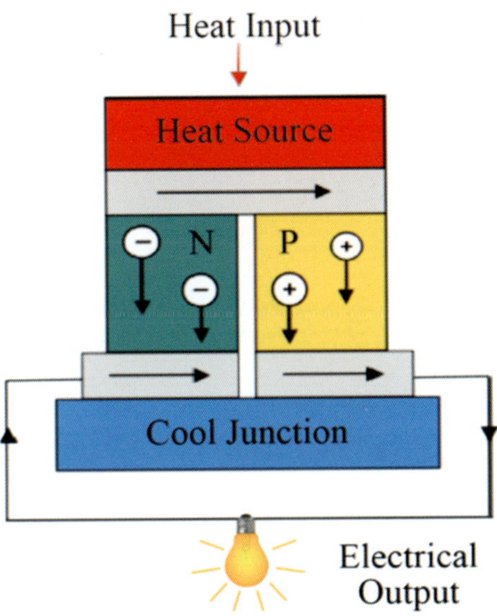

Fig. 7.8. Thermoelectric devices

7.6.5 Fuel cell

The idea of a fuel cell was first conceived in 1839 by Sir William Grove who thought that electricity could be produced by reversing the process of electrolysis, in which hydrogen and oxygen are combined to produce electricity, with water and heat as by-product through an electrochemical reaction. Thus, fuel cell, the electrochemical device will continue to generate power as long as fuel, the hydrogen is supplied. Hydrogen (H_2) can, in principle, be a future environmental-friendly energy carrier when it is produced from renewable energy. The ideal scheme would be to produce hydrogen by splitting water molecules using solar light. Hydrogen could then be used as an energy carrier to provide electricity in our homes, fuel our cars, etc. Hydrogen fuel cells, **Fig. 7.9**, use hydrogen and oxygen as fuel to generate electricity. The fuel molecules in the cell must be ionized to react. The ionization must be catalyzed by the electrodes, and an electrolyte must conduct the formed ionic species so that they can react. The sub-product of this reaction is only water, no CO_2 is produced during the conversion of hydrogen to electricity. The result of the electrochemical process is a maximum of 1.2 V and 1 W cm^{-2} of power.

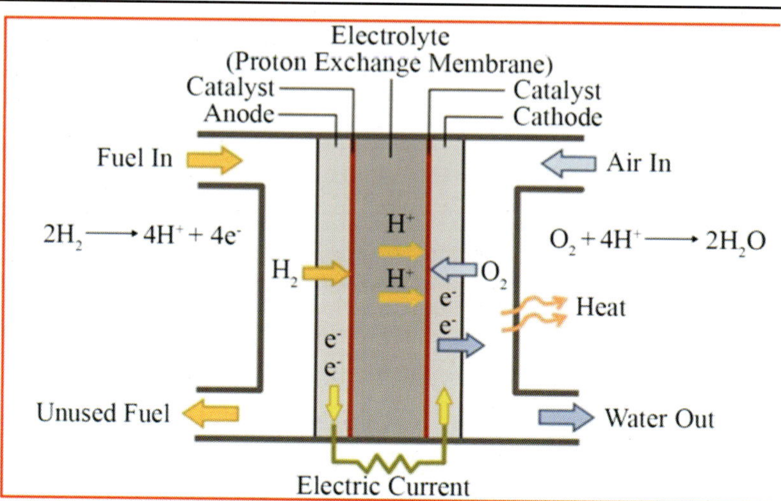

Fig. 7.9. Schematic representation of a fuel cell

The three fundamental components of a hydrogen fuel cell are therefore the fuel (H_2 and O_2), the catalyst and the electrolyte. In order for a hydrogen fuel cells to become an economically viable two main problems need to be addressed include the nature of the catalyst and that of the electrolyte in the cell. Currently, the electrodes in a fuel cell are made of a metal such as platinum (Pt), which is a rare and expensive metal, sensitive to the CO and sulphur species that are dispersed in the atmosphere. These species deactivate the platinum surface (a phenomenon called 'poisoning of the catalyst'). To avoid these problems fuel cells works at high temperatures (>70°C) since the poisoning agents at these temperatures tend to de-adsorb. Nanotechnology is already forthcoming to address some of the issues concerning fuel cell catalysts. Improvements in catalytic activity through nanotechnology is concerned with increasing the material catalytic activity and avoiding the use of rare metals. Since the current generated at an electrode is proportional to the active surface of the catalyst, fuel cells that have higher power density can be formed from nanomaterials, which have a higher surface area to volume ratio. Electro catalytic properties are also proportional to particle size; so nanoparticles and nanomaterials have increased catalytic activity compared to bulk materials. Carbon nanomaterials are particularly suitable since they act both as a support for the platinum nanomaterials and as a conductor. Suitable carbon nanomaterials are carbon foams containing nanopores, different types of carbon nanotubes and single-walled nano horns.

Hydrogen storage and transport

The combustion of hydrogen in air ($2H_2 + O_2 \rightarrow 2H_2O$) is straightforward reaction with no detrimental by-products. However, problem in making use of hydrogen as a fuel are related to its storage and transport. Nanotechnology can contribute in this field by developing new molecules that allow high hydrogen loading capacity, light weight, low in volume, good regeneration kinetics, and low cost. Complex nano metal hydrides, such as $LiBH_4$, $NaBH_4$, $LiAlH_4$ and $NaAlH_4$, which have an intermediate bonding of hydrogen, and carbon nano materials, such as carbon nanotubes and graphene are the potential candidates to meet the requirement of hydrogen storage. Equation 7.1 represent the facile Cycling of Ti doped lithium aluminium hydride ($LiAlH_4$).

$$Li\,AlH_4 \underset{Me_2O\,;\,25^{\circ}C}{\overset{\eta^{oo}}{\rightleftarrows}} LiH + Al + 1.5\,H_2 \tag{7.1}$$

7.6.6 Batteries

A battery is a device that converts chemical energy into electrical energy. There are two types of batteries viz. primary and secondary (rechargeable). In simple terms, a battery is an electrochemical device that generates direct current through a coupled set of reduction-oxidation (redox) reactions. The positive electrode captures electrons and the negative electrode releases electrons. The battery consists of a positive electrode facing a negative electrode divided by a porous separator that prevents the electrodes from touching, and an ionic electrolyte, which is a conducting medium that ensures movement of the ions from one electrode to the other.

Applications of batteries range from mobile phones to personal computers, to electronic vehicles to hybrid vehicles to power tools and backup power systems. In all the cases, high power and light weight are important. Some of commercially available batteries based on nanotechnology are given in **Table 7.3**.

Nanotechnology may offer the following benefits in batteries in the following ways:

- Reducing the possibility of batteries catching fire by providing less flammable electrode material.
- Increasing the available power from a battery and decreasing the time required to recharge a battery. These benefits are achieved by coating the surface of an electrode with nanoparticles. The increases in surface area of the electrode allow more current to flow between the electrode and the chemicals inside the battery. This technique could increase the

efficiency of hybrid vehicles by significantly reducing the weight of the batteries needed to provide adequate power.

- Increasing the shelf life of a battery by using nanomaterials to separate liquids in the battery from the solid electrodes when there is no draw on the battery. This separation prevents the low level discharge that occurs in a conventional battery, which increases the shelf life of the battery dramatically.

Table 7.3. Nanomaterials- based commercially available batteries

Product	Advantages
Battery with chemicals isolated from electrode by "nanograss" when the battery is not in use	Very long shelf life
Silver-zinc battery using nanoparticles in the silver cathode	Higher power density, low combustibility
Lithium-ion battery with the cathode made from a nano-porous phosphate structure	Higher power, quicker recharge, less combustible than standard lithium-ion batteries
Lithium-ion battery with the anode composed of lithium titanate spindle nanoparticles	Higher power, quicker recharge, less combustible than standard lithium-ion batteries

Lithium ion batteries

Intercalation-based batteries using the small lithium (Li^+) ion are the most used electrode material. These batteries have at least one redox-active electrode with an open crystal structure with holes capable of intercalating Li^+. Graphite has been the preferred anode material for Li-ion batteries due to its relatively high capacity, good cyclability, and low redox potential versus lithium metal. Intercalation/de-intercalation of Li^+ in and out of the graphite upon charge/ discharge forms the basis of graphite as the anode of the battery (**Fig. 7.10**). The thermodynamic equilibrium saturation concentration of the graphite is LiC_6, corresponding to a theoretical capacity of 372 mAhg $^{-1}$. To achieve higher energy density for advanced Li-ion batteries, new anode materials with higher capacity is needed. In addition, the graphite anode limits the recharge rates of Li-ion batteries. CNTs are attractive host anode materials for Li^+ intercalation. When the battery is charged, the Li^+ moves from the positive electrode to the negative one via the electrolyte. On discharge, the opposite occurs, releasing energy in the process. Ideally, the structure of the redox-active crystal should be capable of reversibly intercalating the small Li^+ ion.

Various types of nanomaterials as catalyst and anodes/cathode material have been studied to improve the characteristics of lithium ion batteries, some of which are mentioned here.

Catalyst made from nitrogen-doped carbon-nanotubes: This type of catalyst could be used in lithium-air batteries, which can store up to 10 times more energy as compared to lithium-ion batteries containing platinum as catalyst.

Silicon nanoparticles: The use of silicon nanoparticles, rather than solid silicon as anode of a lithium battery, prevents the cracking of the electrode and fast recharge within 10 minutes.

Silicon coated carbon nanotubes for anodes: The use of silicon can increase the capacity of Li-ion batteries by up to 10 times by depositing silicon on nanotubes aligned parallel to each other to prevent damage to the anode when the silicon expands.

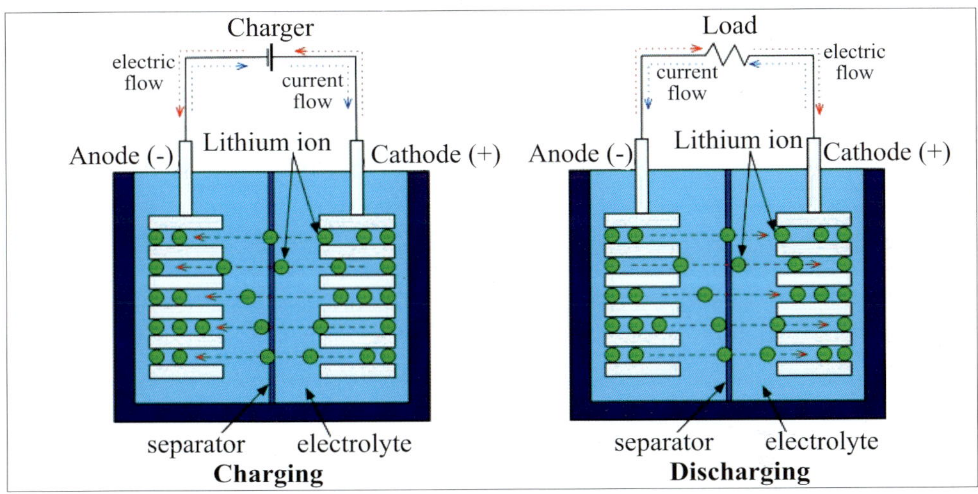

Fig. 7.10. Schematic representation of a lithium-ion battery

Carbon nanotubes grown on graphene have demonstrated as very promising electrode material with high surface area and low electrical resistance. The base of each nanotube is bonded, atom to atom, to the graphene sheet the nanotube-graphene structure is essentially one large molecule with a huge surface area.

Silicon nanowires on a stainless steel substrate: These anodes could have up to 10 times the power density of conventional lithium ion batteries. Silicon nanowires swell as lithium ions are absorbed during discharge of the battery and contract as the lithium ions leave during recharge of the battery the nanowires do not crack, unlike anodes that used bulk silicon

Carbon nanofibers as electrodes: Such electrodes show four times the storage capacity of current lithium ion batteries. Lithium sulfur batteries (the cathode containing sulfur), which have the capability of storing several times the energy of lithium-ion batteries. Such batteries are being developed using cathodes made up of carbon nanofibers encapsulating the sulfur cathodes made up of mesoporous carbon nanoparticles, with the sulfur inside the nanopores.

Aligned carbon nanotubes on a substrate as the anode/cathode: The carbon nanotubes attached to organic molecules that help the nanotubes to align on the electrode substrate, as well as provide many oxygen atoms as centers for attachment of lithium ions to the electrode. This could increase the power density of lithium ion batteries significantly.

Nanocomposites- based cathodes have increased the energy density of Li-ion batteries. Lithium ion batteries with electrodes made from nano-structured lithium titanate has significantly improved the charge/discharge capability at sub-freezing temperatures as well as increasing the upper temperature limit at which the battery remains safe from thermal runaway. These batteries are small enough to be implanted in the eye, to power an artificial retina. Long shelf life battery uses "nanograss" to separate liquid electrolytes from the solid electrode until power is needed. Lithium ion batteries with nanophosphate coated electrodes meet the safety requirements for electric cars while improving the performance.

7.8 FUEL

Nanotechnology can address the growing short fall of fossil fuels such as diesel and gasoline through:

- Making production of fuels from low grade raw materials economical
- Increasing the mileage of engines
- Making the production of fuels from normal raw materials more efficient

Catalysts made from nanoparticles have a greater surface area to interact with the reacting chemicals than catalysts made from larger particles. The larger surface area allows more chemicals to interact with the catalyst simultaneously, which makes the catalyst more effective. This increased effectiveness can make a process such as the production of diesel fuel from coal more economical, and enable the production of fuel from currently unusable raw materials such as low grade crude oil. Use of cerium oxide NPs and tungsten oxide NPs as a nanocatalysts makes the process of refining gasoline more efficient.

Nanotechnology, in the form of genetic engineering, can also improve the performance of enzymes used in the conversion of cellulose into ethanol in high yield. Currently ethanol added to gasoline, is made from corn, molasses, etc. It can allow materials that often go to waste, such as wood chips, grass and other biomass waste to be turned into ethanol. Nanocatalysts with spherical shape are shown to be effective in converting cellulose to sugar and production of enzymes to covert cellulosic materials to ethanol.

7.9 AUTOMOBILES

Today, nanotechnology, like in any other sectors, has opened doors for the automotive sectors. Though may be still in an early stage of R&D, use of nanotechnology in automobiles is promisingly sustainable, safe, comfortable, and eco-friendly; it is also economical. CO_2 free engines, safe driving, quiet cars, self-cleaning body, and windscreens etc. can be the key drivers for the idea of 'nano in cars' as well as heavy automobiles to come alive. Some of the mature application of nanotechnology in automobile sector include:

a. Exhaust Catalysts

Use of catalytic converters has significantly reduced emissions of hazardous air pollutants from automobiles. Application of nanocatalysts is showing further improvements on existing technology.

Most catalytic converters utilize the precious and costly platinum metal. The use of platinum nanoparticles as catalyst show enhanced catalytic activity over existing bulk catalysts due to the fact that a much greater surface area of the metal is exposed. Efficiency increase of 50% or greater can be achieved in some cases. The cost of catalytic converters based on nanoparticulate platinum and platinum nano-composites could be significantly reduced.

b. Shock Absorbers

Shock absorbers provide the comfortable ride in vehicles ranging from sports cars to sport utility vehicles and pickup trucks. Nanotechnology–specifically magnetic nanoparticles are shown to enhance the shock absorber capabilities further than ever before.

Magnetic fluids are comprised of magnetic nanoparticles dispersed in a fluid. Depending on the size of the nanoparticles, the magnetic fluid may be able to change its apparent viscosity in proportion to the strength of the magnetic field applied to it. Therefore, the viscosity can be controlled dynamically, which allows for active damping. Large amounts of mechanical power can be controlled with a small amount of electrical power, making this method of vibration control much more efficient than traditional. Some magnetic fluids can transform themselves into a nearly solid state, making it possible to adjust the stiffness thousands of times per second. Shock absorbers based on magnetic fluids, therefore, provide a very smooth ride and can be adjusted to the desired level.

Shock absorbers based on magnetic fluids are used today in some advanced cars such as Audi Le Mans Quattro. Energy is derived from the electronic control system, and the on-board computer adjusts the shock absorbers based on information provided by sensors that detect the actual driving situation within a few thousandths of a second. The driver can switch between a sporty feel, where the magnetic fluid is at low viscosity, and a more comfortable ride, where the viscosity is set at a higher level.

c. Coatings

Nanotechnology can be best used in automotives in the coatings such as protective and decorative finishes, heat reflective window glasses and water repellant and dust repellent surfaces. However, there still remains a vast area where nanocoatings have to pave its way. Hardcoatings of ceramics improves wear and friction characteristics of components, along with the specialty of detecting even fractional concentrations of gas in vehicle interiors. Self-repairing is yet another concept where materials can refine their original shape under the influence of external temperature. In addition, the electro-chromic coatings are prophetic of the enormous boon for future cars.

d. Coolants

The rising cost of fuel continues to be a matter of concern. Consumers are focused now on purchasing vehicles with increased fuel efficiency. That is why automakers are looking for technology that will improve the fuel efficiency of their vehicles.

Nanoparticles added to heat transfer fluids increase their performance. Solid nanoparticles conduct heat better. They work better because they stay suspended in liquids longer than larger particles. They also have a much greater surface area, which is where heat transfer takes place. The smaller the particle, the greater its ability to enhance heat transfer. Nano-additives, including nanoparticles and nanopowders, could potentially reduce the size of automotive cooling equipment while increasing its heat transfer capabilities. Engines and other components could also be smaller and lighter, providing a lighter weight vehicle. In addition, engines could potentially run at more optimal temperatures. These factors lead to more fuel efficient automobiles. Reduced consumption of fuel also result in reduced emissions to the environment as well.

Numerous other potential applications exist in the automotive market for nanomaterials. Areas currently under investigation include spark plugs using nanoscale metal and ceramic powders, nanocatalysts for octane enhancers, fuel additive for diesel engines, seatbelts, and vehicle levelling sensors.

7.10 DEFENSE AND SECURITY

Nanotechnology opens a broad spectrum of possible military uses that both expand and extend existing systems and define radical new applications. The lesson from history is that the better equipped an army is, the more likely their chances of winning in battle. The idea that nanotechnology could lead to lighter weight platforms smarter devices for soldiers in the field, uniforms that offer ballistic and other protection, and more deadly weapons has proved irresistible.

a. Light weight military platforms

The nanomaterials such as nanocomposites which are lighter and stronger than metals are able to create light weight military platforms and armors for vehicles and soldiers, without functionality compromising on strength and ruggedness. With the addition of new nanocoatings that enhance 'stealth' capabilities, the latest aircraft and missiles can now travel longer distances because of lower fuel consumption whilst reducing the likelihood of detection by an enemy. Nanostructured ceramic coatings exhibit much greater toughness than conventional wear-resistant coatings for machine parts. Such coatings can extend the life of moving parts in power tools to industrial machinery, saving huge maintenance cost.

b. Camouflage/stealth

Comouflage/steath is defined as a means of delay or denial of detection of an military object by detectors such as CCD cameras, thermal cameras, radars or sonars.

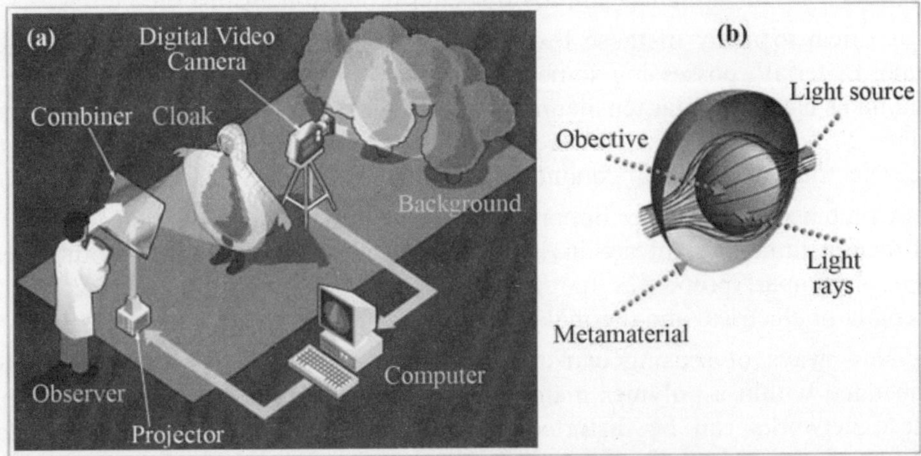

Fig. 7.11. The Complete Invisibility System background projection (a); Schematic representation of metamaterial showing bending of light (b)

Apart from designer nanoparticles to enhance the performance of coatings, which can make surfaces harder, smoother and less liable to corrode, nanotechnology offer the potential for both 'stealth' coatings that minimize the likelihood of radar detection, and coatings that can change color in order to autonomously camouflage an object in different environments. Surfaces with locally variable color, for example, through the use of mobile pigment particles can be used for camouflage. Surfaces with tailored absorption properties for electromagnetic radiation can reduce radar and infrared signatures. It is likely that camouflage paint schemes, of the near future, applied on vehicles as 'smart' coatings, could be altered with an electrical impulse as simply as changing pixel colors on a screen. The idea of 'invisibility cloaks', far from mere fantasy, can be created by simply projecting what is behind a military target onto its surface thus making it effectively camouflaged (**Fig. 7.11a**). The phenomenon is known as e-camouflaging. Another way of creating invisibility is by using metamaterials as illustrated in **Fig. 7.11b**. Metamaterials are basically artificially engineered materials having negative refractive index. Light is neither absorbed nor reflected by the metamaterial; it rather passes "like water flowing around a rock." As a result, only the light from behind the object can be seen. Applications are still very limited at this stage, as the metamaterial can only be produced in small quantities.

c. Protection for soldiers

A military uniform today needs to be multifunctional such as providing ballistic protection; the ability to autonomously gather and transmit information; assist in the healing process if a soldier if wounded; and protect its wearer from extreme environments such as heat and cold and the threat of chemical and biological warfare. It needs to be lightweight yet strong and, of course, cheap to produce. Nanofiber-based garments can provide better protection against projectiles.

Critical to many of these textile applications are carbon nanotubes, a new 'super material', possessing some extraordinary properties. For example, carbon nanotubes have a similar tensile modulus to the current best carbon fibres, but are nearly 20 times stronger than steel. Similarly, remarkable properties are true for their electrical and thermal conductivity.

Combined with another important phenomenon, that nano-particles below the diffraction limit of light are invisible to the naked eye, carbon nanotubes can therefore impart properties to their host matrix whether it is an increase in strength, or electrical and thermal conductivity.

New ways of sensing can also be achieved with networks of nanotubes embedded within a polymer matrix that can also form part of battle uniforms. These networks can be distorted by external stimuli, thus creating flexible sensors. Since most polymers are insulators, any distortion of the conductive network of the nanotubes will cause an increase in electrical resistance. Thus the carbon nanotube network becomes an integrated analogue switch, whose on/off characteristics will be a function of the host polymer. For example, a stimulus could be swelling of the polymer in the presence of water, a solvent or gas, or mechanical strain. In this way, nanotube networks enable the creation of smart, responsive textiles when incorporated in a polymer fiber.

d. Surveillance

Surveillance on an enemy using detectors working or a broad range of an electromagnetic spectrum eg. UV, visible, infrared, Microwave and beyond is an important military task both in peace time and a war.

Nanotchnology has enabled tinier and more efficient listening, imaging and information processing devices than one could ever have imagined in the past. Nanotechnology is enabling new forms of almost 'invisible' surveillance through the incorporation and integration of different technologies with IT, otherwise called ubiquitous computing, including radio frequency identification chips (RFIDs), integrated circuits, quantum dot tags, minute (bio) sensors, intelligent fabrics, films and smart surfaces. Because surveillance techniques can be miniaturized so effectively, the idea of using live insects ('spy' bees) or creating tiny winged robots that emulate insects to be flown into an enemy territory to record data, has formed the basis of new surveillance technologies.

e. Affordable Energy

Lighter energy-storage and conversion systems such as fuel cells using nanoparticle-based membranes or hydrogen tanks based on nanotubes or fullerenes could make all-electric military vehicles practical. For small systems, contracting molecules or deforming materials offer the potential of muscle-like motion. Soldier-worn systems could sense the state of health of the wearer and react by releasing drugs or, using smart materials, by compressing wounds. Energy for communication could be generated by normal body movements using nano piczo electrical materials.

f. Deadly Weapons

Employing advances in nanomaterials and explosives, conventional guns could shoot to larger ranges with high targeting accuracy at reduced mass. The same holds for missiles. With nanotechnology, the size of autonomous mini- and microrobots could decrease, principally down to far below 0.1mm. These could move in all media: on the ground, in water and in air, using propulsion principles known from larger technical systems (wheel, track, propeller, wing, jet, rotor) or biomimetic ones (using legs, wriggling like a snake, hopping, using flapping wings, flagella, etc.). Mini/Microrobots could be designed for a wide variety of purposes such as, reconnaissance, communication, and target designation for big weapons, actuation on a small scale or as weapons. One possible scenario would involve entering into a target object or subject and approaching a central node; there the robot could eavesdrop, manipulate or destroy.

Another possibility would be acting in high numbers, choking air intakes, blocking windows, putting abrasives into mechanics, etc. Biological-technical hybrids, such as insects or small mammals controlled by nerve/brain electrodes, could fulfil similar purposes. Thus, they should be classed with mini-/ microrobots.

For military uses of outer space, nanotechnology offer many possibilities for markedly smaller *satellites*, together with smaller launch vehicles. Small satellites could be used in swarms for radar, communication or intelligence. Small satellites could damage or destroy other satellites–either by a direct hit with high relative velocity or through manipulations after rendezvous and docking. Nanotechnology enabled electromagnetic acceleration could also be used for kinetic-energy space weapons.

7.11 AGRICULTURE

Agriculture is the backbone of most developing countries; more than 60% of the population depend on it for their livelihood. Nanotechnology can help to improve our understanding of the biology of different crops and thus potentially enhance yields or nutritional values. In addition, it can offer routes to added value crops or environmental remediation. Nanotechnology has the potential to revolutionize the agricultural sector with new tools for the molecular treatment of diseases, early detection, enhancing the ability of plants to absorb nutrients etc. Smart sensors and smart delivery systems will help the farmers to combat viruses and other crop pathogens. In the near future, nanostructured catalysts will be available to increase in efficiency of pesticides and herbicides, allowing lower doses to be used thereby saving farmers from life threatening diseases caused by exposure to the excessive use of pesticide and insecticides.

a. Precision farming

Precision farming has been a long-desired goal to maximize output (*i.e.* crop yields) while autonomous sensors linked into a GPS system for real-time monitoring. These nanosensors could be distributed throughout the field where they can monitor soil conditions and crop minimizing input (*i.e.* fertilizers, pesticides, herbicides, etc.) through monitoring environmental variables and applying targeted action. Precision farming makes use of computers, global satellite positioning systems, and remote sensing devices to measure highly localized environmental conditions thus determining whether crops are growing at maximum efficiency or precisely identifying the nature and location of problems. By using centralized data to determine soil conditions and plant development, seeding, fertilizer, chemical and water use can be optimized to lower production costs and potentially increase production all benefiting the farmer. Precision farming can also help to reduce agricultural waste and thus keep environmental pollution to a minimum. Although not fully implemented yet, tiny sensors and monitoring systems enabled by nanotechnology will have a large impact on future precision farming methodologies.

The union of biotechnology and nanotechnology in sensors will create miniaturized equipment of increased sensitivity allowing an earlier response to environmental changes. For example:

- Nanosensors utilizing carbon nanotubes or nano-cantilevers are small enough to trap and measure individual proteins or even small molecules.

- Nanoparticles or nanosurfaces can be engineered to trigger an electrical or chemical signal in the presence of a contaminant such as bacteria.
- Other nano sensors work by triggering an enzymatic reaction or by using nano engineered branching molecules called dendrimers as probes to bind to target chemicals and proteins.

Ultimately, precision farming, with the help of smart sensors, will allow enhanced productivity in agriculture by providing accurate information, thus helping farmers to make better decisions.

b. Smart delivery systems

The use of pesticides has exponentially increased in the second half of the 20th century with DDT becoming one of the most effective and widespread used pesticide throughout the world. However, most of these pesticides were found to be highly toxic, affecting human and animal health and as a result whole ecosystems. As a consequence they were banned in many countries. To maintain crop yields, integrated Pest Management systems, which mix traditional methods of crop rotation with biological pest control methods, are becoming popular and implemented in many countries, such as Tunisia and India.

In future, nanoscale devices with novel properties could be used to make agricultural systems "smart". For example, devices could be used to identify plant health issues before these become visible to the farmer. Such devices may be capable of responding to different situations by taking appropriate remedial action. If not, they will alert the farmer to the problem. In this way, smart devices will act as both a preventive and an early warning system. Such devices could be used to deliver chemicals in a controlled and targeted manner in the same way as nanomedicine has implications for drug delivery in humans.

Technologies such as encapsulation and controlled release methods have revolutionized the use of pesticides and herbicides. Many companies make formulations which contain nanoparticles within the 100-250 nm size range that are able to dissolve in water more effectively than existing ones (thus increasing their activity). Other companies employ suspensions of nanoscale particles (nanoemulsion), which can be either water or oil-based and contain uniform suspensions of pesticidal or herbicidal nanoparticles in the range of 200-400 nm. These can be easily incorporated in various media such as gels; creams, liquids etc., and have multiple applications for preventative measures, treatment or preservation of the harvested product.

7.12. FOOD SECTOR

The benefits of nanotechnology in food sector are many and expected to grow with time. This new, rapidly developing technology impacts every aspect of the food system from cultivation to food production to processing, packaging, transportation, shelf life and bioavailability of nutrients.

a. Food ingredients for nutrients and supplement, color, texture and flavour

Nanomaterials are used as ingredients and additives (*e.g.*, vitamins, antimicrobials, antioxidants) in nutrients and health supplements for enhanced absorption and bioavailability. Public acceptance of food and food products containing nanomaterials depends on their perceived safety. The food industry is also beginning to use nanotechnology to develop nano ingredients to improve color, texture and flavour of food. The nanoparticles TiO_2 and SiO_2 and amorphous silica are used as food additives.

b. Food Processing

Food processing is the conversion of raw ingredients into food and its other forms by making it marketable and with long shelf life. Processing includes toxin removal, prevention from pathogens, preservation, improving the consistency of foods for better marketing and distribution. Processed foods are usually less susceptible to early spoilage than fresh foods and are better suited for long distance transportation from the source to the consumer. Nano capsules delivery systems plays an important role in processing sector and the functional property are maintained by encapsulating simple solutions, colloids, emulsions, biopolymers and others into foods. Nano sized self-assembled structural lipids serves as a liquid carrier of healthy components that are insoluble in water and fats called as nanodrops. They are used to inhibit transportation of cholesterol from the digestive system into the bloodstream.

c. Food Packaging

Nanomaterials used for food packaging provide many benefits such as improved mechanical barriers, detection of microbial contamination and potentially enhanced bioavailability of nutrients. This is perhaps the most versatile application of nanotechnology in food and food-related industries. The packing has a great significance in preserving the food to make it marketable. Innovations in packaging have lead to quality packing and consumer friendly approach in determining the shelf life, biodegradable packing and many more. Nanotechnology in packaging is categorized based on the purpose of the application.

i. Barrier Protection

The food products are preserved by maintaining it in an inert and low oxygen atmosphere for inhibiting microbial growth and spoilage, thus the material used should be impermeable to gases. Polymer nanocomposites containing nano clays dispersed in polymer matrix act as efficient barrier for gases by complicating the path to flow of the gaseous molecules together with favoring the filler-matrix interactions for better performance due to the large surface area of the nanofillers. Nanocomposites prepared from nanoclay fillers and polymer matrix such as nylons, polyolefin, PET, PA, epoxy resin, poly methane are used in food packaging to get higher quality. However, polyamide based nanoclay composites have been developed largely and commercialized and are recognized for their durability and protection. Further, nanocomposites from cellulose based polymers and carbon nanotubes are under development.

ii. Antimicrobial Packaging

Microbial growth in food items is a common problem and leads to its spoiling due to the presence of pathogens. The barriers consisting of natural nanoparticles act to control microbial growth. Some of the nanomaterials effective to control microbial growth includes: Silver nano particles used in all forms including biotexiles, electrical appliances, refrigerators, kitchenware's; Zinc oxide -- its antibacterial nature increases with decreasing particle size, it can be stimulated by visible light, and they are incorporated in a number of polymers including polypropylene; coating of titanium dioxide in packing material to control *E. coli*. Titanium oxide combined with silver nano particles also help to improve the disinfection process. Chiston, a biopolymer derived from chitin, is recently reported to show antimicrobial properties in additional to material for encapsulation.

iii. Biodegradables

Dumping of non-degradable plastics changes the soil nature and leads to accumulation of toxic gases in the atmosphere there by causing a big pollution issue. Hence, more and more use of bio-degradable plastics is a highly desirable solution. However, the poor mechanical strength and higher permeablity to water and gases are limitations for their application in many areas including food packaging. These disadvantages are likely to be resolved by nanotechnology in that incorporation of natural or synthetic nanoparticles in packaging material have the potential to create properties like bio-degradability and high mechanical strength. Nanoparticles to be used for this purpose include proteins, carbohydrates, lipids from animal and plant materials, metal oxides and carbon nanotubes in addition of porous nanofibers of collagen, cellulose synthesized from corn. These nano materials added along with nanoclay also have additional novel properties like sensors, antibacterial action and as biocatalysts.

iv. Smart Packaging

Sensors are used to sense oxygen content, temperature, pathogens and indicators for proper alarming and to regulate the internal environment of the food stuffs. A recent report shows that the current smart packaging segment is dominated by oxygen scavengers, moisture absorbers and barrier packing product, accounting for 80% of the market. Bakery and meat products have attracted most nano-enabled packaging technology to date. Some examples of sensors include gold nanoparticle incorporated enzymes for microbes detection, gas sensing related to condition of food products : nanofibrils of perylene-based fluorophores indicates fish and meat spoilage by detecting gaseous amines. Others include zinc oxide and titanium oxide nanocomposites for the detection of volatile organic compounds. Nanobarcodes are used for tagging and also for security. Thus the use of smart sensors is beneficial to the consumers in terms of better quality identification and producers for rapid distribution and authentication of the food products.

7.13. BUILDING AND CONSTRUCTION MATERIALS

Nanotechnology generated products have many unique characteristics important to fix current construction problems. Some of the advantages of using nanomaterials in construction sectors include:

- Lighter and stronger structural composites
- Low maintenance coating
- Improving pipe joining materials and techniques.
- Better properties of cementitious materials
- Reducing the thermal transfer rate of fire retardant and insulation
- Increasing the sound absorption of acoustic absorber
- Increasing the reflectivity of glass.

The abbreviated list is not an exhaustive list of applications of nanotechnology in construction. Some of these applications are examined in detail below.

a. Concrete

Concrete is one of the most common and widely used construction materials. Its properties have been well studied at macro or structural level without fully understanding the properties of the cementitious materials at the micro level. An understanding has been developed on the hydration process, alkali-silicate reaction (ASR), and fly ash reactivity using nanotechnology. The better understanding of the structure and behavior of concrete at micro/nano-scale is quite helpful to improve concrete properties and prevent the illness, such as ASR.

Addition of nanoscale materials into cement could improve its performance. It is observed that addition of nano SiO_2 could significantly increase the compressive strength of concrete, containing large volume fly ash, at early age and improve pore size distribution by filling the pores between large fly ash and cement particles at nanoscale. The dispersion/slurry of amorphous nano silica is used to improve segregation resistance for self-compacting concrete. It is also reported that adding small amount of carbon nanotube (1%) by weight could increase both compressive and flexural strength.

b. Structural Composites

Steel is a major construction material. Its properties, such as strength, corrosion resistance, and weld ability, are very important for the design and construction. The new steel was developed with higher corrosion-resistance and weld ability by incorporating copper nanoparticles at the steel grain boundaries. Due to its high performance, this type of steel is suitable for application where lightweight and rigid designs are required. For certain applications, the components made out of such a material could be even thinner and lighter than that made from aluminium and titanium due to its ultra-high strength and modulus of elasticity. Its good corrosion and wear resistance can keep life-cycle costs low.

c. Coating

The exterior coatings on buildings incorporating certain nanoparticles or nanolayers have been developed for specific purposes. It is one of the major applications of nanotechnology in construction. For example, TiO_2 is used to coat glazing because of its sterilizing and anti-fouling properties. The TiO_2 will break down and disintegrate organic dirt through powerful catalytic reaction. Furthermore, it is hydrophilic, which allow the water to spread evenly over the surface and wash away dirt previously broken down. Other special coatings based on nanomaterials have also been developed, such as thermal control, energy saving, anti-reflective etc.

d. Nanosensors

Nano and microelectromechanical systems (MEMS) sensors have been developed and used in construction to monitor and/or control the environment condition and the materials/structure performance. One advantage of these sensors is their dimension. Nanosensor ranges from 10^{-9} m to 10^{-5}m in size. The micro sensor ranges from 10^{-4} to 10^{-2} m. These sensors could be embedded into the structure during the construction process. Smart aggregate, a low cost piezoceramic based multi-functional device, has been applied to monitor early age concrete properties such as moisture, temperature, relative humidity and early age strength development. The sensors can also be used to monitor concrete corrosion and cracking. The smart aggregate can also be used for structure health monitoring. The disclosed system can monitor internal stresses, cracks and other physical forces in the structures during the structures' life. It is capable of providing an early indication of the health of the structure before a failure of the structure can occur.

7.14 ENVIRONMENT AND POLLUTION CONTROL

Pollution results from resource production and consumption, which in their current state are very wasteful. Most of the waste products cannot be reintegrated into the environment effectively or cheaply. Thus, processes like petroleum and coal extraction, transportation, and consumption continue to result in photochemical smog, acid-mine drainage, oil slicks, acid rain, and fly ash. Inefficient burning of fuel discards much of its free energy during the conversion of chemical energy into heat and then to mechanical energy. Biological systems, on the other hand, efficiently oxidize fuel through molecular-scale mechanisms without extracting the chemical energy through thermalization. Overcoming the Promethean paradigm requires controlling reactions at the nanoscale.

An increasing variety of nano materials with environmental control applications has been developed over the past several years. For example, nano materials have been used to remediate contaminated soil and groundwater at hazardous waste sites eg., sites contaminated by chlorinated solvents or oil spills. Because of their higher reactivity, chemical or biological surface mediated reactions can be greatly enhanced in comparison to the same material at much larger sizes. These properties allow a nanomaterial in prologned contact with contaminants, thereby resulting in rapid reduction of contaminant concentrations. Furthermore, because of their small size, nanoscale materials may pervade very small spaces in the subsurface and remain suspended in groundwater if appropriate coatings are used. Appropriate coating may allow the particles to travel farther than macro-sized particles, achieve wider distribution, and therefore improve contaminant reduction. Ongoing bench- and pilot-scale research is being performed to investigate environmental pollution control by nanoparticles such as TiO_2, self-assembled monolayers on mesoporous supports (SAMMS), dendrimers, carbon nanotubes, metalloporphyrinogens, and swellable organically modified silica (SOMS). In the following subsections we discuss the application of nanomaterials in the purification of water and air as a step forward in environmental and pollution control together with cleaning of oil spills.

7.14.1 Water purification and Remediation

Currently, millions of people around the world do not have access to safe drinking water; water supplies are under increasing strain from depletion, contamination and pollution. Since, traditional desalination processes are energy intensive, an affordable means of extracting clean water from sea-water or contaminated water is desperately needed. Pollutants in water can be converted into harmless materials through chemical reactions. Studies have shown that nanomaterials could be highly suitable as hydro-dehalogenation and for the remediation of various organic and inorganic groundwater contaminants (**Fig 7.12**).

Nanotechnology eases the water cleansing process because inserting nanopa-rticles into underground water sources is cheaper and more efficient than pum-ping water for treatment. The desalination using nano-sized fibers as electrodes is not only cheaper, but also more energy efficient. Traditional water filtering systems use semi-permeable membranes as electro dialysis or reverse osmosis are costly affair with higher percentage of rejected water. Permeable membranes with sub-nanometer pores filter out ions and impurities while letting water (or other liquids) through are attracting attention for these applications. Membranes that can even filter out viruses are now available.

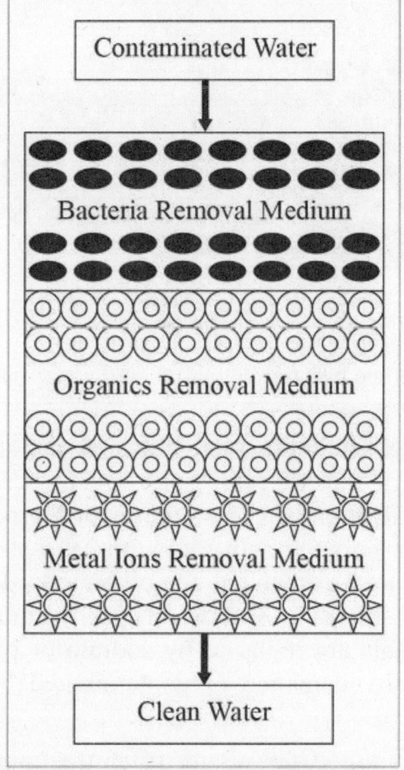

Fig. 7.12. Water purification system based on nanomaterials

Recently, carbon nano materials such as carbon nanotubes and graphene have been hailed as promising candidates materials for membranes. Amazing properties of graphene such as fast water transport, high chemical stability, and the cost-effectiveness of preparation in large quantities make graphene-based membranes superior in comparison to other water purification technologies. The physically confined graphene oxide (PCGO) membranes created with interlayer spacing varying from 6.6–9.8Å (compared to 13.7Å for water-soaked GO) successfully filter out common ions like Na+ and K+, while allowing water to pass through (**Fig. 7.13**). Incorporating graphene flakes into the stacked-layer laminates can also control the swelling of GO membranes. The hydrophobicity of graphene limits water intake into the stack and curtails swelling. The GO-Gr membranes achieved a salt rejection level of 97%. The combination of ion-permeation suppression with fast water transport in PCGO membranes makes them an attractive proposition for water filtration and desalination.

Fig. 7.13. Graphene membrane based desalination and water purification

Decreasing the pore size of a membrane to the nanometer range would increase the selectivity of the molecules allowed to pass through. Also widely used in separation, purification, and decontamination processes are ion exchange resins, which are organic polymer substrate with nano-sized pores on the surface where ions are trapped and exchanged for other ions. Ion exchange resins are mostly used for water softening and water purification. In water, poisonous elements like heavy metals are replaced by sodium or potassium. However, ion exchange resins are easily damaged or contaminated by iron, organic matter, bacteria, and chlorine.

Nanotechnology is applied for waste water treatment and for groundwater remediation. Besides commercialized products and applications, many of the following techniques are still under development or in a testing stage:

- *Filtration/separation:* Nano membranes are state of the art technology. As membrane materials organic polymers and inorganic ceramics are used. Applications of nano filtration can be found, *e.g.* in drinking water treatment, in the food industry as well as in the textile and dyestuff industry. A great potential for nanotechnology is expected in desalination of sea water.
- *Surface treatment* : Nano-based coatings are developed to minimize fouling processes and deposits on membranes, in heat exchangers and reactors and on ship hulls.
- *Sorption*: Nano-based sorbents may be used for different water treatment purposes. Examples are arsenic separation from drinking or

wastewater and a paper like material made from nanowire for absorption from water of hydrophobic liquids such as mineral oil suspended in water.

- *Nanocatalysts:* Nanoscale titanium dioxide is used as a photo catalyst in waste water treatment. During photo-catalysis the illumination of the titanium dioxide results in water and oxygen from the air being converted into reactive hydroxyl radicals, which break down organic pollutants resistant to biological degradation in the water. Other focused area is the development of a catalyst (ferromagnetic nano-sized carriers containing traces of palladium) for treatment of wastewater contaminated with halogenated hydrocarbons.
- *Nano reagents*: Several kinds of nanoparticles such as zero-valent iron are being used for *in-situ* groundwater remediation. Besides pure nano-iron, bimetallic nanoparticles and iron nanoparticles on activated carbon have also been used, in part for groundwater remediation.

7.14.2 Air Purification

Air pollution can be remediated using nanomaterials in several ways. One is through the use of nano-catalysts with increased surface area for gaseous reactions. Catalysts currently in use include a nanofiber catalyst made of manganese oxide to remove volatile organic compounds from industrial smoke-stacks. Nanotechnology products for air pollution control are applied in air filters and in photocatalysis. Some of these materials and methods are discussed below.

- Catalytic converters: In the automotive sector, catalytic converters are state of the art technology. A three-way catalytic converter has a stainless steel body that contains catalytic material as a layer on a substrate (wash coat). The nano particles of the noble metal catalyst in the washcoat minimize the amount of noble metal but with the same catalytic performance.
- *Filtration/separation*: In the automobile industry filter media lined with nanofibers are frequently used in passenger cabin air. Nanofiber-coated filter media are also applied for air filtration (*e.g.* dust removal) at industrial plants and for filtration of the inlet air for gas turbines. Amongst other things, nano-structured membranes have great potential for CO_2 capture from power plant flue gases.
- *Nanocatalysts*: Nanocatalysts for sustainable air purification are mostly ready for practical use. For example, through the use of nanoscale titanium dioxide in cement photo-catalytic active concrete surfaces can be produced for air purification in cities and towns.
- Carbon nanotubes (CNT) based membranes have been developed successfully for trapping greenhouse gas emissions caused by coal

mining and power generation. CNT can trap gases up to a hundred times faster than other methods, allowing integration into large-scale industrial plants and power stations. This new technology both processes and separates large volumes of gas effectively, unlike conventional membranes that can only do one or the other effectively.

7.14.3 Oil spill cleaning

According to the U.S. Environmental Protection Agency (EPA), about 14,000 oil spills are reported each year. Dispersing agents, gelling agents and biological agents are most commonly used for cleaning up oil spills. However, none of these methods can recover the oil lost. Recent developments of nano-wires made of potassium manganese oxide can clean up oil and other organic pollutants while making oil recovery possible. These nanowires form a mesh that absorbs up to twenty times its weight hydrophobic liquids while rejecting water with its water-repelling coating. Since potassium manganese oxide is very stable even at high temperatures, the oil can be boiled off the nanowires and both the oil and the nanowires can then be reused. The Interface Science Corporation successfully launched a new oil remediation and recovery application, which used the water repelling nanowires to clean up the oil spilled by the damaged oil platforms and refineries.

7.15 ARCHAEOLOGICAL PRESERVATION

There are several examples to show that nanotechnology, of course, without any awareness about its nature, was practiced since time immemorial. Like in so many other sectors as discussed above, nanomaterials could be of immense help for the conservation of cultural heritage which is the base of our history and culture. In the past 20 years many, attempts have focused on developing suitable materials for consolidating and protecting heritage documents, statues as well as structures. Many materials have been used with varying degrees of success, but in most of the cases there is increasing decay or the rejection of the intervention.

Actually, preservation of ancient culture available in the form of built environment and art is so important for the present and future generation to learn and get inspired from our traditional past. Though, as we know, the decisions concerning conservation of the built environment in the past have been the domain largely of architectural historians, urban planners, conservation specialists, and related professionals. Sure enough conservation cannot remain a closed and solely self-referential profession and one has to understand the basic causes of deterioration and decay of the heritage articles. In particular, the deterioration of stone and wood with which the ancient structures and monuments are built is a complex physical-chemical process caused by the interaction of several factors: climate of the locations, urban pollution, and the material properties.

In this context, the protection and appreciation of the ancient materials like wood and stone, is of particular interest, especially through new technologies. Every direct method to consolidate and, more particularly, to protect and conserve the ancient structures must have the following characteristics:

(a) To be respectful of the environment;

(b) To be careful not to damage the wood surfaces or stone material of the ancient structures;

(c) Not to affect the structural characteristics of the material that forms the structure to be preserved;

(d) To be well absorbed by capillarity;

(e) Not to produce any change of colour of the treated material; and

(f) To have a good penetration and ensure a high degree of consolidation.

There are numerous studies and experiment relating to technological innovation for the preservation and appreciation of the ancient built heritage. Nanotechnology in particular is believed to offer solution in this area by tuning the properties of nanomaterials to preserve both new construction and cultural heritage.

The nanomaterials, nanostructured inorganic oxides, and more particularly titanium sesquioxide (Ti_2O_3) and silicon (Si_2O_3), have been demonstrated to show a high level of biocompatibility that can be used for the consolidation of archaeological wood and stones. Nanoparticles of silicon oxide are used as precursor to obtain lithium silicate on reacting it with lithium hydroxide/ carbonate. The new form of water-soluble lithium silicate, to be applied in situ show better performance than the conventional material with macro dimensions in preserving stone and wood materials of ancient structures, while perfectly respecting the environment, without changing the physico-chemical and mechanical structures of the treated materials, and all with no lasting effects. The concentration of lithium silicate in the consolidating product ready for use is less than 20% by weight, while the ratio Li_2O : SiO_2 is in the range 1.0: 1.0 to 1.0 : 20.0.

Suggested further readings

1. B. O'Regan and M. Gratzel, A low-cost, high-efficiency solar-cell based on dye sensitized colloidal TiO_2 films, *Nature*, 353 (6346), (1991) 737–740.
2. M. K. Patra, K. Manzoor, S. R. Vadera and N. Kumar, "Functional nanomaterials for sensors and display applications", *Encyclopedia of Nanoscience and Nanotecnology*, Vol. **14** (Ed. H.S. Nalwa), American Scientific Publishers, USA, 2011, pp 385-435.
3. N. Kumar and R. Kumar, *Nanotechnology and nanomaterials in the treatment of life threatening diseases*, Elsevier, 2014.

4. S. Kumbhat, K. Sharma, R. Gehlot, A. Solanki and V. Joshi, Surface plasmon resonance based immunosensor for serological diagnosis of dengue, *J. Pharm. Biomed. Anal.*, 52(2), 2010 255–259.
5. S. Kumbhat, D.R. Shankaran, S.J. Kim, K.V. Gobi, V. Joshi and N. Miura, Surface plasmon resonance biosensor for dopamine using D3 dopamine receptor as a biorecognition Molecule, *Biosens. Bioelectron.*, 23 (2007) 421–427.
6. L. Dai, D.W. Chang, J.B. Baek and W. Lu, Carbon nanomaterials for advanced energy conversion and storage, *Small*, 8 (2012) 1130–1166.
7. S. R. Vadera and N. Kumar, Nanotechnology and nanomaterial for camouflage and stealth application", *Nanotech. Insight.*, 5(3&4) (2014) 16–30.
8. N. Kumar, Nanotechnology for Rural Development in India, Nano Digest Sept' 2014.
9. N. Kumar, "Nanotechnology for Remediation and Purification of Water", *Nanotech Insights*, 6 (1), (2015) 29–39.
10. P. V. Kamat, "Quantum Dot Solar Cells: The Next Big Thing in Photovoltaics", *J. Phys. Chem. Lett.* 4 (2013) 908–918.

REVIEW QUESTIONS

Q 1. Name the consumer products in the field of sports which are being made using nanotechnology? Mention the nano component introduced and the resultant unique feature.

Q 2. How has nanotechnology been successful in introducing following properties in textile goods: i. Antimicrobial action ii. Self-cleaning effect iii. UV protection and iv. Making smart textiles.

Q 3. How has nanotechnology improved the diagnosis for better healthcare? What do you understand Lab-on-a chip (LOC)?

Q 4. What do you understand by nano drug delivery systems (NDDS)? Give the mechanism of targeted drug delivery.

Q 5. Explain how nanotechnology is contributing in developing targeted drugs in terms of composition and delivery system

Q 6. Write short notes on i. Theranostics with two examples, ii. Regenerative medicines, and iii. Tissue engineering.

Q 7. What is the impact of nanotechnology in the field of information technology?

Q 8. What are the differences between spintronics, photonics and quantum electronics? List out the nanomaterials used in each type of the above.

Q 9. Discuss the impact of nanotechnology in the evolution of displays technology'.

Q 10. How nanotechnology is useful in harvesting, generating and storing energy from renewable sources?

Q 11. Discuss the following technologies based on nanomaterials: a. Thermoelectrical device, b. Piezoelectric devices, c. Fuel cell, and d. Lithium ion batteries.

Q 12. ''Nanotechnology is making remarkable contributions in defence sectors'', Justify the statement.

Q13. Discuss the role of nanotechnology in food preservation and packaging.

Q14. Explain decontamination and desalination of water using nanomaterials.

Q 15. How nanotechnology is useful in building construction and preservation of heritage buildings?

■■■■

Subject Index